P9-AFP-045

WITHDRAWN
UTSA LIBRARIES

RENEWALS 458-4574

N

WITHDRAWN
UTSA LIBRARIES

Environmental Tobacco Smoke

Environmental Tobacco Smoke

Edited by
Ronald R. Watson
and
Mark Witten

CRC Press
Boca Raton London New York Washington, D.C.

Library
University of Texas
of San Antonio

Library of Congress Cataloging-in-Publication Data

Environmental tobacco smoke / edited by Ronald R. Watson and Mark Witten.
    p. cm.
Includes bibliographical references and index.
ISBN 0-8493-0311-7 (alk. paper)
1. Tobacco smoke—Toxicology. 2. Tobacco smoke pollution—Health aspects.
3. Passive smoking—Health aspects. I. Watson, Ronald R. (Ronald Ross)
II. Witten, Mark L. (Mark Lee), 1953–

RA1242.T6 E5725 2000
615.9′52395—dc21

00-057190
CIP

This book contains information obtained from authentic and highly regarded sources. Reprinted material is quoted with permission, and sources are indicated. A wide variety of references are listed. Reasonable efforts have been made to publish reliable data and information, but the author and the publisher cannot assume responsibility for the validity of all materials or for the consequences of their use.

Neither this book nor any part may be reproduced or transmitted in any form or by any means, electronic or mechanical, including photocopying, microfilming, and recording, or by any information storage or retrieval system, without prior permission in writing from the publisher.

All rights reserved. Authorization to photocopy items for internal or personal use, or the personal or internal use of specific clients, may be granted by CRC Press LLC, provided that $.50 per page photocopied is paid directly to Copyright Clearance Center, 222 Rosewood Drive, Danvers, MA 01923 USA. The fee code for users of the Transactional Reporting Service is ISBN 0-8493-0311-7/00/$0.00+$.50. The fee is subject to change without notice. For organizations that have been granted a photocopy license by the CCC, a separate system of payment has been arranged.

The consent of CRC Press LLC does not extend to copying for general distribution, for promotion, for creating new works, or for resale. Specific permission must be obtained in writing from CRC Press LLC for such copying.

Direct all inquiries to CRC Press LLC, 2000 N.W. Corporate Blvd., Boca Raton, Florida 33431.

Trademark Notice: Product or corporate names may be trademarks or registered trademarks, and are used only for identification and explanation, without intent to infringe.

Library
University of Texas
at San Antonio

© 2001 by CRC Press LLC

No claim to original U.S. Government works
International Standard Book Number 0-8493-0311-7
Library of Congress Card Number 00-057190
Printed in the United States of America 1 2 3 4 5 6 7 8 9 0
Printed on acid-free paper

Preface

The role of tobacco smoking in causing health problems is well defined. However, the effects of their smoke on non-smokers need further definition. Most non-smokers are exposed to some tobacco smoke each day. While many communities and countries are making certain areas, buildings, and occupations non-smoking and smoke free, the extent of damage due to environmental tobacco smoke (ETS) needs definition and prevention. Which of the many diseases, cancers, and pathologies clearly associated with smoking are tobacco smoke–induced in non-smokers? As public exposure to environmental smoke needs to be reduced, a better understanding of its consequences must be developed. In addition, the effects of smoking bans in the workplace and changes in advertising need to be assessed on the exposure of non-smokers to environmental smoke. Animal models, epidemiology studies, and medical reports help define such problems and their consequences.

The objective of this book will be to bring into one place the key observations on the nature and effects of ETS exposure. This book will focus on the pathological effects of ETS in pregnant women, newborns, youths, adults, and the elderly. It delves into developing associations with asthma, tobacco allergy, heart disease, and cancer due to ETS. It investigates other pathological effects such as DNA damage, gene activation, and immunosuppression, just beginning to be studied. In addition, this book probes the role of the political system and its laws in modifying behaviors, exposure risk, and health consequences of environmental smoke. The authors summarize the ability of antioxidant supplements to lower ETS damage. The effects of restrictions on tobacco smoking in public places are vital to reduce ETS exposure and improve health. In addition, the use of animal models strengthens the precision of studies. The concerns of the tobacco industry and potential problems in defining relationships between ETS and health problems are also included. A major focus relates to obvious and major effects of ETS on lung function promoting asthma, which has increased dramatically in children. Clearly, ETS is a significant health risk that can be eliminated.

Editors

Ronald R. Watson, Ph.D. initiated and directed the NIH Specialized Alcohol Research Center at the University of Arizona College of Medicine, Tucson, for 6 years. Dr. Watson has edited 50 books, including 8 on drugs of abuse.

Dr. Watson attended the University of Idaho, but graduated from Brigham Young University in Provo, Utah with a degree in Chemistry in 1966. He completed his Ph.D. degree in 1971 in Biochemistry at Michigan State University in East Lansing. His postdoctoral schooling was completed at the Harvard School of Public Health in Nutrition and Microbiology in Cambridge, Massachusetts, including a 2-year postdoctoral research experience in immunology. He was an Assistant Professor of Immunology and did research at the University of Mississippi Medical Center in Jackson from 1973 to 1974. He was an Assistant Professor of Microbiology and Immunology at the Indiana University Medical School from 1974 to 1978 and an Associate Professor at Purdue University, in Lafayette, Indiana, in the Department of Food and Nutrition from 1978 to 1982. In 1982, he joined the faculty at the University of Arizona, Tucson, in the Department of Family and Community Medicine. He is also a Research Professor in the University of Arizona's newly formed College of Public Health. He has published 450 research papers and review chapters.

Dr. Watson is a member of several national and international nutrition, immunology, cancer, and research societies. Dr. Watson and Dr. Mark Witten are funded by the Arizona Disease Control Research Commission to assess the role of vitamin E on environmental tobacco smoke's modulation of lung function. Dr. Watson has just set up the first murine model to measure the actions of environmental tobacco smoke on heart function and during mouse AIDS.

Mark L. Witten, Ph.D. is currently a Research Professor and Director of the Joan B. and Donald R. Diamond Lung Inquiry Laboratory in the Department of Pediatrics at the University of Arizona College of Medicine. Prior to this appointment, he was an Instructor in Medicine at Harvard Medical School in Boston, Massachusetts.

Dr. Witten received his Ph.D. degree with a double major in Physiology and Exercise Physiology at Indiana University in 1983. He has authored over 80 peer-reviewed manuscripts and book chapters. In 1992, he published the first animal model of sidestream cigarette smoke exposure.

Contributors

Ann Aschengrau
School of Public Health
Boston University
Boston, MA

Edward G. Barrett
LRRI
Albuquerque, NM

David E. Bice
LRRI
Albuquerque, NM

Paul D. Blanc
Division of Occupational and
  Environmental Medicine
Department of Medicine and
  Cardiovascular Research Institute
University of California
San Francisco, CA

Carlos Blanco
Allergy Unit
Hospital NTRA SRA del Pino
Canary Islands, Spain

Michael Brauer
Occupational Hygiene Program
School of Occupational and
  Environmental Hygiene
University of British Columbia
Vancouver, BC, Canada

D. Jeff Burton
IVE, Inc.
Bountiful, UT

Teresa Carrillo
Allergy Unit
Hospital NTRA SRA del Pino
Canary Islands, Spain

Anna Maria Castellazzi
Laboratorio di Immunologia
Clinica Pediatrica dell'Universita
  di Pavia
IRCCS Policlinico San Matteo
Pavia, Italy

Rodolfo Castillo
Allergy Unit
Hospital NTRA SRA del Pino
Canary Islands, Spain

Daphne Chan
Division of Clinical Pharmacology
  and Toxicology
The Hospital for Sick Children
Toronto, Ontario, Canada

Yinhong Chen
College of Public Health
  and School of Medicine
University of Arizona
Tucson, AZ

Helen Dimich-Ward
Department of Medicine
  — Respiratory Division
University of British Columbia
Vancouver, BC, Canada

Mark D. Eisner
Division of Occupational and
  Environmental Medicine
Department of Medicine and
  Cardiovascular Research Institute
University of California
San Francisco, CA

Samuel S. Gidding
Nemours Cardiac Center
Alfred I. duPont Institute
Wilmington, DE

Melbourne F. Hovell
Center for Behavioral Epidemiology
  and Community Health
Graduate School of Public Health
San Diego State University
San Diego, CA

Julia Klein
Division of Clinical Pharmacology
  and Toxicology
The Hospital for Sick Children
Toronto, Ontario, Canada

Elizabeth A. Klonoff
Department of Psychology
San Diego State University
San Diego, CA

Gideon Koren
Division of Clinical Pharmacology
  and Toxicology
The Hospital for Sick Children
Toronto, Ontario, Canada

Hope Landrine
Department of Psychology
San Diego State University
San Diego, CA

Timothy Lash
School of Public Health
Boston University
Boston, MA

Peter N. Lee
Independent Consultant in Statistics
  and Epidemiology
Hamilton House
Sutton, Surrey, U.K.

Georg E. Matt
Center for Behavioral Epidemiology
  and Community Health
Graduate School of Public Health
  and Department of Psychology
San Diego State University
San Diego, CA

Murray D. Meek
St. Michael's Hospital
University of Toronto
Toronto, Ontario, Canada

Alfredo Morabia
Division of Clinical Epidemiology
University Hospitals
Geneva, Switzerland

Nancy Ortega
Allergy Unit
Hospital NTRA SRA del Pino
Canary Islands, Spain

Julian Rodriquez
College of Public Health
  and School of Medicine
University of Arizona
Tucson, AZ

Walter K. Schlage
INBIFO Institut für biologische
  Forschung GmbH
Köln, Germany

Ashok Teredesai
INBIFO Institut für biologische
  Forschung GmbH
Köln, Germany

Dennis R. Wahlgren
Center for Behavioral Epidemiology
  and Community Health
Graduate School of Public Health
San Diego State University
San Diego, CA

Shengjun Wang
Lung Injury Laboratory
Department of Pediatrics
School of Medicine
University of Arizona
Tucson, AZ

Ronald R. Watson
College of Public Health
  and School of Medicine
University of Arizona
Tucson, AZ

Gayle C. Windham
Reproductive Epidemiology Section
Division of Environmental and
  Occupational Disease Control
Department of Health Services
Oakland, CA

Hanspeter Witschi
ITEH and Department of Molecular
  Bioscience
School of Veterinary Medicine
University of California
Davis, CA

Mark L. Witten
Lung Injury Laboratory
Department of Pediatrics
School of Medicine
University of Arizona
Tucson, AZ

Joy M. Zakarian
Center for Behavioral Epidemiology
  and Community Health
Graduate School of Public Health
San Diego State University
San Diego, CA

Jin Zhang
College of Public Health
  and School of Medicine
University of Arizona
Tucson, AZ

Table of Contents

1 Difficulties in Determining Health Effects Related to Environmental Tobacco Smoke Exposure

Peter N. Lee

CONTENTS

0-8493-0311-7/00/$0.00+$.50
© 2001 by CRC Press LLC

1

RODUCTION

The possibility that exposure to environmental tobacco smoke (ETS) might cause the nonsmoker serious health effects first attracted major attention in 1981 when Hirayama, in Japan,[1] and Trichopoulos, in Greece,[2] both reported an increased risk of lung cancer related to marriage to a smoker. Since then, numerous individual studies and reviews of the evidence have been conducted. Earlier reviews, in 1986, by the U.S. Surgeon General[3] and the U.S. National Research Council[4] concluded that ETS exposure was causally related to lung cancer and was associated with an increased incidence of respiratory symptoms and infections in children and referred to other possible health effects where more data were needed. The claims against ETS widened in 1993, when the U.S. EPA[5] stated that the association with respiratory symptoms and infections in children was causal, as was that with middle ear disease, exacerbation of asthma and reduced lung function in children, and with respiratory symptoms and reduced lung function in adults. By 1997, the California EPA[6] extended the list of claimed effects to include cardiovascular disease, nasal sinus cancer, asthma induction in children, Sudden Infant Death Syndrome (SIDS), and reduced birthweight. Similar conclusions were reached in the U.K. in 1998 by the Scientific Committee on Tobacco and Health (SCOTH),[7] based on a series of papers reviewing the evidence on lung cancer,[8] ischaemic heart disease,[9] SIDS,[10] and middle ear disease[11] and on respiratory illness, asthma, allergy, and bronchial reactivity in children.[12–16]

Although the above reflects what might be regarded as the "conventional medical view" on ETS as expressed by authoritative committees, it is important to realize that over the same 20-year period many researchers have expressed doubts about such an interpretation of the evidence (e.g., References 17 to 30). Underlying this difference of opinion are the particular difficulties involved in determining health effects related to ETS. The purpose of this chapter is to discuss these difficulties, first in relation to the interpretation of results from a single study and second in relation to the combined evidence from multiple studies, and then briefly to review the validity of the conventional medical view in light of whether these difficulties have adequately been taken into account.

1.2 PROBLEMS WITH INDIVIDUAL STUDIES

1.2.1 LACK OF POWER

When designing a study, one should consider the magnitude of effect one might plausibly expect, and then ensure the study is large enough to actually detect such an effect as statistically significant. When studying ETS, the likely extent of exposure is so low that only very modest effects on health, if any, are to be expected, so that extremely large studies are indicated. In terms of particulate matter or nicotine, average ETS exposure is probably less than 0.5% of that of a typical smoker of 20 cigarettes a day,[31] suggesting that if lung cancer risk is linearly related to exposure, then the relative risk associated with marriage to a smoker is less than 1.1, perhaps as low as 1.01 or 1.02.[32] It would take a very large study indeed to detect such a small relative risk (RR). In this context, it is interesting to note that the International Agency of Research on Cancer (IARC) took 10 years to design and implement a multicenter case-control study aimed at having an 80% power to detect an RR of 1.3 with 95% confidence. In the event, despite accumulating data on 650 lung cancer cases in nonsmokers — the largest number in almost any such study — the estimated RR for spousal ETS exposure was 1.16, with 95% confidence limits as wide as 0.93 to 1.44.[33] In other words, although at first sight large, the study did not allow one to distinguish between the various key possibilities — no association (RR = 1.0), the very weak association suggested by dose extrapolation (RR < 1.1), the somewhat larger association suggested by recent meta-analyses (RR $\simeq$ 1.2),[8] and the even larger association they hoped to detect (RR = 1.3). Bearing in mind that this is almost the largest study of ETS and lung cancer, it follows that the lack of power is a general problem in interpreting results of individual studies, especially when one realizes that the majority of studies involve under 100 lung cancer cases. The same limitation applies to much of the evidence for other health effects.

1.2.2 Susceptibility to Bias

In an ideal world, effects of ETS exposure would be studied by a double blind, randomized, intervention trial in which exposure and disease incidence were precisely known. In practice, such studies are not feasible, and one must make do with non-randomized prospective and case-control epidemiological studies, which are open to various possibilities of bias. To infer causality from an association observed between ETS and a health effect, one must be able to rule out the possibility that the association did not in fact arise as an artefact from one or more of these biases. Such biases may arise from errors in determining the health effect, errors in determining ETS exposure, confounding by other relevant variables, or by erroneously including smokers in what is intended to be a study of nonsmokers ("smoker misclassification bias"). These sources of bias are discussed below.

1.2.2.1 Diagnostic Inaccuracy

Evidence from numerous studies[34] shows that 15% or more of positive clinical diagnoses of lung cancer are false, with no lung cancer being seen at autopsy, and that there are an even larger number of false negative diagnoses, where lung cancers seen at autopsy were not discovered in-life. Inasmuch as classification of patients as having lung cancer in epidemiological studies is very rarely based on autopsy evidence, diagnostic inaccuracy will certainly exist and is likely to be more common in those studies that do not insist on histological confirmation and accept diagnosis based on X-ray or cytology. It is difficult to assess the extent to which diagnostic inaccuracy might bias the reported association of ETS with lung cancer. Random misclassification with diseases unassociated with ETS would lead to underestimation of any true relationship, but it is not clear whether misclassification is random (misdiagnosis might be correlated with ETS exposure itself or with factors associated with ETS) or whether ETS is unassociated with all diseases that might be confused with lung cancer (e.g., other cancers or other lung diseases). The extent and direction of bias arising from diagnostic inaccuracy have been studied little in the context of either lung cancer or other diseases associated with ETS.

Another related issue is the use of endpoints which are surrogates for the disease in question. One example is a study relating smoking by the spouse to the presence of "epithelial, possibly precancerous lesions" (EPPL) in the lungs of nonsmokers sampled at autopsy.[35,36] Higher EPPL scores were seen where the spouse smoked, but this excess seems uninterpretable for a number of reasons. Notably, EPPL scores declined with age and were no higher in current heavy smokers than in nonsmokers, in direct contradiction to what one would expect from a valid marker of lung cancer risk.[37]

The use of surrogate endpoints is particularly relevant when considering the evidence linking ETS exposure to heart disease. Glantz and Parmley,[38] in a review paper, cited results from experimental animal studies and from experimental and clinical studies in humans and put forward a wide range of mechanisms as support for their view that ETS exposure was an important contributor to the risk of heart disease death in nonsmokers. Lee and Roe[25] reviewed these results in detail and, apart from pointing out other limitations in many of the studies, cast grave doubts on the relevance of many of the endpoints used. They noted that there is no obviously appropriate animal model for coronary artery atherosclerosis and criticized investigators for overreadiness to reach conclusions about the effects of long-term ETS exposure on this endpoint from ultra short-term studies in humans or laboratory animals. They also expressed reservations about the relevance of effects of carbon monoxide exposure on exercise tolerance and cardiac muscle performance to the possible development of progressive heart disease following ETS exposure.

1.2.2.2 Errors in Determining ETS Exposure

There are enormous problems in accurately quantifying the extent of ETS exposure, especially over a lifetime. In an ideal world, one would carry out a prospective study in which ETS exposure is measured at regular intervals by means of air measurements and/or uptake of relevant smoke constituents in biological fluids. In practice, with one minor exception,[39] studies of lung cancer and heart disease have assessed ETS exposure based on questionnaire responses relating to a limited number of potential major sources of ETS exposure. Smoking by the spouse has been the most commonly used index of exposure, but a number of studies have attempted to determine whether (and, in some cases, how much) the subjects are exposed from other household members, in the workplace, socially, in travel, and/or in childhood. While studies have shown that cotinine levels in nonsmokers are increased if the spouse smokes and, to a lesser extent on average, in relation to workplace ETS exposure,[40] it does not necessarily follow that having a spouse or workplace colleague who smokes always implies an increased average lifetime total ETS exposure. Also, the extent of ETS exposure associated with whether or not a specific other individual smokes depends on various relevant factors, including room size, frequency of contact with the individual, and the number of years the subject has known the individual. Studies of ETS should, strictly, take such considerations into account, but rarely do so in practice.

In case-control studies, reliance on data reported by the subject leads to the possibility of recall bias, with answers to questions depending on awareness (or presence) of disease. That recall bias might be important is indicated by a study conducted in Scotland[41] in which the presence of respiratory symptoms in nonsmokers was found to be strongly associated with self-reported ETS exposure (classified as "none at all," "a little," "some," or "a lot"), but not with serum cotinine level. The authors suggested that the presence of respiratory disease might bias self-reported exposure, and it is certainly plausible that the presence of diseases such as lung cancer or coronary heart disease might lead to the subjects being more likely to recall occasions when they were exposed to ETS and to overestimate exposure relative to healthy controls.

Another factor affecting the validity of data collected on ETS exposure is the use of proxy respondents, such as next-of-kin, when the lung cancer or heart disease case is already dead or too ill to be interviewed. Proxy respondents were used in about one third of the 50 or so studies of ETS and lung cancer, and in some of these, including three of the largest U.S. case-control studies,[42-44] proxy responses were used for a relatively large proportion of cases, but not at all for controls, creating an obviously unsatisfactory imbalance with a propensity for bias.

It might seem to be a definite advantage to use an objective marker of ETS exposure, as it avoids the problems of recall bias, and indeed cotinine has been determined in urine, serum, or saliva in quite a number of studies of diseases in infants or children and of birthweight.[40] However, some problems with the use of cotinine should be noted. First, having a half-life of only about a day or so, it is only a marker of recent ETS exposure and may be of limited value in studies of a chronic disease, especially where the subject is a hospitalized patient. Second, cotinine is only one metabolite of nicotine, and the speed and extent to which cotinine is metabolized from nicotine may vary for a number of reasons, including genetic and environmental factors. Third, cotinine determination is subject to analytical error, particularly at the low levels resulting from ETS exposure.[40]

In studies of infants and children, care should be taken not to assume that associations of adverse health effects with smoking by the parents necessarily indicate ETS exposure. There is a strong correlation of maternal smoking after pregnancy with maternal smoking during pregnancy, and unless the study is sufficiently large and collects detailed information on smoking at the different time points, it may be impossible to distinguish between potential effects of maternal smoking on the fetus and of ETS exposure from the mother's smoke during early life. Similarly, associations with paternal smoking may, in theory, result from effects on sperm as well as from direct effects of ETS exposure.

Effects on birth outcome of maternal smoking in pregnancy are not normally considered as effects of ETS exposure, inasmuch as ETS exposure is defined to involve the **inhalation** of tobacco smoke other than by puffing on a cigarette, cigar, or pipe.[45] However, effects on birth outcome of the exposure of nonsmoking mothers during pregnancy to the tobacco smoke of others are usually included under the ETS heading, being sometimes referred to as "tertiary smoking."

1.2.2.3 Confounding

Because epidemiological studies do not involve random allocation of subjects to ETS exposure, the possibility always arises that associations observed between ETS exposure and disease may partly or wholly arise from the confounding effects of other variables which are themselves correlated both with ETS exposure and with risk of the disease in question. Studies of ETS vary tremendously in the extent to which potential confounding variables are taken into account, many paying little or no attention to the problem. In the following paragraphs, some of the more important potential confounders are discussed.

One source of potential confounding, which is often ignored completely, arises from the failure to restrict analysis to subjects satisfying the precondition for potential exposure. Where exposure is defined as marriage to a smoker, marriage is the precondition for potential exposure, and the risk in the exposed group should properly be compared with that in those who are married to a nonsmoker. However, unmarried subjects are not excluded in many studies, so one ends up comparing an exposed group, all of whom are married, with a comparison group, many of whom are not married. This leads to an obvious potential for confounding by marital status or the numerous variables with which it is correlated. The same problem arises with regard to workplace exposure, where being employed is the precondition and the unemployed should be excluded from analysis.

In view of the strong relationship of age to risk of many diseases, it is of obvious importance to adjust for age in analysis, and in practice the majority of studies do so. However, a number of studies, typically case-control studies, in which cases and controls were matched on age at the design stage, do not. In some of these studies (about one in four of all the studies of ETS and lung cancer), the matching on age did not apply to the nonsmoking subjects analyzed for potential ETS effects, but to the total cases and controls including smokers. Just because the total cases and controls are age matched, it does not follow that the nonsmoking cases and controls also are, and it can easily be demonstrated that moderate bias can result from failure to adjust for age in these circumstances.[32]

Recent reviews[46,47] have clearly shown that increased fruit and vegetable consumption is associated with a reduced risk of lung cancer and less clearly shown that diets high in fat, saturated fat, and cholesterol may independently increase the risk of lung cancer. While most of these data relate to smokers, there are also a number of studies suggesting similar relationships apply in nonsmokers.[48–56] Another recent review[57] has clearly demonstrated that smokers have unhealthier patterns of nut.ient intake than nonsmokers, and various studies[58–63] have also shown similar, but generally smaller, differences in relation to ETS exposure, with marriage to a smoker typically associated with reduced fruit and vegetable consumption and increased saturated fat intake. This implies that failure to take diet into account in studies of ETS and lung cancer will cause confounding.

It is notable that, in their recent review of the evidence on lung cancer and ETS, Hackshaw et al.[8] consider fruit and vegetable consumption as the **only** potential source of confounding, estimating a bias of 1.02 which they consider to be negligible in contrast to their overall RR estimate for smoking by the husband of 1.24. Based on our own analysis of more data sets than considered by Hackshaw et al., we estimate (details available on request) a somewhat larger bias of 1.04 from confounding due to fruit and vegetable consumption and a bias of 1.02 from confounding due to dietary fat consumption.

Diet is not, of course, the only potential confounding variable associated with ETS exposure. ETS exposure has been associated with a greater likelihood of working in jobs with an increased risk of lung cancer and with lower social class, education, and income.[51,58,63–67] Again, adjustment for these factors was rarely applied in the ETS and lung cancer studies.

In view of the wide range of potential risk factors, adjustment for confounding might be expected to be even more important for heart disease than it is for lung cancer. This is to some extent reflected by the fact that the proportion of studies taking into account a wide range of variables in analysis is much greater in the heart disease studies[25] than in the lung cancer studies.[32] However, even for heart disease, many of the studies paid only limited attention to confounding variables, with about half ignoring the classical coronary risk factors of blood pressure, cholesterol, and body mass index and very few taking into account such factors as exercise, race, diet, alcohol, diabetes, and family history of heart disease or hypertension. Estimating the bias, due to uncontrolled confounding from the data published in those studies that had reported results of adjusted analyses, is not easy, inasmuch as researchers typically only present two RRs, one totally unadjusted and the other adjusted for age and a variety of other variables. However, when an attempt was made to study the available information,[25] it was found that many of the effects of adjustment were strikingly large, increasing or decreasing the RR estimate by 20% or more, with adjustment for classical risk factors tending to decrease the RR estimate (though in some cases increasing it). Given that the effect of adjustment is often as large as or larger than the magnitude of the claimed association of heart disease risk with ETS exposure, it is surprising that researchers do not present fuller details making clear the extent of adjustment resulting from each individual factor considered.

The possibility of confounding is also a potential problem with regard to the evidence relating ETS exposure to effects in infants and children. In this context, it is of interest to consider the evidence reviewed by Strachan and Cook[12] on lower respiratory illness in infancy and early childhood. Here, unadjusted and adjusted odds ratios were available for only 15 of the 40 outcome/study combinations considered, with adjustment decreasing the risk estimate in 11 cases and increasing it in 4, the magnitude of adjustment usually only being quite small. However, although a total of almost 20 factors were considered in at least one of the studies, the number of risk factors actually taken into account in any particular study was typically quite small, averaging about four and never exceeding seven. Some factors, including presence of pets, damp/cold housing, or any index of parental neglect or of nutrition, were never considered at all. Another factor that was not considered was solid feeding; a later paper[68] reported that a significant association between parental smoking and respiratory illness disappeared on adjustment for solid feeding. Also, in view of the statement in the 1986 U.S. National Research Council Report[4] that the mechanism of the increased risk of respiratory symptoms in the children of smoking parents may "either be a direct effect of ETS or due to a higher risk of

cross-infection in such homes," it is remarkable that so few of the studies have
collected data on the presence of infections or symptoms in parents, siblings, and
other household members.

That adjustment for potential confounding variables can have a dramatic effect
is illustrated clearly by the data for SIDS. Recent reviews of the evidence[10,28] clearly
show a very strong crude association, with maternal prenatal and postnatal smoking
typically associated with about a threefold increased risk of SIDS. The great majority
of the studies present RR estimates that are unadjusted or adjusted for quite a short
list of potential confounding factors. However, it was noticeable that those few
studies that adjusted for a long list found that adjustment reduced the RR dramatically. Thus, for example, a study in New Zealand[69,70] reported RRs that dropped
after adjustment from 4.09 to 1.65 for maternal prenatal smoking and from 4.24 to
1.79 for maternal postnatal smoking, while a study in the U.K.[71] reported a drop
from 4.84 to 1.78 for maternal prenatal smoking. The authors of these papers seem
to regard the adjusted RRs, which are statistically significant, as being indicative of
a causal relationship. However, this interpretation is highly questionable since part,
if not all, of the excess RR remaining after adjustment may be due to **residual
confounding**, resulting from the long list of confounders considered not being long
enough and from errors in determining the confounding factors that have been
considered.[28] Golding[72] takes an opposite view, arguing that the true relationship
may be underestimated in studies that control for numerous factors associated with
SIDS, if some of the factors adjusted for are markers of the consequences of parental
smoking that is on the pathway to SIDS. However, as discussed elsewhere in detail,[28]
there seems little basis for this view.

1.2.2.4 Misclassification of Smoking Status

Virtually all the epidemiological evidence on ETS relates to lifelong never smokers.
This is partly because possible effects of ETS exposure on smokers are viewed as
of relatively little concern, but mainly because it is likely to be extremely difficult
to detect reliably any effect of ETS on a smoking-associated disease in the presence
of a history of smoking. This difficulty arises as the total extent of a smoker's
exposure to smoke constituents will be largely determined by his own smoking habits
and less by his much smaller exposure from ETS. Since smoking and ETS exposure
are correlated, any errors in the assessment of active smoking history are likely to
cause a residual confounding effect substantially larger than any plausible possible
effect of ETS.

In studies relating smoking by the spouse to risk of lung cancer in never smokers,
bias may arise if in fact a proportion of the subjects are current or ex-smokers. While
random errors in ETS exposure or in diagnosis of lung cancer will tend to reduce
any true positive association, it is important to note that random errors in misclas-
sifying ever smokers as never smokers will tend to increase it or to create a spurious
positive association if marriage to a smoker is actually unrelated to the risk of lung
cancer. This positive bias (smoking misclassification bias) arises because smokers
tend to marry smokers, so that misclassified smokers are likely to be more frequent

among those with a spouse who smokes. Elsewhere,[73] Lee and Forey discuss the mathematics of this bias in detail. They show that the size of the bias depends on the misclassification rate, the excess risk associated with active smoking, the degree of concordance between husband and wife, the proportion of true smokers in the population, and the proportion of spouses who smoke. It is little affected by misclassification of nonsmokers as smokers, or of smoking by spouses, which can be ignored for practical purposes. In Reference 73, the authors present a method for correcting individual study RR estimates for misclassification bias and show that other published methods[4,5,74] have errors.

Lee and Forey[75] also summarize evidence from 42 studies relating to denial of current smoking (based on high cotinine levels in urine, serum, or saliva in self-reported nonsmokers) or denial of past smoking (based on inconsistencies between self-reports made at different times). The estimated extent of misclassification of current or past smoking was found to vary markedly from study to study. Differences in the circumstances in which questions are asked are a major reason for this. A smoker's readiness to admit his habit may depend on a variety of factors, including the social acceptability of smoking (which itself varies by country and time period), whether the person asking the questions disapproves of or has advised against smoking (e.g., a doctor), whether the subject considers smoking in the past or at a low level is of interest to the interviewer, and whether the subject can remember past smoking. Variation in the cotinine cutoffs used to detect unreported smoking also contributes to the variation in estimated misclassification rates. Lee and Forey[75] clearly showed that occasional smokers deny current smoking more often than regular smokers and that current smokers are more frequently detected among ex-smokers than among never smokers. Although misclassification rates vary widely by study, they estimated that for the U.S. population it was reasonable to carry out bias correction assuming that the equivalent of 2.5% of average risk ever smokers report being never smokers, though values in the range of 1 to 4% are plausible. Using these estimates, and appropriate published estimates of husband and wife smoking concordance,[21] they concluded that misclassification bias is an important determinant of the apparent slight excess lung cancer risk observed in nonsmokers married to smokers. Indeed, their central estimates of a 2.5% misclassification rate and a concordance ratio of 3.0 reduced an overall RR estimate of 1.13 based on data for 13 U.S. studies of spousal smoking in females to 1.01.

For Asian studies, which form a substantial proportion of the evidence on ETS and lung cancer, the proportion of women who report smoking is much lower than in Western populations, and the estimated misclassification bias is much lower using the misclassification rate of 2.5%. However, there is evidence from three studies of Asian women[76] indicating much higher misclassification rates than in Western populations, and that misclassification bias is an important issue there, too. In their review, Hackshaw et al.[8] estimate that misclassification bias would reduce an overall RR estimate for lung cancer and spousal smoking from 1.24 to 1.18. However, this is an underestimate of the bias because they use misclassification rate estimates that are somewhat lower than are appropriate for U.S. and European populations and because they totally ignore the evidence of very high misclassification rates in

Asia. Surprisingly, Hackshaw et al.[8] reject such evidence out-of-hand,[77] although there has long been speculation that, as smoking is regarded as socially unacceptable by Japanese and other Asian women, misclassification rates are likely to be very high there.

Misclassification of smoking habits causes bias when studying the relationship of spousal smoking to lung cancer because of the concordance of smoking habits between spouses. Given that there is also likely to be some concordance of smoking habits between workplace colleagues, misclassification bias may also be relevant when studying workplace ETS exposure. However, this author is not aware that anyone has so far attempted formal bias correction here.

Misclassification bias correction has also not been attempted when studying the relationship of heart disease to ETS exposure. This is presumably because the excess risk associated with active smoking is much lower for heart disease than it is for lung cancer, so that, all other things being equal, the bias would consequently be expected to be much lower. However, other things may not be equal. Misclassification rates are particularly high in patients advised by their doctors to give up smoking following a diagnosis of heart disease,[78] so subjects claiming to be nonsmokers may include some smokers particularly at risk of subsequent death from heart disease. Support for this comes from a prospective study in Denmark[79] in which serum was taken for cotinine determination at baseline. Cumulative heart disease incidence in self-reported nonsmokers with cotinine levels inconsistent with nonsmoking was 17.9% (based on 5 cases), an incidence substantially **higher** than that in self-reported current smokers (4.3% based on 72 cases). This contrasts with the usual calculations for lung cancer which assume misclassified smokers have a risk **lower** than that of average ever smokers (taking into account evidence showing they smoke less and, if ex-smokers, tended to give up smoking a long time ago). Clearly, the possibility that misclassification bias is important in studies of heart disease and ETS cannot be rejected.

1.2.3   INADEQUATE CONTROLS

When estimating RRs (via odds ratios) from case-control studies, it is assumed that the controls are representative, as regards ETS exposure, of the population from which the cases were derived. If this is not so, bias may occur. Nonrepresentativeness may arise from using hospital controls suffering from a disease associated with ETS exposure, leading to underestimation of the true RR. It may also arise from using population controls derived from an area not equivalent to the catchment area of the hospital(s) from where the cases came or taken over a time period that is not the same as that for the cases. Bias here may be in either direction.

1.2.4   CASE-ONLY STUDIES

A recent technique, used, for example, when studying genetic effects, is the case-only study, which looks for possible interactions between mutations in a gene and environmental factors. A recent U.S. lung cancer case-only study[80] reported that, after adjusting for age, radon exposure, saturated fat intake, and vegetable intake, there was a significant tendency for glutathione transferases class mu (GSTM1) mutations (polymorphisms) to be more common in nonsmokers if they were exposed to ETS in the household (odds ratio 2.6, 95% CI 1.1 to 6.1) and argued that their result "suggests that the observed excess lung cancer risk among never-smoking women results from cancers in two distinct groups: one that is genetically at high risk and one that is genetically at lower risk of lung cancer from exposure to ETS." There are a number of reasons why their suggestion may be invalid, including the marginal nature of the statistical significance and the doubtful representativeness of their sample, with useful tissue samples only being available for 106 out of the original 618 lung cancer cases among lifelong nonsmoking women.

The most important reservation, however, lies in the problems of interpreting results from a case-only study, such as this. Such a study **cannot** determine the RR of lung cancer due to household ETS exposure or GSTM1 mutations as there are no controls. It can only determine whether there is an interaction between the two relationships, and then only assuming that household ETS exposure and GSTM1 mutations are unrelated in the population at large. This assumption seems open to question. Since genetic susceptibility markers are not randomly distributed in the population — for example, they will certainly vary by ethnic group — an association with ETS in the population at large is far from impossible.

In any case, even if one accepts this assumption, the results do not imply that the population divides into two groups, one with high lung cancer risk from household exposure and one with low risk. Taking into account that the original case-control study from which these samples were derived[42] found only a very weak association of household ETS exposure with lung cancer (RR = 1.1), the results of the later study would imply that in the absence of GSTM1 mutations, household ETS exposure would reduce the risk of lung cancer, not increase it.

1.2.5 ERRORS IN STATISTICAL ANALYSIS

Clearly, in any study incorrect statistical analysis may result in overestimation or underestimation of the true relationship of ETS exposure to the disease in question, and it would not be practical here to consider all the possible ways in which errors may arise. However, some relevant problems will be referred to.

One point to make is that there are quite a number of examples in the ETS literature where gross errors in published data have appeared (and have been reproduced by other authors). One example is a study in China[81] that reported RRs (95% CIs) for smoking by the husband of 1.40 (1.12 to 1.76) for 1 to 9 cigarettes per day, 1.97 (1.42 to 2.72) for 10 to 19 cigarettes per day, and 2.76 (1.85 to 4.10) for 20+ cigarettes per day based on a study involving a total of 54 lung cancer cases and 93 controls; these results were reproduced in a review[8] as an example of a dose-response relationship. In fact, the widths of the CIs are vastly too narrow for such a small study, implying a minimum sample size far greater than was actually known to exist. This is demonstrated in a recent paper[82] which provides a number of simple methods for checking for possible errors in reported odds ratios, RRs, and CIs and cites other examples of errors in the ETS literature.

Another problem arises from the uses of multiple endpoints, for example, in studies of respiratory symptoms or lung function. Care should be taken not to overinterpret the existence of one or two marginally significant associations. The problem of multiple testing is heightened when there are also multiple exposure indices, further increasing the chance of a false positive relationship being seen among the legions of statistical analyses conducted. In theory, the proper use of p-values applies only to hypotheses defined in advance, not those suggested from such exploratory analyses.

A related problem results from looking for associations in subsets of the data, commonly known as "data-dredging." Again, the chance of a false positive relationship will be increased. It should be noted that when one does check for variation in the ETS/disease association in subsets of the population (e.g., by age, sex, race, or occupation), it is appropriate to carry out formal tests of interaction. Just because a significant association is seen in one age group and not in others does **not** imply that an interaction exists. The data may in fact be more simply, and often better, explained by a model in which the association is invariant over age, happening by chance to be more evident in a specific age group. Failure to carry out tests of interaction is a common weakness in epidemiological analyses, not only when investigating subsets, but also when carrying out multiple regression analyses. If one does not do so one is assuming, but never confirming, that the effects of the various risk factors studied are independent.

While it is valuable to investigate the dose-response relationship, a problem with many studies is that they limit attention to a simple test for trend. It is important to remember that such a test simply investigates whether a linear dose-response relationship with a positive slope fits the data significantly better than does a horizontal line. It does not test whether a linear dose response actually fits the data at all, nor whether there is any tendency at all for risk to rise with increasing dose among the ETS exposed subjects. One might, for example, find a significant positive trend for a study with RRs of 1.0, 2.2, 2.0, and 1.8 for ETS doses of 0, 1, 2, and 3 units, despite the fact that response actually declines over the positive dose range and the likelihood that a linear relationship may be totally inappropriate. More detailed modelling is often necessary, but is rarely conducted.

It should also be realized that biases described previously, such as confounding, smoker misclassification bias, and recall bias, affect not only simple comparisons between ETS exposed and ETS unexposed groups, but also analyses relating risk to dose of ETS. For example, since the degree of concordance between the smoking habits of spouses increases with the amount smoked, the magnitude of smoker misclassification bias will increase with the amount smoked, thus creating a spurious dose-response relationship.[32]

1.3   PROBLEMS IN ASSESSING THE OVERALL EVIDENCE

Some of the problems described in Section 1.2 in relation to individual studies will also apply when attempting to assess the overall evidence. However, there are also a number of additional problems.

1.3.1 PLAUSIBILITY

Some would argue[5,8,74] that ETS must cause lung cancer because of the known relationship of active smoking with lung cancer, because ETS is effectively diluted mainstream smoke, and because no zero threshold exists. In considering this argument, it should be noted that while the evidence for active smoking indicates an increased risk in relation to the lowest levels studied, these are typically about five to ten cigarettes a day, and it is actually unknown whether one cigarette a day carries an excess risk, let alone whether the lower dose level implied by the relatively small uptake of tar and nicotine from ETS exposure does so. It should also be noted that although ETS and mainstream smoke have many chemicals in common, there are important differences in the chemical and physical composition of the two types of smoke[83,84] and that concentrations of chemicals in ETS are typically much lower than permissible exposure limits approved by regulators.[85] Although some scientists believe in a zero threshold for carcinogenesis, this belief is far from universally held. In a survey of toxicologists,[86] 75% disagreed with the statement that "there is no safe level of exposure to cancer-causing agents" (and this percentage would no doubt be much higher for effects other than cancer). Furthermore, the U.S. EPA has recently accepted that low doses of the carcinogen chloroform are without risk to humans.[87] That exposure may not in fact imply risk is strongly suggested by the fact that the huge 1979 U.S. Surgeon General's Report[88] did not even mention the possibility that ETS might cause lung cancer, although it concluded that smoking did and exposure to smoke constituents from ETS was evidently not zero.

Even if one accepts that a risk from ETS is probable — perhaps bearing in mind interindividual variation in ETS exposure and in susceptibility — the actual magnitude of expected risk is impossible to predict. Observed approximate linearity of dose response in the active smoking dose range does not imply low dose linearity, and the true risk could be very small indeed. To assess whether this is a meaningful risk, and how much that risk is, requires assessment of the epidemiological evidence.

Some studies have reported associations of ETS exposure with diseases that are not reported to be associated with active smoking. Such results seem *a priori* highly implausible, partly because of the similarity of ETS and mainstream smoke (though, as noted above, this is not complete), but mainly because it is clear that smokers have substantially more ETS exposure than do nonsmokers, not only from their own smoking, but also from the greater likelihood that they will spend time in smoky environments. Since there is no evidence for the existence of any counterbalancing protective effect from active smoking, it is extremely unlikely that a true effect of ETS exposure could exist in the absence of an observed effect in active smokers.

A question of plausibility also arises if the observed association with ETS exposure is of similar magnitude to that from active smoking. This would seem to imply either that effects arise solely from ETS exposure, the association with active smoking being due to the increased ETS exposure of smokers, or that the dose-response relationship for the effect under consideration is not at all linear, but more like one in which any exposure above a given amount increases risk by a similar amount.

In their review of the evidence relating ETS exposure to heart disease, Law et al.[9] attempt to explain why the estimated RR associated with ETS exposure is as high as 1.30 when that from smoking 20 cigarettes a day is only 1.78. They postulate a mechanism based on platelet aggregation involving a markedly nonlinear dose-response relationship, with risk rising steeply up to 0.2 cigarettes a day (the dose that they believe average ETS exposure is equivalent to) and then rising much less steeply. There are, however, several limitations to the evidence cited.[25] These include a lack of unanimity in the literature regarding the role of enhanced platelet aggregation in heart disease, doubts about the interpretation of the study Law et al. cited[89] as evidence that platelet aggregation increased risk of heart disease, conflicting data on platelet aggregation from acute smoking experiments, and lack of confirmatory animal evidence. It is also notable that Law et al. never actually looked at the ETS dose-response data from the epidemiological studies they reviewed. Had they done so, they would have noted the highly heterogeneous nature of the data, with the largest studies consistently showing no association at all between ETS exposure and heart disease risk and the smaller studies consistently showing a significant trend with the highest exposure level often associated with a risk higher than that seen for active smoking. These data hardly fit in with the explanations Law et al. put forward.

1.3.2 PUBLICATION BIAS

There is an increasing tendency to assess the overall relationship of an exposure to a disease by carrying out meta-analyses of published data. While such analyses are valuable as a guide to the overall magnitude of the association, two important points should be remembered. First, if some or all of the individual studies are subject to the biases described in Section 1.2, then the overall meta-analysis estimates will also be subject to the same biases, unless corrections for bias to the individual study estimates are made before the results are meta-analyzed. Second, even if such bias correction is made, the overall estimate may still be biassed if the literature taken into account is not representative of the total data available. It is well documented[90] that in many situations scientists tend not to submit, and journals tend not to publish, results from studies, particularly small studies which find no effect. Hence, the published literature may give a biassed impression. On the basis that large studies are more likely than small studies to publish their findings anyway regardless of their results, the possibility of publication bias can be investigated by studying whether smaller studies tend to give higher RR estimates than do large studies. There are also a number of additional techniques that can be used, but all depend on assumptions that are difficult or impossible to verify, rendering estimation of the extent of publication bias uncertain.[90]

Although it seems unlikely, but not impossible, that large studies of ETS and lung cancer have been conducted but not reported, there is very clear evidence that publication bias has severely affected the literature on ETS and heart disease. In 1981, results relating spousal smoking to lung cancer were published from the huge American Cancer Society (ACS) Cancer Prevention Study (CPS-I),[91] but although it was apparent that this study would allow similar analyses based on many deaths in nonsmokers, and although this author was told by the ACS in the 1980s that such analyses had been conducted, even now they have not published their findings. Only when, years later, LeVois and Layard obtained the data and published their own analyses[92] did it become apparent in the literature that there was a huge study showing no association whatsoever with ETS exposure.

One would have thought that results from a study involving more heart disease deaths than the rest of the published literature put together would then have been included in meta-analyses, but this has not always been the case. Thus, in their meta-analyses, Law et al.[9] did not include them. The reasons cited included the fact that LeVois and Layard were consultants to the tobacco industry, which they believe rendered their analysis open to suspicion, and their claim that their analyses of data from the later CPS-II study[92] gave different results from those reported by the ACS themselves.[93] In fact, the two sets of analyses really gave quite similar answers. In any case, if Law et al. distrusted LeVois and Layard's analyses, why did they not do their own analysis of the data?

Law et al.[9] also claim that the results of the CPS-I study were an outlier, statistically inconsistent with the results from other studies, but this is not sufficient justification for omitting such important data. As Lee and Roe[25] show elsewhere, there is huge between-study heterogeneity of the RR for heart disease and spousal smoking, with estimates from studies involving over 1000 heart disease deaths (or cases) on average showing little evidence of an association (RR = 1.04, CI = 1.00 to 1.07), studies involving 100 to 999 deaths showing a modest association (RR = 1.33, CI = 1.18 to 1.50), and studies involving <100 deaths showing a very strong association (RR = 1.54, CI = 1.21 to 1.95). Many of the studies showing a very strong association were reported only as abstracts or dissertations and not published in peer-reviewed journals, and some had obvious weaknesses in study design.[25] The same heterogeneity was evident for the dose-response analyses, with significant trends never seen in the large, better quality studies, but frequently seen in the other studies.

In view of the fragmentary nature of the available evidence relating ETS to cancers other than of the lung and to chronic obstructive lung disease, it is also worth noting that no one has ever published results for these diseases based on either CPS-I or CPS-II!

Another form of publication bias applies when reviewers unfairly select evidence to present or highlight. One example of this is seen in recent papers by the IARC summarizing the evidence on ETS and lung cancer. In earlier reviews,[94] they presented the data in a common form in a table, with RRs and CIs being given for each study in relation to spousal smoking. Inasmuch as there was, at that time, little evidence in relation to other ETS exposure endpoints, this was an unobjectionable procedure. However, later, when data on a variety of ETS exposure indices became available from a number of studies, it was expected that IARC reviewers would have produced tables summarizing evidence for each index separately. Instead, they[95,96] presented one (or sometimes two) RR for each study, with no consistency whatsoever about which index of ETS exposure results were presented for. It can be demonstrated[24] that there was a clear tendency to distort the evidence by presenting in the tables only results for those indices where the RR happened to be high. For example, in a Japanese study,[97] RRs were presented in relation to eight sources of ETS exposure. Six were in the range 0.8 to 1.2 and were not mentioned in the IARC reviews, with reference only made to two (4.00 for mother smoking, 3.20 for husband's father smoking) that gave high RR values.

1.3.3 CONSISTENCY OF FINDINGS

When conducting meta-analyses, a common approach is to use the "fixed-effects" model, with RR estimates from the individual studies weighted by the inverse of their variance, thus making large studies contribute more than small studies to the overall RR estimate. However, this method implicitly assumes that a common RR applies to all studies, and it is often the situation that there will be evidence of statistically significant heterogeneity between studies. If this is the situation, some researchers use the "random-effects" model to provide an overall estimate. This method assumes that the true RR for each study varies according to a log normal distribution. The weight attached to each study is now derived from the between-study variation as well as from the within-study variation and results in an overall estimate which gives more relative weight to small studies than does the fixed-effects model.

There are a number of objections to the random-effects model. One is that there is no good reason why the study means should be log normally distributed. More seriously, the approach fails to try to explain the heterogeneity. If the truth is that in one situation the true RR is 1.2 and in another it is 1.8, then that is how the data should be presented. There is little point in presenting an average estimate of 1.4, 1.5, or 1.6 which may not apply in any situation.

Investigating potential sources of heterogeneity should be an important part of any meta-analysis, but it is frequently ignored. As already noted, the evidence on ETS and heart disease shows a strong relationship of study size and study quality to the magnitude of the observed RR. Detailed analysis of data on lung cancer and the husband's smoking also shows a number of sources of between-study variation.[23] For example, RR estimates for 12 studies where no age adjustment or proper matching had been carried out averaged 1.53 (CI = 1.30 to 1.81) against only 1.10 (CI = 1.02 to 1.18) for studies where it had been. Though all these associations may not represent independent relationships, there were also significant variations in RR by study location, period, size, quality, and type, by histological type, by whether histological confirmation was required, by whether confounding variables had been taken into account, and by whether dose-response relationships had been studied. For some of these factors, the differences in RRs seen were larger than the overall RR associated with the husband's smoking.

1.4 INTERPRETATION OF THE EVIDENCE

It is not possible, in the space available, to discuss fully all the data and arguments involved when attempting to assess the evidence between ETS and the various diseases it is claimed to cause. Rather, conclusions are summarized briefly for some of the main endpoints (those for which this author has scrutinized the evidence in detail), with reference sometimes made to other papers which enlarge on the reasons for these views.

1.4.1 LUNG CANCER

As is shown elsewhere,[23,24,32,76] the strength of the evidence linking passive smoking to lung cancer has clearly been overstated by the conventional medical view. The aggregated data on spousal smoking suggest that the small risk of lung cancer in nonsmokers is almost 20% higher in those who are married to a smoker than in those who are not. However, this excess could wholly or in substantial part be due to the combined effects of bias, most importantly that resulting from smoker mis-classification bias, from confounding by diet and other variables, and from other sources of bias. The precise extent of all the biases remains uncertain, but it seems impossible to conclude with any certainty either that ETS causes lung cancer or that it does not do so.

1.4.2 HEART DISEASE

The detailed critique of the claims of Glantz and Parmley by Lee and Roe[25] shows clearly that the available experimental and epidemiological evidence does not justify a conclusion that ETS exposure actually increases the risk of heart disease. The epidemiological evidence of an association of risk with spousal smoking is weakened by the existence of various forms of bias and is undermined by recent reports from three large studies that find essentially no relationship. The experimental clinical and animal evidence is difficult to interpret for a combination of reasons, including unrealistic exposure levels, inappropriate endpoints, and poor study designs.

1.4.3 NASAL SINUS CANCER

Three studies[98–100] have reported that nonsmokers exposed to ETS have an increased risk of nasal sinus cancer. Though the RRs are relatively high (2.55, 3.00, and 5.73), they are extremely variable, being based on 35 cases at most, with no evidence of a dose-response relationship presented and a lack of control for potential confounding variables. The magnitude of the reported association also seems implausibly high, bearing in mind the much weaker relationships reported with active smoking. One should regard these fragmentary data as far from conclusive.

1.4.4 SIDS

The overall data clearly show a strong association of both prenatal and postnatal maternal ETS exposure, with some evidence also of a relationship with paternal smoking.[10,28] However, as noted in Section 1.2, those studies that adjusted for large numbers of potential confounding variables found that the association was massively weakened, suggesting that residual confounding may explain part, if not all, of the association remaining after adjustment.[28]

1.4.5 BIRTHWEIGHT

A meta-analysis[101] estimated that ETS exposure is, on average, associated with a 23-g decrement in birthweight. This difference does not necessarily imply harm to the infant and can be compared with an estimate of 102 g for the reduction in birthweight relating to birth at an elevation of 1000 m.[102] According to this author's own (unpublished) review of the evidence, there are considerable difficulties in interpreting the weak association between ETS and reduced birthweight. Thus, of the over 40 studies reviewed, none that adjusted for ten or more potential confounding variables reported a statistically significant relationship, some reported associations adjusted for no potential confounding variables at all, while other studies found that adjustment markedly weakened the strength of the reported association.

1.4.6 LOWER RESPIRATORY TRACT ILLNESSES

The most common cause of such illnesses is exposure to infectious agents such as viruses. If ETS increases their incidence, the mechanism by which this occurs remains to be determined. Though the overall evidence clearly shows an association between parental smoking and increased incidence of lower respiratory tract illnesses,[12] and although attempted adjustment for potential confounding variables has generally had little effect, there is reason to believe (see Section 1.2) that such adjustment has been inadequate. Also if, in fact, parental smoking does affect incidence, the evidence is not at all clear whether this is a postnatal ETS effect.

1.4.7 MIDDLE EAR DISEASE

Thornton and Lee[29] have discussed the evidence in detail based on a review of almost 60 studies. The overall data suggest a weak association of ETS exposure to recurrent otitis media, otitis media with effusion, and unspecified middle ear disease, but not with acute otitis media or with persistent otitis media with effusion. However, there are a number of problems in interpreting the data, including the difficulty of distinguishing prenatal and postnatal ETS effects and the generally inadequate control for confounding, and the evidence for a causal role of ETS is not convincing.

1.5 CONCLUSIONS

There are considerable difficulties in conducting adequate studies of the relationship of ETS to disease and in assessing the overall evidence. When these are taken properly into account, it is clear that many, if not all, of the health claims made against ETS cannot be justified.

ACKNOWLEDGMENTS

I thank my colleagues Barbara Forey and Dr. John Fry for assistance with some of the statistical analyses referred to and for helpful discussions, Dr. Francis Roe and Jan Hamling for comments on the draft, and Pauline Wassell for typing the manuscript. Philip Morris provided financial support, for which I am grateful. I alone bear responsibility for any views expressed.

REFERENCES

1. Hirayama, T., Non-smoking wives of heavy smokers have a higher risk of lung cancer: a study from Japan, *Br. Med. J.*, 282, 183, 1981.
2. Trichopoulos, D., Kalandidi, A., Sparros, L., and MacMahon, B., Lung cancer and passive smoking, *Int. J. Cancer*, 27, 1, 1981.
3. U.S. Surgeon General, The Health Consequences of Involuntary Smoking, a Report of the Surgeon General, DHHS (CDC) 87-8398, U.S. Department of Health and Human Services, Rockville, MD, 1986, 359.
4. Committee on Passive Smoking, Board on Environmental Studies and Toxicology and National Research Council, *Environmental Tobacco Smoke. Measuring Exposures and Assessing Health Effects,* National Academy Press, Washington, D.C., 1986, 337.
5. U.S. Environmental Protection Agency, *Smoking and Tobacco Control, Monograph 4: Respiratory Health Effects of Passive Smoking: Lung Cancer and Other Disorders,* NIH Publication No 93-3605, National Institutes of Health, Washington, D.C., 1993, 364.
6. California Environmental Protection Agency (EPA), *Health Effects of Exposure to Environmental Tobacco Smoke,* CaliEPA Office of Environmental Health Hazard Assessment, California Environmental Protection Agency, Sacramento, 1997.
7. Scientific Committee on Tobacco and Health, *Report of the Scientific Committee on Tobacco and Health,* Her Majesty's Stationery Office, London, 1998, 141.
8. Hackshaw, A.K., Law, M.R., and Wald, N.J., The accumulated evidence on lung cancer and environmental tobacco smoke, *Br. Med. J.*, 315, 980, 1997.
9. Law, M.R., Morris, J.K., and Wald, N.J., Environmental tobacco smoke exposure and ischaemic heart disease: an evaluation of the evidence, *Br. Med. J.*, 315, 973, 1997.
10. Anderson, H.R. and Cook, D.G., Passive smoking and sudden infant death syndrome: review of the epidemiological evidence, *Thorax*, 52, 1003, 1997.
11. Strachan, D.P. and Cook, D.G., Parental smoking, middle ear disease and adenotonsillectomy in children, *Thorax*, 53, 50, 1998.
12. Strachan, D.P. and Cook, D.G., Parental smoking and lower respiratory illness in infancy and early childhood, *Thorax*, 52, 905, 1997.
13. Cook, D.G. and Strachan, D.P., Parental smoking and prevalence of respiratory symptoms and asthma in school age children, *Thorax*, 52, 1081, 1997.
14. Strachan, D.P. and Cook, D.G., Parental smoking and allergic sensitisation in children, *Thorax*, 53, 117, 1998.
15. Strachan, D.P. and Cook, D.G., Parental smoking and childhood asthma: longitudinal and case-control studies, *Thorax*, 53, 204, 1998.
16. Cook, D.G. and Strachan, D.P., Parental smoking, bronchial reactivity and peak flow variability in children, *Thorax*, 53, 295, 1998.
17. Armitage, A.K., Ashford, J.R., Gorrod, J.W., and Sullivan, F.M., Environmental tobacco smoke — is it really a carcinogen?, *Med. Sci. Res.*, 25, 3, 1997.
18. Gori, G.B., Science, policy and ethics: the case of environmental tobacco smoke, *J. Clin. Epidemiol.*, 47, 325, 1994.
19. Hood, R.D., Wu, J.M., Witorsch, R.J., and Witorsch, P., Environmental tobacco smoke exposure and respiratory health in children: an updated critical review and analysis of the epidemiological literature, *Indoor Environ.*, 1, 19, 1992.
20. Huber, G.L., Brockie, R.E., and Mahajan, V.K., Smoke and mirrors. The EPA's flawed study of environmental tobacco smoke and lung cancer, *Regulation*, 3, 44, 1993.
21. Lee, P.N., *Environmental Tobacco Smoke and Mortality,* S. Karger, Basel, 1992, 224.

22. Lee, P.N., Lung cancer and ETS: is the epidemiologic evidence conclusive?, in *American Statistical Association, Proceedings of the Section on Statistics and the Environment*, American Statistical Association, Alexandria, VA, 1994, 65.

23. Lee, P.N., Difficulties in assessing the relationship between passive smoking and lung cancer, *Stat. Methods Med. Res.*, 7, 137, 1998.

24. Lee, P.N. and Thornton, A.J., A critical commentary on views expressed by IARC in relation to environmental tobacco smoke and lung cancer, *Indoor Built Environ.*, 7, 129, 1998.

25. Lee, P.N. and Roe, F.J.C., Environmental tobacco smoke exposure and heart disease: a critique of the claims of Glantz and Parmley, *Hum. Ecol. Risk Assess.*, 5, 171, 1999.

26. Mantel, N., What is the epidemiologic evidence for a passive smoking–lung cancer association, in *Indoor Air Quality*, Kasuga, H., Ed., Springer-Verlag, Berlin, 1990, 341.

27. Nilsson, R., Environmental tobacco smoke and lung cancer: a reappraisal, *Ecotoxicol. Environ. Saf.*, 34, 2, 1996.

28. Thornton, A.J. and Lee, P.N., Parental smoking and sudden infant death syndrome: a review of the evidence, *Indoor Built Environ.*, 7, 87, 1998.

29. Thornton, A.J. and Lee, P.N., Parental smoking and middle ear disease in children: a review of the evidence, *Indoor Built Environ.*, 8, 21, 1999.

30. Witorsch, R.J., Parental smoking and respiratory health and pulmonary function in children: a review of the literature and suggestions for future research, in *Environmental Tobacco Smoke, Proceedings of the International Symposium at McGill University 1989*, Ecobichon, D.J. and Wu, J.M., Eds., Lexington Books, Canada, Lexington, MA, 1990, 205.

31. Phillips, K., Howard, D.A., Browne, D., and Lewsley, J.M., Assessment of personal exposures to environmental tobacco smoke in British nonsmokers, *Environ. Int.*, 20, 693, 1994.

32. Lee, P.N., A Review of the Epidemiology of ETS and Lung Cancer, Unpublished, 1997, 56. Available on request from P.N. Lee Statistics and Computing Ltd., 17 Cedar Road, Sutton, Surrey, SM2 5DA, U.K.

33. Boffetta, P., Agudo, A., Ahrens, W., Benhamou, E., Benhamou, S., Darby, S.C., Ferro, G., Fortes, C., Gonzalez, C.A., Jöckel, K.-H., Krauss, M., Kreienbrock, L., Kreuzer, M., Mendes, A., Merletti, F., Nyberg, F., Pershagen, G., Pohlabeln, H., Riboli, E., Schmid, G., Simonato, L., Trédaniel, J., Whitley, E., Wichmann, H.-E., Winck, C., Zambon, P., and Saracci, R., Multicenter case-control study of exposure to environmental tobacco smoke and lung cancer in Europe, *J. Natl. Cancer Inst.*, 90, 1440, 1998.

34. Lee, P.N., Comparison of autopsy, clinical and death certificate diagnosis with particular reference to lung cancer. A review of the published data, *APMIS*, 102(Suppl. 45), 1, 1994.

35. Trichopoulos, D., Mollo, F., Tomatis, L., Agapitos, E., Delsedime, L., Zavitsanos, X., Kalandidi, A., Katsouyanni, K., Riboli, E., and Saracci, R., Active and passive smoking and pathological indicators of lung cancer risk in an autopsy study, *J. Am. Med. Assoc.*, 268, 1697, 1992.

36. Agapitos, E., Mollo, F., Tomatis, L., Katsouyanni, K., Lipworth, L., Delsedime, L., Kalandidi, A., Karakatsani, A., Riboli, E., Saracci, R., and Trichopoulos, D., Epithelial, possibly precancerous, lesions of the lung in relation to smoking, passive smoking, and socio-demographic variables, *Scand. J. Soc. Med.*, 24, 259, 1996.

37. Lee, P.N., Active and passive smoking and pathological indicators of lung cancer. A report of limited value? [Letter], *J. Am. Med. Assoc.*, 270, 1690, 1993.

38. Glantz, S.A. and Parmley, W.W., Passive smoking and heart disease. Mechanisms and risk, *J. Am. Med. Assoc.*, 273, 1047, 1995.

39. de Waard, F., Kemmeren, J.M., van Ginkel, L.A., and Stolker, A.A.M., Urinary cotinine and lung cancer risk in a female cohort, *Br. J. Cancer*, 72, 784, 1995.

40. Lee, P.N., Uses and abuses of cotinine as a marker of tobacco smoke exposure. in *Analytical Determination of Nicotine and Related Compounds and Their Metabolites*, Gorrod, J.W. and Jacob, P., III, Eds., Elsevier, Amsterdam, 1999, 669.

41. Tunstall-Pedoe, H., Brown, C.A., Woodward, M., and Tavendale, R., Passive smoking by self report and serum cotinine and the prevalence of respiratory and coronary heart disease in the Scottish heart health study, *J. Epidemiol. Community Health*, 49, 139, 1995.

42. Brownson, R.C., Alavanja, M.C.R., Hock, E.T., and Loy, T.S., Passive smoking and lung cancer in nonsmoking women, *Am. J. Public Health*, 82, 1525, 1992.

43. Stockwell, H.G., Goldman, A.L., Lyman, G.H., Noss, C.I., Armstrong, A.W., Pinkham, P.A., Candelora, E.C., and Brusa, M.R., Environmental tobacco smoke and lung cancer risk in nonsmoking women, *J. Natl. Cancer Inst.*, 84, 1417, 1992.

44. Fontham, E.T.H., Correa, P., Reynolds, P., Wu-Williams, A., Buffler, P.A., Greenberg, R.S., Chen, V.W., Alterman, T., Boyd, P., Austin, D.F., and Liff, J., Environmental tobacco smoke and lung cancer in nonsmoking women. A multicenter study, *J. Am. Med. Assoc.*, 271, 1752, 1994.

45. Lee, P.N., Passive smoking, *Food Chem. Toxicol.*, 20, 223, 1982.

46. Block, G., Patterson, B., and Subar, A., Fruit, vegetables, and cancer prevention: a review of the epidemiological evidence, *Nutr. Cancer*, 18, 1, 1992.

47. Ziegler, R.G., Mayne, S.T., and Swanson, C.A., Nutrition and lung cancer, *Cancer Causes Control*, 7, 157, 1996.

48. Candelora, E.C., Stockwell, H.G., Armstrong, A.W., and Pinkham, P.A., Dietary intake and risk of lung cancer in women who never smoked, *Nutr. Cancer*, 17, 263, 1992.

49. Koo, L.C., Kabat, G.C., Rylander, R., Tominaga, S., Kato, I., and Ho, J.H.-C., Dietary and lifestyle correlates of passive smoking in Hong Kong, Japan, Sweden, and the U.S.A., *Soc. Sci. Med.*, 45, 159, 1997.

50. Mayne, S.T., Janerich, D.T., Greenwald, P., Chorost, S., Tucci, C., Zaman, M.B., Kiely, M., and McKneally, M.F., Dietary beta carotene and lung cancer risk in U.S. nonsmokers, *J. Natl. Cancer Inst.*, 86, 33, 1994.

51. Cardenas, V.M., Environmental Tobacco Smoke and Lung Cancer Mortality in the American Cancer Society's Cancer Prevention Study II, Thesis, Emory University, Atlanta, GA, 1994.

52. Knekt, P., Järvinen, R., Seppänen, R., Rissanen, A., Aromaa, A., Heinonen, O.P., Albanes, D., Heinonen, M., Pukkala, E., and Teppo, L., Dietary antioxidants and the risk of lung cancer, *Am. J. Epidemiol.*, 134, 471, 1991.

53. Ko, Y.-C., Lee, C.-H., Chen, M.-J., Huang, C.-C., Chang, W.-Y., Lin, H.-J., Wang, H.-Z., and Chang, P.-Y., Risk factors for primary lung cancer among non-smoking women in Taiwan, *Int. J. Epidemiol.*, 26, 24, 1997.

54. Alavanja, M.C.R., Brown, C.C., Swanson, C., and Brownson, R.C., Saturated fat intake and lung cancer risk among nonsmoking women in Missouri, *J. Natl. Cancer Inst.*, 85, 1906, 1993.

55. Alavanja, M.C.R., Brownson, R.C., and Benichou, J., Estimating the effect of dietary fat on the risk of lung cancer in nonsmoking women, *Lung Cancer*, 14(Suppl. 1), S63, 1996.

56. de Stefani, E., Fontham, E.T.H., Chen, V., Correa, P., Deneo-Pellegrini, H., Ronco, A., and Mendilaharsu, M., Fatty foods and the risk of lung cancer: a case-control study from Uruguay, *Int. J. Cancer*, 71, 760, 1997.

57. Dallongeville, J., Marécaux, N., Fruchart, J.-C., and Amouyel, P., Cigarette smoking is associated with unhealthy patterns of nutrient intake: a meta-analysis, *J. Nutr.*, 128, 1450, 1998.

58. Thornton, A., Lee, P., and Fry, J., Differences between smokers, ex-smokers, passive smokers and non-smokers, *J. Clin. Epidemiol.*, 47, 1143, 1994.

59. Le Marchand, L., Wilkens, L.R., Hankin, J.H., and Haley, N.J., Dietary patterns of female nonsmokers with and without exposure to environmental tobacco smoke, *Cancer Causes Control*, 2, 11, 1991.

60. Shibata, A., Paganini-Hill, A., Ross, R.K., Yu, M.C., and Henderson, B.E., Dietary β-carotene, cigarette smoking, and lung cancer in men, *Cancer Causes Control*, 3, 207, 1992.

61. Sidney, S., Caan, B.J., and Friedman, G.D., Dietary intake of carotene in nonsmokers with and without passive smoking at home, *Am. J. Epidemiol.*, 129, 1305, 1989.

62. Tribble, D.L., Giuliano, L.J., and Fortmann, S.P., Reduced plasma ascorbic acid concentrations in nonsmokers regularly exposed to environmental tobacco smoke, *Am. J. Clin. Nutr.*, 58, 886, 1993.

63. Matanoski, G., Kanchanaraksa, S., Lantry, D., and Chang, Y., Characteristics of nonsmoking women in NHANES I and NHANES II epidemiologic follow-up study with exposure to spouses who smoke, *Am. J. Epidemiol.*, 142, 149, 1995.

64. Friedman, G.D., Petitti, D.B., and Bawol, R.D., Prevalence and correlates of passive smoking, *Am. J. Public Health*, 73, 401, 1983.

65. Cress, R.D., Holly, E.A., Ahn, D.K., Kristiansen, J.J., and Aston, D.A., Contraceptive use among women smokers and nonsmokers in the San Francisco Bay area, *Prev. Med.*, 23, 181, 1994.

66. Osler, M., The food intake of smokers and nonsmokers: the role of partner's smoking behavior, *Prev. Med.*, 27, 438, 1998.

67. Whitlock, G., MacMahon, S., Vander Hoorn, S., Davis, P., Jackson, R., and Norton, R., Association of environmental tobacco smoke exposure with socio-economic status in a population of 7725 New Zealanders, *Tob. Control*, 7, 276, 1998.

68. Wilson, A.C., Forsyth, J.S., Greene, S.A., Irvine, L., Hau, C., and Howie, P.W., Relation of infant diet to childhood health: seven year follow up of cohort of children in Dundee infant feeding study, *Br. Med. J.*, 316, 21, 1998.

69. Mitchell, E.A., Ford, R.P.K., Stewart, A.W., Taylor, B.J., Becroft, D.M.O., Thompson, J.M.D., Scragg, R., Hassall, I.B., Barry, D.M.J., Allen, E.M., and Roberts, A.P., Smoking and the sudden infant death syndrome, *Pediatrics*, 91, 893, 1993.

70. Mitchell, E. A., Taylor, B.J., Ford, R.P.K., Stewart, A.W., Becroft, D.M.O., Thompson, J.M.D., Scragg, R., Hassall, I.B., Barry, D.M.J., Allen, E.M., and Roberts, A.P., Four modifiable and other major risk factors for cot death: the New Zealand study, *J. Paediatr. Child Health*, 28(Suppl. 1), S3, 1992.

71. Blair, P.S., Fleming, P.J., Bensley, D., Smith, I., Bacon, C., Taylor, E., Berry, J., Golding, J., and Tripp, J., Smoking and the sudden infant death syndrome: results from 1993-5 case-control study for confidential inquiry into stillbirths and deaths in infancy, *Br. Med. J.*, 313, 195, 1996.

72. Golding, J., Sudden infant death syndrome and parental smoking — a literature review, *Paediatr. Perinat. Epidemiol.*, 11, 67, 1997.

73. Lee, P.N. and Forey, B.A., Misclassification of smoking habits as a source of bias in the study of environmental tobacco smoke and lung cancer, *Stat. Med.*, 15, 581, 1996.

74. Wald, N.J., Nanchahal, K., Thompson, S.G., and Cuckle, H.S., Does breathing other people's tobacco smoke cause lung cancer?, *Br. Med. J.*, 293, 1217, 1986.

75. Lee, P.N. and Forey, B.A., Misclassification of smoking habits as determined by cotinine or by repeated self-report — a summary of evidence from 42 studies, *J. Smoking-Related Dis.*, 6, 109, 1995.

76. Lee, P.N., Passive smoking and lung cancer: strength of evidence on passive smoking and lung cancer is overstated [Letter], *Br. Med. J.*, 317, 346, 1998.

77. Hackshaw, A.K., Law, M.R., and Wald, N.J., Passive smoking and lung cancer: authors' reply [Letter], *Br. Med. J.*, 317, 348, 1998.

78. Lee, P.N., *Misclassification of Smoking Habits and Passive Smoking. A Review of the Evidence*, Springer-Verlag, Heidelberg, 1988, 103 (International Archives of Occupational and Environmental Health Supplement).

79. Suadicani, P., Hein, H.O., and Gyntelberg, F., Mortality and morbidity of potentially misclassified smokers, *Int. J. Epidemiol.*, 26, 321, 1997.

80. Bennett, W.P., Alavanja, M.C.R., Blomeke, B., Vähäkangas, K., Castrén, K., Welsh, J.A., Bowman, E.D., Khan, M.A., Flieder, D.B., and Harris, C.C., Environmental tobacco smoke, genetic susceptibility, and risk of lung cancer in never-smoking women, *J. Natl. Cancer Inst.*, 91, 2009, 1999.

81. Geng, G.-Y., Liang, Z. H., Zhang, A.Y., and Wu, G.L., On the relationship between cigarette smoking and female lung cancer, in *Smoking and Health 1987, Proceedings of the 6th World Conference on Smoking and Health, Tokyo, 9-12 November 1987,* International Congress Series No. 780, Aoki, M., Hisamichi, S., and Tominaga, S., Eds., Elsevier Science Publishers B.V., Amsterdam, 1988, 483.

82. Lee, P.N., Simple methods for checking for possible errors in reported odds ratios, relative risks and confidence intervals, *Stat. Med.*, 18, 1973, 1999.

83. Guerin, M., Jenkins, R.A., and Tomkins, B.A., *The Chemistry of Environmental Tobacco Smoke: Composition and Measurement*, Lewis Publishers, Chelsea, MI, 1992.

84. Redhead, C.S. and Rowberg, R.E., *Environmental Tobacco Smoke and Lung Cancer Risk*, U.S. Congressional Research Service, Washington, D.C., 1995, 75.

85. Gori, G.B. and Mantel, N., Mainstream and environmental tobacco smoke, *Regul. Toxicol. Pharmacol.*, 14, 88, 1991.

86. Kraus, N., Malmfors, T., and Slovic, P., Intuitive toxicology: expert and lay judgments of chemical risks, *Risk Anal.*, 12, 215, 1992.

87. Butterworth, B.E., Dorman, D.C., Gaido, K.W., Sumner, S.J., Corton, J.C., Borghoff, S.J., and Conolly, R.B., Research at CIIT on the risks to human health from exposure to chemicals, *CIIT Activities*, 19, 7, 1999.

88. U.S. Surgeon General, Smoking and Health, a Report of the Surgeon General, (PHS)79-50066, Office on Smoking and Health, Rockville, MD, 1979.

89. Elwood, P.C., Renaud, S., Sharp, D.S., Beswick, A.D., O'Brien, J.R., and Yarnell, J.W.G., Ischemic heart disease and platelet aggregation. The Caerphilly Collaborative Heart Disease Study, *Circulation*, 83, 38, 1991.

90. Thornton, A. and Lee, P., Publication bias in meta-analysis, *J. Clin. Epidemiol.*, 53, 207, 2000.

91. Garfinkel, L., Time trends in lung cancer mortality among nonsmokers and a note on passive smoking, *J. Natl. Cancer Inst.*, 66, 1061, 1981.

92. LeVois, M.E. and Layard, M.W., Publication bias in the environmental tobacco smoke/coronary heart disease epidemiologic literature, *Regul. Toxicol. Pharmacol.*, 21, 184, 1995.

93. Steenland, K., Thun, M., Lally, C., and Heath, C., Jr., Environmental tobacco smoke and coronary heart disease in the American Cancer Society CPS-II cohort, *Circulation*, 94, 622, 1996.

94. Saracci, R. and Riboli, R., Passive smoking and lung cancer: current evidence and ongoing studies at the International Agency for Research on Cancer, *Mutat. Res.*, 222, 117, 1989.

95. Trédaniel, J., Boffetta, P., Saracci, R., and Hirsch, A., Environmental tobacco smoke and the risk of cancer in adults, *Eur. J. Cancer*, 29A, 2058, 1993.

96. Trédaniel, J., Boffetta, P., Saracci, R., and Hirsch, A., Exposure to environmental tobacco smoke and risk of lung cancer: the epidemiological evidence, *Eur. Respir. J.*, 7, 1877, 1994.

97. Shimizu, H., Morishita, M., Mizuno, K., Masuda, T., Ogura, Y., Santo, M., Nishimura, M., Kunishima, K., Karasawa, K., Nishiwaki, K., Yamamoto, M., Hisamichi, S., and Tominaga, S., A case-control study of lung cancer in nonsmoking women, *Tohoku J. Exp. Med.*, 154, 389, 1988.

98. Hirayama, T., Cancer mortality in nonsmoking women with smoking husbands based on a large-scale cohort study in Japan, *Prev. Med.*, 13, 680, 1984.

99. Fukuda, K. and Shibata, A., Exposure-response relationships between woodworking, smoking or passive smoking, and squamous cell neoplasms of the maxillary sinus, *Cancer Causes Control*, 1, 165, 1990.

100. Zheng, W., McLaughlin, J.K., Chow, W.-H., Chien, H.T.C., and Blot, W.J., Risk factors for cancers of the nasal cavity and paranasal sinuses among white men in the United States, *Am. J. Epidemiol.*, 138, 965, 1993.

101. Windham, G.C., Eaton, A., and Waller, K., Is environmental tobacco smoke exposure related to low birthweight?, *Epidemiology*, 6(Abstr.), 41S, 1995.

102. Jensen, G.M. and Moore, L.G., The effect of high altitude and other risk factors on birthweight: independent or interactive effects?, *Am. J. Public Health*, 87, 1003, 1997.

2 Birthweight and Gestational Age in Relation to Prenatal Environmental Tobacco Smoke Exposure

Gayle C. Windham

CONTENTS

2.1 INTRODUCTION

Active smoking has been causally associated with a number of developmental and reproductive endpoints; the largest body of work concerns low birthweight, showing consistent, strong associations for years.[1] Infants of women who smoke during pregnancy are estimated to have twice the risk of low birthweight or, put another way, a weight decrement of 150 to 200 g at birth, compared to infants of non-smokers. Persons exposed to environmental tobacco smoke (ETS) are subjected to most of the same constituents as those contained in mainstream smoke, although the pattern and amounts of exposure differ. Therefore, exposure of non-smoking women to ETS during pregnancy is also of concern with respect to the growth and development of the fetus.

0-8493-0311-7/00/$0.00+$.50
© 2001 by CRC Press LLC

The purpose of this chapter is to summarize the evidence regarding the association between maternal exposure to ETS and fetal growth, based on published reviews as well as more recent study results in the epidemiologic literature. In 1992, the U.S. Environmental Protection Agency (EPA)[2] published a report on the health effects of ETS, but it did not include reproductive and developmental outcomes. The California EPA (CalEPA) and the Department of Health and Human Services have conducted the most thorough agency review to date on these outcomes.[3] Data from the studies on fetal growth in the California report were included in a pooled (or meta) analysis.[4] The World Health Organization (WHO), on behalf of the G-8 countries, held a consultation of international experts in early 1999 to evaluate the evidence regarding the health effects of ETS exposure on children.[5] Much of this chapter was originally written as a background paper for that meeting. In addition to studies included in the meta-analysis, those published since then (by 1999) are also noted here. Some background on the design of epidemiologic studies of this topic is summarized first.

2.1.1 Assessing Exposure to ETS in Health Effects Studies

In this chapter, ETS exposure of the pregnant woman is the concern, not "passive" exposure of the fetus to tobacco smoke products from an actively smoking mother. Some studies have attempted to examine effects of ETS exposure among groups of women that include smokers by adjusting for maternal smoking, but the effects attributable to each will be difficult to untangle. The most appropriate women to examine may be never smokers, in order to avoid issues of quitting early in pregnancy and denial of smoking during pregnancy due to the social stigma. In epidemiologic studies, exposure to ETS has generally been determined in three ways: ascertainment of spousal smoking status, estimation of the number of hours exposed (at home, work, or elsewhere), and measurement of biomarkers. Animal studies of tobacco smoke exposure face the challenges of simulating the human experience; animal models for ETS exposure have only recently come into use, and animal data will not be discussed here.

Ascertaining spousal smoking status was common in early studies, but it is the crudest method of determining exposure to ETS. Because women may be exposed to ETS in places other than the home, this method can lead to misclassification of the unexposed group, which if non-differential with respect to outcome usually makes an association more difficult to detect. Furthermore, dose-response relationships are difficult to assess because the amount smoked may not reflect comparable exposures between couples, depending on several elements of the home environment. Most such studies are based on an interview of the pregnant woman; reporting of spouse's smoking status appears valid and reliable, but information on the amount smoked is less accurate.[6,7]

Studies that base exposure assessment on all sources of exposure should provide more useful information. Nevertheless, misclassification may still occur as respondents vary in their awareness of and ability to quantify their exposure in multiple locations,[7] which tends to be underestimated.[8] Furthermore, ETS exposure is ubiquitous in many areas, making it difficult to identify an unexposed comparison group.[9,10] More recently, there has been an effort to quantify exposures by measuring biomarkers, with the preferred marker being cotinine, a metabolite of nicotine, due to its specificity for tobacco smoke. Measurement of cotinine has been shown to distinguish ETS-exposed and unexposed non-smokers,[11] but not consistently. Because of the relatively short half-life of cotinine (15 to 25 h in plasma), it reflects integrated exposures of the past few days only, so questionnaire data are important for assessing chronic ETS exposure.

Because cotinine levels in non-smokers exposed to ETS are generally one to two orders of magnitude lower than in smokers,[9,11] the effect of ETS exposure would naturally be expected to be much lower than that of active smoking. Thus, the pronounced effect of ETS exposure seen in some studies has been called "biologically implausible."[12] However, this presupposes a linear relationship, which may not be appropriate. Furthermore, compounds other than nicotine (e.g., carbon monoxide, toluene, cadmium, and polyaromatic hydrocarbons [PAHs]) are also likely toxicants of interest. Because the ratio of these constituents in mainstream vs. sidestream smoke varies and the constituents are metabolized at different rates,[13,14] it is difficult to extrapolate dose-response relationships for the unidentified active component(s) based on cotinine levels. There is also a suggestion that cotinine levels measured during pregnancy may not be comparable to measurements when not pregnant,[15] further complicating interpretation.

2.1.2 MEASURES OF EFFECT ON FETAL GROWTH

Generally, fetal growth has been assessed at birth based on weight and gestational age. The measures commonly used include low birthweight (LBW) (<2500 g) and intrauterine growth retardation (IUGR) or small for gestational age, which is typically defined as less than the tenth percentile of weight for gestational age based on a standard population. The LBW category includes infants that are growth retarded as well as infants who grew appropriately, but were born prematurely. These may result from different etiologies; therefore, some investigators examine LBW in term births only (although IUGR is preferred to increase study power). Preterm births (<37 weeks) are also examined as a separate category. Mean birthweight or gestational age may be examined as endpoints as well. Other measures of fetal size, such as length or head circumference, may be important for identifying mechanisms of effect, but few studies have measured these with respect to ETS exposure.

Several factors are important in interpreting the results of an epidemiologic study, including the power of the study to detect an effect (or sample size), quality of the exposure assessment (discussed above), and consideration of potential confounders. When examining fetal growth, a number of covariables should be considered as potential confounders, including maternal age, race, parity or prior reproductive history, socioeconomic status, and/or access to prenatal care. Few studies have information on maternal stature or weight gain, but these are also important determinants of fetal weight, as are certain illnesses, complications of pregnancy, and infant gender. However, these may not all be related to ETS exposure and, thus, would not be confounders in individual studies. Gestational age at delivery is the strongest predictor of birthweight.

When evaluating a body of epidemiologic literature, basing interpretation only on the tallying of statistically significant findings can be misleading,[16,17] as epidemiologic data seldom satisfy the criteria of randomized experimental trials. Furthermore, statistical significance is influenced by sample size. Finally, such tallying does not take into account possible sources of bias in the studies. Meta-analyses, which combine results of multiple studies, provide more statistical power, but can also incorporate weaknesses of the original studies and so require care to assess biases and sources of heterogeneity across studies.

2.2 REVIEW OF THE LITERATURE

2.2.1 MEAN BIRTHWEIGHT

More than 25 studies have examined mean birthweight in relation to ETS exposure, with all but a few finding a decrement in weight, ranging from 5 to over 200 g, as shown in Figure 2.1, for studies that provided data for calculating error or confidence limits.[4,18-38] The CalEPA review[3] and subsequent meta-analysis[4] concluded there was a consistent, slight effect of ETS exposure on birthweight. Qualitatively, the better studies (adjusted for confounders, exposure from multiple sources, etc.) showed a mean weight decrement ranging from 25 to 100 g. In the meta-analysis,[4] the 19 studies conducted among non-smoking mothers which provided sufficient data for statistical inclusion yielded a pooled weight decrement of 31 g (95% confidence limits [CL] –42, –20). Pooling was done with different subsets of the studies to examine heterogeneity among studies and the effect of including only adjusted measures or studies with better exposure assessment. The pooled weight decrements varied between 25 and 44 g. There were three studies[32-34] that based exposure on measurement of cotinine in non-smokers, although using different assay methods and fluids (e.g., saliva and serum), and adjusted for at least some confounders. These studies yielded a pooled weight decrement of 82 g (CL –126, –37) with ETS exposure based on cotinine levels greater than 1 to 2 ng/ml (in non-smokers).

Of the studies that examined dose-response effects, several found evidence for such trends,[21,25,31] but not all.[26,32] Another study found a weight decrement only with high exposure, which the authors interpreted as a threshold effect.[35] Two studies noted that the birthweight decrements in infants of women highly exposed to ETS (≥5 h/day or ≥20 cigarettes/day by others) were similar to those of infants of light smokers.[30,31] There were insufficient data to distinguish effects of home vs. workplace exposure in the studies. Studies were conducted throughout the world, including the U.S., Canada, Western Europe, England, Scandinavia, China, and Japan. Region was not a significant predictor in the meta-analysis, but studies from Asia tended to find slightly lower mean weight decrements.

Some criticisms of the studies in early reviews included lack of control for confounders and crude exposure assessment,[12] which were addressed to some extent in studies that appeared subsequently. In a brief review which included only nine of the studies in the meta-analysis,[39] the author suggested a possible influence of ETS exposure on fetal growth, but felt studies were limited by methodological shortcomings. The importance of considering interactions of ETS exposure with other exposures in

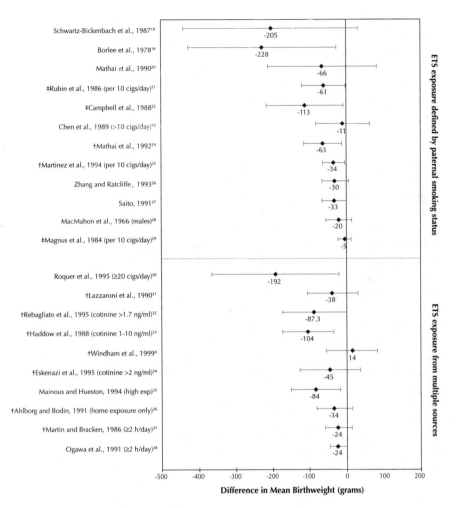

FIGURE 2.1 Summary of differences in mean birthweight and 95% confidence intervals between ETS-exposed and unexposed pregnancies by ETS definition and study size. (Adapted from Windham, G.C. et al., *Pediatr. Perinat. Epidemiol.*, 13, 35–57, 1999. With permission.)

increasing risks of adverse fetal outcome was noted. In a more recent paper, the authors pooled their results on mean birthweight with those of ten other studies (all included in the CalEPA[3] report and the Windham et al.[4] meta-analysis) to yield a weight decrement of 31 g (95% CL –44, –19),[40] entirely consistent with the larger meta-analysis.

Several additional studies (Table 2.1) have been published that were not included in the previous meta-analyses, and their results generally support the findings of a decrement in birthweight with ETS exposure. One of these was conducted in a region not previously represented, namely, Eastern Europe, and found a weight decrement of 58 g among infants of non-smokers who reported passive exposure to cigarette smoke at home or work.[41] Ascertainment of smoking and ETS exposure was conducted when the children were 9 years old, so is subject to recall error, and only a few potential confounders were ascertained. In a study from Brazil,[42] another region not previously represented, women whose partners smoked had infants with a statistically significant decrement in birthweight (30 g), but the analysis included (and adjusted for) maternal smokers.

The other studies were from the U.S. (Table 2.1). A very large study based on state surveillance data of low income, non-smoking women examined exposure to the cigarette smoke of a household member, stratified by age.[43] Adjusting for numerous factors, the decrement in mean birthweight associated with ETS exposure was 90 g in infants of women over 30 years old, with little difference among those of younger non-smokers. (In these data, birthweight decrements of 150 to 300 g were found in infants of active smokers, stratified by the same maternal age categories). The large California study[44] ascertained hours of ETS exposure prospectively during the first trimester, which may have led to some misclassification if exposure changed later in pregnancy. Weight decrements were not found, except among the few women exposed 12 or more hours per week (Table 2.1). Stratifying, mean birthweight was decreased among infants of highly ETS-exposed (≥7 h/day), non-white women, but not among whites. The study from Connecticut[45] enrolled women during pregnancy as well, but ascertained ETS exposure postpartum based on detailed questions of those who reported any exposure. The authors found little evidence for a reduction in birthweight with any of the exposure measures. However, the duration, amount, and intensity variables were analyzed as quasi-continuous, which presumes a linear relationship. Exposure at home and socially was only ascertained for the week prior to delivery, which may not be representative of exposure during the critical period. Another U.S. study showed that non-smoking women who reported smokers in the home had a significantly greater mean urinary cotinine level (40 ng/mg creatinine) than non-exposed women (20 ng/mg creatinine).[46] Birthweight was examined in relation to the entire range of cotinine values, including among smokers, so effects of ETS are difficult to untangle. At the lowest cotinine category presented (31 to 100 ng/mg creatinine), which overlapped the mean of the ETS-exposed women, there was an adjusted birthweight decrement of 57 g and a length decrement of 0.52 cm (95% CL –1.05, 0.01). (Mean birthweight declined in a dose-response manner with higher cotinine levels.)

Thus, among these newer studies, which all adjusted for at least some confounders and were conducted in different regions of the world, most found birthweight decrements consistent with findings from the previous studies of high quality. Of the two that attempted to quantify an ETS exposure gradient, one found potential effects only with high exposure[44] and the other found no effect.[45]

Few measures of infant size, other than birthweight, have been assessed in relation to ETS exposure. Four previous studies were identified with data on infant length, but only one provided adjusted differences.[3,31] The studies generally showed decrements on the order of 0.25 to 1 cm with ETS exposure, most of which were borderline statistically significant, consistent with the one new study noted above.[46]

TABLE 2.1
Recent Studies of Mean Birthweight and ETS Exposure

Authors (year)/Country (ref)	Study design (N)	ETS level[a] (% exposed)	Difference in mean weight (g) (95% CL)
Jedrychowski and Flak (1996) Poland (41)	Retrospective study of (1,165) school-age kids	ETS at home or work during pregnancy (52%)	−73 crude −58 (−119, 3) Adj[b] for 1–3
Horta et al. (1997) Brazil (42)	Interview postpartum (5,166)	Partner smoking (49%), adjusted for maternal smoking	−30 (p <0.05 in ANOVA) but "CL includes unity" Adj[b] for 2, 4, 5, 7, 8
Ahluwalia et al (1997) U.S. – Arizona and N. Dakota (43)	Pregnancy surveillance data; low income (17,412; 13,497 non-smokers)	Other household smokers (21%)	<30 y.o.: 9 (−26, 44) ≥30 y.o.: −90 (−181, 1) Adj[b] for 2, 4–6, 9–12
Wang et al. (1997) U.S. – Boston, MA (46)	Prospective study of pregnant women (740; 483 non-smokers)	Cotinine level (31–100 ng/mg creatinine) (15% reported home ETS)	−57 (−143, 29) Adj[b] for 1–5, 8, 9, 11, 13
Sadler et al. (1999) U.S. – Connecticut (45)	Prospective nested cohort Postpartum interview for ETS (2,714; 2,283 non-smokers)	Regular exposure to other's smoke >1 h/week in 3rd trimester (27%) (duration, intensity, amount)	10 crude 1 (−43, 41) Adj[b] for 1–6, 8–11, 13–17
Windham et al. (2000) U.S. – California (44)	Prospective study of pregnant women (1st trimester) (4,454 singletons; 3,646 non-smokers)	Hours of ETS at home or work 1st trimester (20% any) ≥7 h/day (4%) ≥12 h/day (1%) Non-whites ≥7 h/day	8 (−86, 103) −88 (−290, 114) −119 (−272, 35) Adj[b] for 4, 5, 8–11, 13, 18, 19

[a] In non-smokers unless otherwise stated.

[b] Adjusted for gestational age,[1] parity,[2] infant gender,[3] education and/or social class,[4] maternal stature[5] (height, weight, or BMI), weight gain,[6] prenatal care,[7] pregnancy outcome history,[8] race or skin color,[9] marital status,[10] alcohol consumption,[11] altitude,[12] age,[13] religion,[14] hypertension,[15] preeclampsia,[16] past smoking,[17] stressful life events,[18] caffeine consumption.[19]

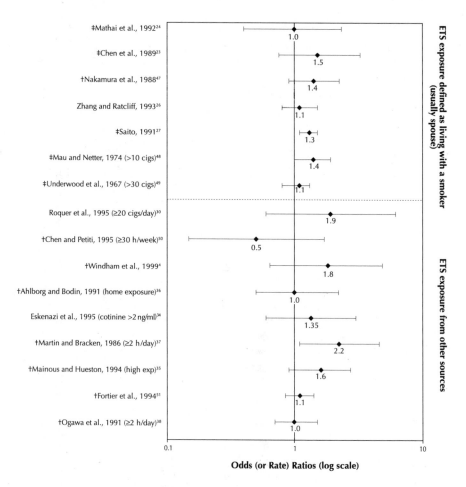

* Studies listed by size—smallest studies at top of each section.
† Adjusted.
‡ OR and CI calculated from data, sometimes estimated. In a few studies, rate ratios were calculated
(Zhang and Ratcliff, Ahlborg and Bodin).

FIGURE 2.2 Odds ratios and 95% confidence intervals for the association of LBW (or IUGR) and ETS by ETS definition and study size. (Adapted from Windham, G.C. et al., *Pediatr. Perinat. Epidemiol.*, 13, 35, 1999. With permission.)

2.2.2 INTRAUTERINE GROWTH RETARDATION OR LOW BIRTHWEIGHT

Sixteen studies of LBW or IUGR were reviewed in the CalEPA document[3] and previously included in the meta-analysis,[4] many of which showed an increased risk, but few statistically significantly so (Figure 2.2). The 11 studies which accounted for gestational age by examining either IUGR or LBW at term were fairly homogeneous and yielded a pooled odds ratio (OR) of 1.2 (95% CL 1.1, 1.3). Only six of these studies provided estimates that were adjusted for confounders; pooling them yielded an OR of 1.1. The eight studies that provided an estimate of effect for LBW yielded a pooled OR of 1.0. This was strongly influenced by a large, but methodologically weak study;[49] removing it yielded an OR of 1.4 (95% CL 1.1, 1.7), which was very similar to the pooled OR for the few LBW studies which adjusted for confounders. Two studies measured cotinine,[33,34] and both reported that the risk of LBW was increased 30%, but one did not assess significance,[33] so the results cannot be pooled.

Several of these studies indicated greater effects with higher levels of exposure.[27,47] In the entire set of these studies, those from Europe and Asia tended to have higher effect estimates than those from North America. This finding was again influenced by the large early study[49] which ascertained exposure only as paternal smoking status from medical records. Based on these studies and supporting data, the CalEPA[3] document stated that there was sufficient evidence that ETS exposure adversely affects fetal growth, with a slight increase in the risk of LBW or IUGR, as did the WHO consensus documents.

Studies published subsequent to those included in the meta-analysis (see Section 2.2.1 for other details) generally found some increase in LBW or IUGR (Table 2.2). In the Polish study,[41] ETS exposure at home or work was associated with a 50% increase in LBW after adjustment. In the Brazilian study,[42] partner smoking was associated with a slight increase in LBW after adjustment for a number of factors including maternal smoking (OR 1.18; 95% CL 0.94, 1.48). The adjusted OR for ETS and IUGR was 1.33 (Table 2.2) (compared to an OR of 2.07 for any active smoking). A study from Sweden[53] reported over a doubled rate of IUGR (defined as more than two standard deviations below the age-related means) with ETS exposure among all women. Among non-smokers, the adjusted OR was nearly 4, but pregnancy history was not included in the models. Active smoking was associated with an OR of about 6. The large U.S. study[43] reported a risk of LBW with ETS exposure at home that was more than doubled in infants of women aged 30 or older, but not increased among infants of younger women (Table 2.2). The increased risk among older women was not as apparent for IUGR (adjusted OR 1.3; 95% CL 0.8, 2.2). The California study[44] reported increased risks of LBW, but not of IUGR, which was the endpoint most strongly related to active smoking. Stratifying the associations with LBW by age and race showed an effect only among non-whites (OR = 3.8; 95% CL 1.5, 9.8), but there was no difference in effect size by maternal age. In contrast to those studies, the Connecticut study[45] found no evidence for increased risk of IUGR with any of several ETS exposure variables (duration, intensity, and amount). This study may have overadjusted as prior LBW and past smoking were included in models, but crude measures showed little association either.

One new study included measurement of nicotine levels in the hair of a small sample of newborns and their mothers.[52] By self-reported ETS exposure at home and/or work, a slightly elevated OR for IUGR was found only among the 12% of non-smokers who were exposed in both places (OR 1.4; 95% CL 0.4, 4.4). Maternal nicotine levels were examined by quartiles in the entire sample that included smokers. The risks of IUGR for infants of women in the second through fourth quartiles were increased three to four times that of the lowest quartile. When smokers were excluded, the same nicotine categories yielded ORs of 2 to 3, which the authors interpreted as effects of ETS exposure. The authors did not discuss why non-smoking mothers might have nicotine levels in the highest quartile, including the possibility

TABLE 2.2
Recent Studies of Fetal Growth[a] and Preterm Delivery by ETS Exposure

Authors (year)/Country	Study design (N)	ETS level[a] (% exposed)	Odds ratio (95% CL)	
			LBW/IUGR[b]	Preterm
Jedrychowski and Flak (1996) Poland (41)	Retrospective study of (1,165) school-age kids	ETS at home or work during pregnancy (52%)	1.46 (0.83–2.6) Adj for 1–3[c]	—
Horta et al. (1997) Brazil (42)	Interview postpartum (5,166)	Partner smoking (49%), adjusted for maternal smoking	1.33 (1.05, 1.68) Adj for 2, 4, 5, 7, 8, 19, 13[c]	1.25 (0.99–1.57) Adj for 4, 5, 7, 8[c]
Ahluwalia et al. (1997) U.S. – Arizona and N. Dakota (43)	Pregnancy surveillance data; low income (17,412; 13,497 non-smokers)	Other household smokers (21%)	<30 y.o.: 0.97 (0.76, 1.23) ≥30 y.o.: 2.42 (1.51, 3.87) Adj for 2, 4–6, 9–12[c]	0.92 (0.76, 1.13) 1.88 (1.22, 2.88) Adj for 2, 4, 5, 6, 9–12[c]
Nafstad et al. (1998) Norway (52)	Case control of IUGR with prospective interview (163; 122 non-smokers)	Maternal hair nicotine (2nd–4th quartile, 66%)	0.96 (0.45, 2.07) 3.17 (1.25, 8.02) "Not materially changed by 2, 4, 5, 3, weakened by 10[c]"	—
Dejin-Karlsson et al. (1998) Sweden (53)	Prospective study from 1st prenatal visit (826 singletons, 563 non-smokers)	Any ETS at home or work (57%)	2.4 (1.0,5.8) 3.9 (1.4, 10.7) Adj for 4, 5, 9, 13	1.2 (0.7,2.3) crude (includes maternal smokers)
Sadler et al. (1999) U.S. – Connecticut (45)	Prospective nested cohort Postpartum interview for ETS (2,714; 2,283 non-smokers)	Regular exposure in 3rd trimester to other's smoke >1 h/week (27%) Duration, h/week	0.82 (0.51, 1.33) 0.99 (0.97, 1.01) Adj for 2–6, 8, 9, 13–17	0.67 (0.34, 1.8) crude, from data presented
Windham et al. (2000) U.S. – California (44)	Prospective study of pregnant women (1st trimester) (4,454 singletons; 3,646 non-smokers)	Hours ETS at home/work (20% any ETS) ≥7 h/day (4%)	1.8 (0.8, 4.1) Adj for 4, 5, 8, 9, 18	1.6 (0.87, 2.9) 2.4 (1.0.5.3) ≤35 weeks Adj for 4, 5, 8, 9, 18 (others examined)

[a] In non-smokers unless otherwise stated.

[b] Odds ratio presented is for LBW in Jedrychowski and Flak, Horta et al., Ahluwalia et al., and Windham et al. Others are for IUGR.

[c] Adjusted for gestational age,[1] parity,[2] infant gender,[3] education and/or social class,[4] maternal stature[5] (height, weight, or BMI), weight gain,[6] prenatal care,[7] pregnancy outcome history,[8] race or skin color,[9] marital status,[10] alcohol consumption,[11] altitude,[12] age,[13] religion,[14] hypertension,[15] pre-eclampsia,[16] past-smoking,[17] stressful life events.[18]

of misreporting of active smoking status. One explanation may be that women absorb ETS directly into the hair so that nicotine reaches relatively higher levels than in serum or urine. This study had some other limitations including small sample size and lack of control for confounders, although the authors appear to have examined some potential confounders.

2.2.3 GESTATIONAL AGE AND PRETERM DELIVERY

Fewer studies have been published which examined gestational age or preterm delivery (<37 weeks) in relation to ETS exposure. Some of these were reviewed previously with the conclusion that there was little evidence of an effect,[3] but were not included in the meta-analysis conducted for fetal growth. Most of these reported ORs centered around the null value[27,34,37,47–49,51] or stated that duration of pregnancy was not related to ETS exposure.[25] A study of East Indian births[24] found a crude OR of 1.6 for preterm delivery and exposure at home (95% CL 0.82, 2.9). A much more thorough prospective study of Swedish women[36] examined ETS exposure at home (live with smoker) or work (most of time at work spent around smokers). Preterm birth was less common with home ETS exposure, but more common with exposure at work (adjusted OR 1.3; 95% CL 0.7, 2.3) compared to no exposure. This risk ratio was increased to 1.5 among women working full time (with presumably greater exposure) and was 1.9 (95% CL 1.0, 3.5) in women only exposed in the workplace (less likelihood of misclassification).

Of the recent studies not previously reviewed, four presented data on preterm delivery (Table 2.2). The study from Brazil[42] reported an adjusted OR of 1.25 associated with partner's smoking, but maternal smoking showed little association with preterm delivery overall, only if the preterm births were also LBW and/or IUGR. Gestational age was determined by clinicians trained using the Dubowitz method, rather than by mother's report of last menstrual period or birth certificate data. Similarly, the Swedish study[53] reported a risk of 1.2 with ETS exposure, but did not adjust for any confounders. One of the U.S. studies reported an almost doubled risk of preterm birth with ETS exposure (at home) among women greater than 30 years old.[43] The California study[44] also found an elevated adjusted risk of preterm delivery at greater hours (≥7 h/day) of ETS exposure, which was stronger and showed a dose-response trend for even earlier deliveries (<35 weeks) (Table 2.2). By race, the association appeared limited to non-whites in this study. There was also a suggestion that infants of older women were more strongly affected.

2.2.4 SUPPORTING DATA

Smoking during pregnancy has long been considered a risk factor for LBW.[1] The literature is extremely consistent, with infants of smokers on average weighing 150 to 200 g less than those of non-smokers and bearing a doubled rate of LBW.[1,54] The association is independent of other variables such as maternal age, education or parity, alcohol and drug use, and prenatal care.[54–56] Furthermore, the effects appear to be consistent with a dose-response relationship.[56] A recent meta-analysis that included 23 studies of active smoking found a pooled relative risk of 1.8 for smoking and LBW.[55] Based on the prevalences of smoking during pregnancy of 18 to 27%

in the U.S. (which may be high now), the authors estimated that 11 to 21% of LBW could be attributable to maternal smoking.

There is evidence to suggest that maternal active smoking has a greater effect on IUGR than on overall LBW.[57,58] Kramer reported a pooled relative risk of 2.4 for IUGR, which yielded an attributable fraction of 22 to 36% for smoking.[57] In developed countries where the smoking prevalence in pregnant women is higher than in developing countries, Kramer[59] found cigarette smoking to be the greatest determinant of IUGR, with the largest etiologic fraction when compared to a number of socioeconomic factors. In developing countries, other factors associated with lower weight gain and maternal size play a comparatively greater role than smoking. Other anthropometric measures have indicated reductions in fetal length, head circumference, and chest circumference with maternal active smoking,[58] suggesting symmetric growth retardation. Although the data on the relationship of maternal smoking and preterm delivery are less consistent, numerous studies have shown an association.[57,62–64] Kramer estimated a pooled relative risk of 1.4, considering a few of the best studies, yielding an attributable fraction of 8 to 14%.[57]

In general, the results of animal studies support an effect of prenatal mainstream and sidestream smoke exposure on fetal weight at term.[3] "Preterm" delivery is not commonly examined in animal studies. Tobacco smoke contains thousands of chemicals, many of which (e.g., nicotine, carbon monoxide, toluene, cadmium, and PAHs) are considered reproductive toxicants and have been associated with adverse effects on fetal growth in animals.[3] A primary mechanism of effect is thought to be via fetal hypoxia: carbon monoxide impairs oxygen availability to the fetus by binding to hemoglobin, and nicotine has strong vasoconstrictive properties, potentially reducing placental blood flow.[3,56] Premature delivery may arise as a result of placenta previa or abruptio placenta, also found associated with smoking. Constituents of smoke may also be related to other biochemical changes, such as lowering estrogen production or metabolism, or increasing catecholamine levels.[56]

2.2.5 MODIFYING FACTORS

Only a few of the studies that examined fetal growth in relation to ETS exposure have reported results of analyses to identify effect modifiers or more susceptible subgroups. A U.S. study[43] had a primary goal of examining effects by maternal age and generally found associations to be much greater among older women (30 or more) than younger women (see above). However, two other studies[31,42] stated that maternal age did not modify the effects of ETS exposure, without presenting the data. Similarly, a study from California[44] found little evidence of effect modification of ETS exposure by maternal age, except perhaps for preterm delivery. In those data, associations of active smoking with LBW and a weight decrement were greater in older women. Several other studies of active smoking have shown stronger effects on LBW or IUGR in older mothers.[60,61] A possible mechanism for such an effect might be a decline in cardiovascular reserve or decreased placental efficiency with age, so that the hypoxic effects of smoke are more difficult to counter.

A study conducted in the U.S.[35] reported that ETS and LBW were significantly associated in non-whites, but not in whites, with a higher OR among the non-whites.

The large California study also examined associations by these categories and found much greater associations in non-whites for decreased birthweight, as well as increased risks of LBW and preterm delivery.[44] Similar modification was generally seen with active smoking as well in that study. However, some studies of active smoking have found no racial differences.[65] There is some evidence that blacks may metabolize nicotine differently than whites, as they generally have higher serum cotinine levels per cigarette consumed.[65–67] Aside from genetic variations in metabolism, such differences may arise from varying behavior patterns regarding types of cigarettes smoked and the style of smoking.

A few ETS studies examined socioeconomic factors as modifiers, but did not yield consistent results. A greater effect among lower social classes might be attributed to poorer housing conditions and ventilation, leading to increased exposure. Another contributing factor might be maternal nutrition, as maternal weight and gestational weight gain are predictors of infant birthweight and poorer prepregnancy nutritional status may make it more difficult to compensate for the effects of tobacco smoke. Studies have shown greater effects of active smoking on fetal growth retardation or birthweight among women with lower prepregnancy weight[68] or lower tricep skinfold thickness.[69]

2.3   IMPACT

Considering the biologic plausibility and numerous consistent epidemiologic studies, it appears that ETS exposure is associated with a slight decrement in birthweight (perhaps 25 to 45 g) or increment (20 to 40%) in fetal growth retardation. This would represent a decrement of about 1% in mean birthweight, compared with a 5 to 6% decrement associated with active smoking. Although it is difficult to separate out the possibility of uncontrolled confounding or misclassification in an individual study with an effect measure of low magnitude, the consistency of the association with decreased birthweight found in numerous studies of varying designs and populations strengthens the evidence. This association appears even stronger in studies using a biomarker of exposure, which should reduce exposure misclassification. Furthermore, some studies have found evidence for dose-response trends. Thus, greater effects may occur in pregnancies where women have heavier exposures. The data on the risks of preterm delivery show less consistent results, but several studies indicate the possibility of increased risk, perhaps more so in older women.

Infants of lower birthweight and gestational age are at increased risk for neonatal mortality and morbidity.[56,70,71] LBW is a major determinant of infant mortality, particularly during the first month of life (or neonatal period). The increased risk of mortality associated with being born with a LBW appears to continue through childhood.[73,74] The longer the gestation for a given birthweight, the lower the mortality.[70,72] Infants born preterm are also subject to a number of complications associated with physiological immaturity, including cerebral palsy, hyaline membrane disease, sepsis, and seizure disorders. Subtle neurological deficiencies also appear more common, and this risk increases with decreasing birthweight.[70] Some data suggest that infants who experience symmetrical growth retardation (in weight,

TABLE 2.3
Attributable Fraction (%) of IUGR or LBW Due to
ETS Exposure in Non-Smokers

Prevalence of ETS exposure	Relative risk = 1.2	Relative risk = 1.4
20%	3.8%	7.4%
40%	7.4%	13.8%
60%	10.7%	19.4%

length, and head circumference) are less likely to exhibit later "catch-up" growth and also appear more likely to have cognitive deficits and difficulties in school.[70]

Thus, if prenatal ETS exposure is causally associated with lowered birthweight, a number of long-lasting effects on child health may result. The overall small magnitude of effect of ETS exposure on mean birthweight may not be clinically significant for an individual infant at low risk. Yet, if the entire birthweight distribution is shifted lower with ETS exposure, as it appears to be with active smoking, infants who are already compromised may be pushed into even higher risk categories. In addition, particularly in developing countries where the spouse is the primary source of prenatal ETS exposure, the child may continue to be exposed to tobacco smoke in the home as critical development continues.

Unlike active smoking among women, prenatal exposure to ETS is about twice as frequent in developing countries (40 to 60%) compared to developed countries (20 to 25%), particularly where the smoking frequency among men is high and there are few restrictions on smoking in public places. A study of a large U.S. sample (NHANES III) found that 88% of non-smokers had detectable levels of cotinine,[9] using an assay with a very low level of detection. The sample, however, was not restricted to pregnant women who may try to avoid smoke exposure due to increased public attention to this potential hazard. The prevalence of reported exposure appears greater at work than at home in this and other recent U.S. surveys; national data indicate that some two thirds of women of reproductive age work. As smoking becomes regulated in the work place, social settings may become more important sources of exposure.[75]

Table 2.3 indicates the range of LBW or IUGR that may be attributable to ETS exposure in non-smokers, based on an increase in risk of 20 to 40% and different rates of exposure. Thus, as the proportion exposed increases, the greater the portion of LBW in a population that can be potentially attributed to ETS exposure. Because ETS exposure of pregnant women is more common than active smoking, it may contribute almost as much to the burden of LBW or IUGR worldwide, despite lower risks to an individual pregnant woman.

2.4   FUTURE EFFORTS

Despite some study weaknesses, there are data from several high quality studies, as well as contributing data from many others, indicating a likely effect of ETS exposure on birthweight. Length of gestation has not been as frequently studied, nor have other anthropometric measures such as infant length. Future studies of these end-points should be large in order to detect a relatively small effect and should collect information on the many potential confounders. To address other methodological weaknesses, ETS exposure assessment must be thorough, with, at a minimum, ascertainment of multiple sources of exposure (home, work, and elsewhere). Some quantification of length or intensity of exposure is also important in order to measure a dose-response trend or threshold effect. As ETS exposure may be ubiquitous in many areas, measurement of a biomarker of exposure is preferred, but it is costly and usually requires prospective studies. Cotinine may be measured in other fluids besides blood, including urine and saliva, as well as nicotine in hair, making it less invasive. Development or modification of bioassays to measure other compounds in tobacco smoke may also be useful. These recommendations suggest the need to standardize study quality.

A next step would be to identify modifying factors; these may help in clarifying mechanisms of effect, as well as defining populations that are more susceptible or at higher risk, in order to target interventions. Another area of investigation will be effective intervention strategies. The U.S., and California in particular, have enacted legislation limiting smoking in public and work places. Data have shown such regulations to be effective in reducing work place exposure,[76] but other environments may be more difficult to control. Community campaigns to educate the public about health concerns for pregnant women and children will be necessary and may require different designs to reach specific target groups at greater risk or more likely to be exposed. Education of health care providers to recommend that their patients avoid exposure may also be useful. Some prevention strategies are discussed in other chapters. Such issues necessitate a multifaceted approach in order to reduce the burden of disease that may result from ETS exposure.

REFERENCES

1. U.S. Department of Health and Human Services (USDHHS), *The Health Consequences of Smoking for Women: A Report of the Surgeon General*, U.S. Department of Health and Human Services, Public Health Service, Office of the Assistant Secretary for Health, Office of Smoking and Health, Atlanta, GA, 1980.
2. U.S. Environmental Protection Agency, Respiratory Health Effects of Passive Smoking: Lung Cancer and Other Disorders, EPA Publication No. EPA/600/6-90/006F, Sections 7.7 and 8.4.3, U.S. EPA, Washington, D.C., 1992.
3. National Cancer Institute, Health Effects of Exposure to Environmental Tobacco Smoke: The Report of the California Environmental Protection Agency, *Smoking and Tobacco Control Monograph No. 10*, NCI, NIH Pub. No. 99-4645, U.S. Department of Health and Human Services, National Institutes of Health, Bethesda, MD, 1999, chap. 2 and 3.
4. Windham, G.C., Eaton, A., and Hopkins, B., Evidence for an association between environmental tobacco smoke exposure and birth weight: a meta-analysis and new data, *Pediatr. Perinat. Epidemiol.*, 13, 35–57, 1999.
5. World Health Organization, Tobacco Free Initiative, *International Consultation on Environmental Tobacco Smoke (ETS) and Child Health*, WHO, Switzerland, 1999.

6. McLaughlin, J.K., Dietz, M.S., Mehl, E.S., and Blot, W.J., Reliability of surrogate information on cigarette smoking by type of informant, *Am. J. Epidemiol.*, 126, 144–146, 1987.
7. Coultas, D.B., Peake, G.T., and Samet, J.M., Questionnaire assessment of lifetime and recent exposure to environmental tobacco smoke, *Am. J. Epidemiol.*, 130, 338–347, 1989.
8. Emmons, K.M., Abrams, D.B., Marshall, R.J., Etzel, R.A., Novotny, T.E., Marcus, B.H., and Kane, M.E., Exposure to environmental tobacco smoke in naturalistic settings, *Am. J. Public Health*, 82, 24–27, 1992.
9. Pirkle, J.L., Flegal, K.M., Bernert, J.T., Brody, D.J., Etzel, R.A., and Maurer, K.R., Exposure of the U.S. population to environmental tobacco smoke, *J. Am. Med. Assoc.*, 275, 1233–1240, 1996.
10. Kawachi, I. and Colditz, G.A., Invited commentary: confounding, measurement error, and publication bias in studies of passive smoking, *Am. J. Epidemiol.*, 144, 909–914, 1996.
11. Jarvis, M.J. and Russell, M.A.H., Measurement and estimation of smoke dosage to non-smokers from environmental tobacco smoke, *Eur. J. Respir. Dis. (Suppl.)*, 133, 68–75, 1984.
12. Hood, R.D., An assessment of potential effects of environmental tobacco smoke on prenatal development and reproductive capacity, in *Environmental Tobacco Smoke. Proceedings of the International Symposium at McGill University 1989*, Ecobichon, D.J. and Wu, J.M., Eds., Lexington Books, Lexington, MA, 1990, 241–268.
13. Guerin, M.R., Jenkins, R.A., and Tomkins, B.A., *The Chemistry of Environmental Tobacco Smoke: Composition and Measurement*, Lewis Publishers, Boca Raton, FL, 1992.
14. U.S. Department of Health and Human Services (USDHHS), *The Health Consequences of Involuntary Smoking: A Report of the Surgeon General*, DHHS Publication No. (CDC) 87-8398, U.S. DHHS, Public Health Service, Centers for Disease Control, Atlanta, GA, 1986.
15. Rebagliato, M., Bolumar, F., du B Florey, C., Jarvis, M.J., Perez-Hoyos, S., Hernandez-Aguado, I., and Avino, M.J., Variations in cotinine levels in smokers during and after pregnancy, *Am. J. Obstet. Gynecol.*, 178, 568–571, 1998.
16. Frieman, J., Chalmers, T.C., Smith, H., Jr., and Kuebler, R.R., The importance of beta, the type II error and sample size in the design and interpretation of the randomized control trial: survey of 71 "negative" trials, *N. Engl. J. Med.*, 299, 690–694, 1978.
17. Greenland, S., Quantitative methods in the review of epidemiologic literature, *Epidemiol. Rev.*, 9, 1–30, 1987.
18. Schwartz-Bickenbach, D., Schulte-Hobein, B., Abt, S., Plum, C., and Nau, H., Smoking and passive smoking during pregnancy and early infancy: effects on birth weight, lactation period, and cotinine concentrations in mother's milk and infant's urine, *Toxicol. Lett.*, 35(1), 73–81, 1987.
19. Borlee, I., Bouckaert, A., Lechat, M.F., and Misson, C.B., Smoking patterns during and before pregnancy: weight, length and head circumference of progeny, *Eur. J. Obstet. Gynecol. Reprod. Biol.*, 8(4), 171–177, 1978.
20. Mathai, M., Skinner, A., Lawton, K., and Weindling, A.M., Maternal smoking, urinary cotinine levels and birth weight, *Aust. N.Z. J. Obstet. Gynaecol.*, 30, 33–36, 1990.
21. Rubin, D.H., Krasilnikoff, P.A., Leventhal, J.M., Weile, B., and Berget, A., Effect of passive smoking on birth weight, *Lancet*, 2, 415–417, 1986.
22. Campbell, M.J., Lewry, J., and Wailoo, M., Further evidence for the effect of passive smoking on neonates, *Postgrad. Med. J.*, 64(755), 663–665, 1988.
23. Chen, Y., Pederson, L.L., and Lefcoe, N.M., Passive smoking and low birth weight [Letter], *Lancet*, 2(8653), 54–55, 1989.

24. Mathai, M., Vijayasri, R., Babu, S., and Jeyaseelan, L., Passive maternal smoking and birth weight in a South Indian population, *Br. J. Obstet. Gynaecol.*, 99(4), 342–343, 1992.
25. Martinez, F.D., Wright, A.L., Taussing, L.M., and the Group Health Medical Associates, The effect of paternal smoking on the birth weight of newborns whose mothers did not smoke, *Am. J. Public Health*, 84, 1489–1491, 1994.
26. Zhang, J. and Ratcliffe, J.M., Paternal smoking and birth weight in Shanghai, *Am. J. Public Health*, 83, 207–210, 1993.
27. Saito, R., The smoking habits of pregnant women and their husbands, and the effect on their infants, *Nippon Koshu Eisei Zasshi*, 38, 124–131, 1991.
28. MacMahon, B., Alpert, M., and Salber, E., Infant weight and parental smoking habits, *Am. J. Epidemiol.*, 82, 247–261, 1966.
29. Magnus, P., Berg, K., Bjerkedal, T., and Nance, W.E., Parental determinants of birth weight, *Clin. Genet.*, 26, 397–405, 1984.
30. Roquer, J.M., Figueras, J., Botet, F., and Jimenez, R., Influence on fetal growth of exposure to tobacco smoke during pregnancy, *Acta Paediatr.*, 84, 118–121, 1995.
31. Lazzaroni, F., Bonassi, S., Manniello, E., Morcaldi, L., Repetto, E., Ruocco, A., Calvi, A., and Cotellessa, G., Effect of passive smoking during pregnancy on selected perinatal parameters, *Int. J. Epidemiol.*, 19, 960–966, 1990.
32. Rebagliato, M., du V Florey, C., and Bulumar, F., Exposure to environmental tobacco smoke in nonsmoking pregnant women in relation to birth weight, *Am. J. Epidemiol.*, 142, 568–571, 1995.
33. Haddow, J.E., Knight, G.J., Palomaki, G.E., and McCarthy, J.E., Second-trimester serum cotinine levels in nonsmokers in relation to birth weight, *Am. J. Obstet. Gynecol.*, 159, 481–484, 1988.
34. Eskenazi, B., Prehn, A.W., and Christianson, R.E., Passive and active maternal smoking as measured by serum cotinine: the effect on birth weight, *Am. J. Public Health*, 85, 395–398, 1995.
35. Mainous, A.G. and Heuston, W.J., Passive smoke and low birth weight, *Arch. Fam. Med.*, 3, 875–878, 1994.
36. Ahlborg, G., Jr. and Bodin, L., Tobacco smoke exposure and pregnancy outcome among working women, a prospective study at prenatal care centers in Orebro County, Sweden, *Am. J. Epidemiol.*, 133, 338–347, 1991.
37. Martin, T.R. and Bracken, M.B., Association of low birth weight with passive smoke exposure in pregnancy, *Am. J. Epidemiol.*, 124, 633–642, 1986.
38. Ogawa, H., Tominaga, S., Hori, K., Noguchi, K., Kanou, I., and Matsubara, M., Passive smoking by pregnant women and fetal growth, *J. Epidemiol. Comm. Health*, 45, 164–168, 1991.
39. Ahlborg, G., Jr., Health effects of environmental tobacco smoke on the offspring of non-smoking women, *J. Smoking-Related Dis.*, 5, 107–112, 1994.
40. Peacock, J.L., Cook, D.G., Carey, I.M., Jarvis, M.J., Bryant, A.E., Anderson, H.R., and Bland, J.M., Maternal cotinine level during pregnancy and birthweight for gestational age, *Int. J. Epidemiol.*, 27, 647–656, 1998.
41. Jedrychowski, W. and Flak, E., Confronting the prenatal effects of active and passive tobacco smoking on the birth weight of children, *Centr. Eur. J. Public Health*, 4, 201–205, 1996.
42. Horta, B.L., Victora, C.G., Menezes, A.M., Halpern, R., and Barros, F.C., Low birth weight, preterm births and intrauterine growth retardation in relation to maternal smoking, *Paediatr. Perinat. Epidemiol.*, 11, 140–151, 1997.

43. Ahluwalia, I.B., Grummer-Strawn, L., and Scanlon, K.S., Exposure to environmental tobacco smoke and birth outcome: increased effects on pregnant women aged 30 years or older, *Am. J. Epidemiol.*, 146, 42–47, 1997.
44. Windham, G.C., Hopkins, B., Fenster, L., and Swan, S.H., Active or passive tobacco smoke exposure and the risk of preterm delivery or low birthweight, *Epidemiology*, 11, 427–433, 2000.
45. Sadler, L., Belanger, K., Saftlas, A., Leaderer, B., Hellenbrand, K., McSharry, J.-E., and Bracken, M.B., Environmental toabcco smoke exposure and small-for-gestational-age births, *Am. J. Epidemiol.*, 150, 695–705, 1999.
46. Wang, X., Tager, I.B., van Vunakis, H., Speizer, F.E., and Hanrahan, J.P., Maternal smoking during pregnancy, urine cotinine concentrations, and birth outcomes, a prospective cohort study, *Int. J. Epidemiol.*, 26, 978–987, 1997.
47. Nakamura, M., Oshima, A., Hiyama, T., Kubota, N., Wada, K., and Yano, K., Effect of passive smoking during pregnancy on birth weight and gestation: a population-based prospective study in Japan, in *Smoking and Health, 1987*, Aoki, M., Hisamichi, S., and Tominaga, S., Eds., Proc. of the 6th World Conf. on Smoking and Health, Tokyo, November 9–12, 1987, Excerpta Medica, Amsterdam, International Congress Series, 780, 267–269, 1988.
48. Mau, G. and Netter, P., The effects of paternal cigarette smoking on perinatal mortality and the incidence of malformations, *Dtsch. Med. Wochenschr.*, 99, 1113–1118, 1974.
49. Underwood, P.B., Kesler, K.F., O'Lane, J.M., and Callagan, D.A., Parental smoking empirically related to pregnancy outcome, *Obstet. Gynecol.*, 29, 1–8, 1967.
50. Chen, L.H. and Petiti, D.B., Case-control study of passive smoking and the risk of small-for-gestational-age at term, *Am. J. Epidemiol.*, 142, 158–165, 1995.
51. Fortier, I., Marcoux, S., and Brisson, J., Passive smoking during pregnancy and the risk of delivering a small-for-gestational-age infant, *Am. J. Epidemiol.*, 139, 294–301, 1994.
52. Nafstad, P., Fugelseth, D., Qvigstad, E., Zahlsen, K., Magnus, P., and Lindemann, R., Nicotine concentration in the hair of nonsmoking mothers and size of offspring, *Am. J. Public Health*, 88, 120–124, 1998.
53. Dejin-Karlsson, E., Hanson, B.S., Ostergren, P.-O,. Sjoberg, N.-O., and Marsal, K., Does passive smoking in early pregnancy increase the risk of small-for-gestational-age infants?, *Am. J. Public Health*, 88, 1523–1527, 1998.
54. American College of Obstetrics and Gynecology (ACOG) educational bulletin. Smoking and women's health, *Int. J. Gynecol. Obstet.*, 60, 71–82, 1997.
55. DiFranza, J.R. and Lew, R.A., Effect of maternal cigarette smoking on pregnancy complications and sudden infant death syndrome, *J. Fam. Pract.*, 40, 385–394, 1995.
56. Walsh, R.A., Effects of maternal smoking on adverse pregnancy outcomes: examination of the criteria of causation, *Hum. Biol.*, 66, 1059–1092, 1994.
57. Kramer, M.S., Determinants of low birth weight: methodological assessment and meta-analysis, *WHO Bull.*, 65, 663–737, 1987.
58. Cliver, S., Maternal smoking and fetal growth, Association of Reproductive Health Professionals (ARHP), *Clin. Proc.*, 17–18, 1996.
59. Kramer, M.S., Socioeconomic determinants of intrauterine growth retardation, *Eur. J. Clin. Nutr.*, 52, S29–S33, 1998.
60. Wen, S.W., Goldenberg, R.L., Cutter, G.R., Hoffman, H.J., Cliver, S.P., Davis, R.O., and DuBard, M.B., Smoking, maternal age, fetal growth, and gestational age at delivery, *Am. J. Obstet. Gynecol.*, 162, 53–58, 1990.
61. Wisborg, K., Henriksen, T.B., Hedegaard, M., and Secher, N.J., Smoking during pregnancy and preterm birth, *Br. J. Obstet. Gynaecol.*, 103, 800–805, 1996.

62. Wright, S.P., Mitchell, E.A., Thompson, J.M.D., Clements, M.S., Ford, R.P.K., and Stewart, A.W., Risk factors for preterm birth: a New Zealand study, *N.Z. Med. J.*, 111, 14–16, 1998.

63. Fox, S.H., Koepsell, T.D., and Daling, J.R., Birth weight and smoking during pregnancy — effect modification by maternal age, *Am. J. Epidemiol.*, 139, 1008–1015, 1994.

64. Cnattingius, S., Axelsson, O., Eklund, G., and Lindmark, G., Smoking, maternal age, and fetal growth, *Obstet. Gynecol.*, 66, 449–452, 1985.

65. U.S. Department of Health and Human Services (USDHHS), *Tobacco Use Among U.S. Racial/Ethnic Minority Groups — African Americans, American Indians and Alaska Natives, Asian Americans and Pacific Islanders, and Hispanics: A Report of the Surgeon General*, U.S. Department of Health and Human Services, Centers for Disease Control and Prevention, National Center for Chronic Disease Prevention and Health Promotion, Office on Smoking and Health, Atlanta, GA, 1998.

66. English, P.B., Eskenazi, B., and Christianson, R.E., Black-white differences in serum cotinine levels among pregnant women and subsequent effects on infant birth weight, *Am. J. Public Health*, 84, 1439–1443, 1994.

67. Klebanoff, M.A., Levine, R.J., Clemens, J.D., DerSimonian, R., and Wilkins, D.G., Serum cotinine concentration and self-reported smoking during pregnancy, *Am. J. Epidmiol.*, 148, 259–262, 1998.

68. Spinillo, A., Capuzzo, E., Nicola, S.E., Colonna, L., Egbe, T.O., and Zara, C., Factors potentiating the smoking-related risk of fetal growth retardation, *Br. J. Obstet. Gynaecol.*, 101, 954–958, 1994.

69. Neggers, Y., Goldenberg, R.L., Cliver, S.P., Hoffman, H.J., and Copper, R.L., The relationship between maternal skinfold thickness, smoking and birth weight in black and white women, *Paediatr. Perinat. Epidemiol.*, 8, 216–221, 1994.

70. McCormick, M.C., The contribution of low birth weight to infant mortality and childhood morbidity, *N. Engl. J. Med.*, 82–90, 1985.

71. Paneth, N.S., The problem of low birth weight, in *The Future of Children: Low Birth Weight*, Behrman, R.E., Ed., The David and Lucile Packard Foundation, Los Altos, CA, 5, 19–34, 1995.

72. Wilcox, A.J. and Skjaerven, R., Birth weight and perinatal mortality: the effect of gestational age, *Am. J. Public Health*, 82, 378–382, 1992.

73. Samuelson, S.O., Magnus, P., and Bakketeig, L.S., Birth weight and mortality in childhood in Norway, *Am. J. Epidemiol.*, 148, 983–991, 1998.

74. Xu, B., Rantakallio, P., and Jarvelin, M.-R., Mortality and hospitalizations of 24-year old members of the low-birthweight cohort in Northern Finland, *Epidemiology*, 9, 662–665, 1998.

75. O'Connor, T.Z., Holford, T.R., Leaderer, B.P., Hammond, S.K., and Bracken, M.B., Measurement of exposure to environmental tobacco smoke in pregnant women, *Am. J. Epidemiol.*, 142, 1315–1321, 1995.

76. Pierce, J.P., Evan, N., Farkas, A.J., Cavin, S.W., Berry, C., Kramer, M., Kealey, S., Rosbrook, B., Choi, W., and Kaplan, R.M., Tobacco Use in California, An Evaluation of the Tobacco Control Program, 1989–1993, La Jolla, CA, University of California, San Diego, 1994.

3 Antioxidant Vitamins in Reducing the Toxicity of Environmental Tobacco Smoke

Ronald R. Watson and Jin Zhang

CONTENTS

3.1 INTRODUCTION

Environmental tobacco smoking (ETS), passive smoke, or second-hand smoke is a pervasive contaminant in public places. Exposure to ETS results in oxidative stress and damage.[1] Oxidative stress, a disturbance in the prooxidant–antioxidant balance in favor of the former,[2] leads to lipid peroxidation; alteration of antioxidant enzymes in the heart, liver, and lung; as well as potential damage to DNA. The accumulated oxidative damage is believed to play an important role in the etiology of a number of chronic diseases induced by tobacco smoke, such as chronic bronchitis and emphysema, cancer, and cardiovascular disease.

Antioxidants are small molecules that act as scavengers of reactive oxygen species and prevent them from causing further cellular damage. A cell may defend itself against oxidative stress through the use of antioxidants. Furthermore, antioxidants can modulate inflammation by interacting with cytokines and reactive oxygen species biology.

With the process of studies on reactive oxygen species in various kinds of diseases, especially in smoking-related heart and pulmonary dysfunction, the role

0-8493-0311-7/00/$0.00+$.50
© 2001 by CRC Press LLC

of the antioxidants is becoming more and more important. This chapter provides an overview of possible associations between antioxidant vitamins and toxicity of ETS.

3.2 TOXICITY OF ENVIRONMENTAL TOBACCO SMOKE

Tobacco smoke (TS) is a complex mixture of particulate and gaseous compounds. It is estimated that cigarette smoke contains over 4000 different chemical constituents.[3–6] The toxicity of TS is due to a large variety of compounds, including nicotine, cadmium, benzapyrene, oxidants, and free radicals that initiate, promote, or amplify oxidative damage.[7,8] The combustion of tobacco during cigarette smoking promotes the oxidation of polycyclic aromatic hydrocarbons and generates free radicals (or reactive oxygen metabolites). Free radicals are generated both in the particulate (tar) phase and gas phase of cigarette smoke. Approximately 10^{14} free radicals per puff have been estimated in the tar phase. More free radicals are in the gas phase, estimated at 10^{15} free radicals per puff.[9] Free radicals and the oxygen singlet have been implicated in the pathogenesis of arteriosclerosis, coronary disease, pulmonary and other cancers, and chronic bronchitis and emphysema.[10]

While the risks associated with smoking have been fairly well documented, the risks from ETS are not as well known. ETS consists of exhaled TS and sidestream cigarette smoke (SSCS) which contains many free radicals and reactive compounds. The composition of mainstream cigarette smoke (smoke drawn into the mouth during puffs) is quantitatively and, to some extent, qualitatively different from that of SSCS (smoke formed from the lit end of a burning cigarette).[11–13] It is known that SSCS is unfiltered, since it does not pass through the column of tobacco or the filter of cigarette, and it is generated from tobacco combustion at a lower temperature with limited availability of oxygen inside the burning cone. These differences make SSCS more toxic than mainstream cigarette smoke on an equal molar basis. Some toxic, carcinogenic, and cocarcinogenic substances such as nicotine, carbon monoxide, carbon dioxide, catechol, benzopyrenes, N-nitrosamines [N-nitrosodimethylamine, N'-nitrosoanabasine, and 4-(methylnitrosamino)-1-(3-pyridyl)-1-butanone], and ammonia are enriched in SSCS.[14,15]

Over the past several years, ETS has come to the forefront of the social debate pitting smokers against nonsmokers. ETS is probably the most important contamination of indoor air. A national survey reported that 37.4% of adult nonsmokers reported exposure to ETS at home or in the workplace, and 25% of nonsmoking adults report exposure at work.[16] A report released by the California Environmental Protection Agency in September of 1997 attributed 35,000 to 62,000 ischemic heart disease deaths every year in the U.S. to ETS. Additionally, over 2000 childhood deaths from both bronchitis and Sudden Infant Death Syndrome as well as 3000 lung cancer deaths have been linked to ETS.[17]

In 1986, the U.S. Surgeon General concluded that ETS can cause lung cancer in adult nonsmokers and that children of parents who smoke have increased frequency of respiratory symptoms, acute lower respiratory tract infection, and reduced lung function. In 1992, the Office of Health and Environmental Assessment of the U.S. EPA classified ETS as a Group A carcinogen, in the same group as benzene and vinyl chloride.[18] Consequently, there has been a focus of public concern on the

$O^=_2$ Superoxide Anion Radical

$^1O_2$ Singlet Molecular Oxygen

H_2O_2 Hydrogen Peroxide

HO^- Hydroxyl Radical

HO_2 Perhydroxyl Radical

Membrane Lipid Oxidation

Peroxidation of Polyunsaturated Phospholipids

Cellular Membrane Damage

DNA, Carbohydrates, Proteins

Atherosclerosis

Coronary Heart Disease

Peripheral Vascular Occlusive Disease

Pulmonary Cancer

Pulmonary Emphysema

FIGURE 3.1 Tobacco smoke and oxidative stress. (From Diana, J.N., *Ann. N.Y. Acad. Sci.*, 686, 1, 1993. With permission.)

hazards posed by ETS. This concern has been reflected by recent federal legislation prohibiting smoking on airlines, in public buildings, and in restaurants.

3.3 ENVIRONMENTAL TOBACCO SMOKE AND OXIDATIVE STRESS

Cigarette smoke contains two very different populations of free radicals, relatively stable free radicals in the tar phase and more reactive small oxygen- and carbon-centered radicals in the gas phase. Thus, it is likely that free-radical processes play a significant role in cigarette smoke toxicity. Such radicals initiate and enhance lipid peroxidation in biological membranes,[9,19,20] which could be related to the incidence of many chronic diseases (Figure 3.1).

Exposure to workplace ETS results in increased oxidative stress and damage, as measured by increased levels of the antioxidant enzymes superoxide dismutase, catalase, glutathione reductase, and glutathione peroxidase.[1] This oxidative damage occurs when the production of reactive oxygen species in a cell exceeds the cell's natural antioxidant defenses. Injury can occur to cellular constituents, membrane lipids, proteins, DNA, and RNA, leading to cell damage or death.[21] The process of cellular oxidative damage has been linked to the etiology of several chronic degenerative conditions, including cancer and coronary disease. These are both closely associated with smoking and exposure to ETS.[10] An increasing oxidative stress in mice resulted from SSCS exposure. The oxidative stress was indicated by an alteration of antioxidant enzymes in heart, liver, and lung tissues.[21,22] Exposure to ETS at work increased the level of certain antioxidant enzymes in the blood of employees.[23] Oxidative stress is believed to play a role in carcinogenesis by oxidative damage to DNA. Nuclear DNA is one of the cellular targets of reactive oxygen species. Reactive oxygen species can react with nuclear DNA, resulting in a number of damaged DNA products. Many of these DNA lesions can be repaired. However, both the cellular defense and repair systems can be overwhelmed to such a degree that the total extent of the damage cannot be repaired, particularly during exposure to reactive oxygen species-generating xenobiotics. As DNA damage accumulates within a cell, the likelihood of a cytotoxic or mutagenic event is believed to be part of the etiology of a number of chronic diseases[22–24] such as chronic bronchitis and emphysema, cancer, and cardiovascular disease.

3.4 OXIDATIVE STRESS AND PULMONARY AND HEART DISEASE

As the number of smokers goes up in the U.S., ETS is more and more pervasive. For a nonsmoker living with a smoker the exposure is equivalent to about 1% of that from actively smoking 20 cigarettes a day (based on plasma cotinine).[25] Passive smoking is associated with an increase in risk of chronic respiratory disease in adults of 25% (10 to 43%) and an increase the risk of acute respiratory illness in children by 50 to 100%. It is likely that passive smoking increases the risk of ischaemic heart disease and that exposure in pregnancy lowers birthweight.[25] The most common health risks associated with cigarette smoke are listed in Table 3.1.

Cigarette smoke is known to contain numerous oxidants and prooxidants that are capable of producing free radicals and possibly initiating lipid peroxidation.[26] Lung function can be altered by both free radical and oxidant exposure. Increased lipid peroxidation, as an indicator of oxidative stress, is associated with pulmonary airway narrowing in the general population.[27] On this point, airway thickening and narrowing with expiratory airflow obstruction can be a result of oxidative stress in passive smoking subjects. Nitrogen dioxide (NO_2), one of the major oxidant air pollutants present in environmental smoke, for example, is found as high as 250 ppm in cigarette smoke.[28] NO_2 is a common indoor air pollutant that is a proinflammatory air pollutant under conditions of repeated exposure. The data indicate that four sequential exposures to NO_2 result in a persistent neutrophilic inflammation in the airway.[29]

The cumulative smoking history is highly correlated with both leukocytosis and elevation of acute phase reactions.[30–33] This reflects a smoking-induced inflammatory

---

TABLE 3.1
Health Risks Associated with Cigarette Smoke

Heart disease
    Cardiovascular disease
    Atherosclerosis
    Coronary heart disease
    Peripheral vascular occlusive disease
    Cerebrovascular disease
Pulmonary disease
    Lung cancer
    Chronic obstructive lung disease (COLD)
    Chronic mucus hypersecretion (cough and phlegm)
    Airway thickening and narrowing with expiratory airflow obstruction
    Emphysema

Source: From Diana, J.N., *Ann. N.Y. Acad. Sci.*, 686, 1, 1993. With permission.)

---

response with increasing accumulation of alveolar macrophages and neutrophils in the lungs. Smoking causes an increase in oxidative metabolism of macrophages and neutrophils. The increased oxidative metabolism of phagocytes is accompanied by increased generation of reactive radicals. Smokers have higher neutrophil myeloperoxidase activity than nonsmokers.[34]

Also, passive smoking is a dominant risk factor for atherosclerotic disease.[35–37] An important pathologic mechanism by which cigarette smoke could potentially induce the development of vascular disease is by the impaired production of nitric oxide (NO), a molecule that significantly influences the physiologic balance in the vasculature and appears to inhibit monocyte adhesion as well as thrombosis.[38] Cigarette smoke contains a large amount of free radicals and prooxidants, NO, NO_2, peroxynitrite, and nitrosamines.[26] The particulate phase of smoke contains high concentrations of substances leading to the production of O_2^- and H_2O_2. These two compounds lead to the production of OH, which is highly cytotoxic. Furthermore, O_2^- can reach the vascular endothelium and reduces vascular NO levels significantly.[39] Long-term exposure to passive smoking impairs the arterial endothelial function, probably through impaired endothelial NO activity,[40] and increases the thickness of the carotid wall.[41] Furthermore, accelerated atherosclerotic plaque development has been found in animals subjected to cigarette smoking.[42] The free radicals entering the body are first trapped by serum aqueous and lipophilic antioxidants, which interact and provide greater protection against lipid peroxidation than antioxidant on its own. After failure of this antioxidant barrier, LDL lipid peroxidation can take place.[43] Oxidized LDL is readily taken up by scavenger receptors on the macrophage. The outcome is LDL cholesterol accumulation in macrophages. This is generally accepted as a key event of atherosclerosis.[44] There is ample evidence that oxidative stress is involved in the pathogenesis of atherosclerosis, endothelial dysfunction, platelet aggregation, and heart failure.[45]

TABLE 3.2
Antioxidant Mechanisms in Human Biology

Antioxidant enzymes
Superoxide dismutase
   Catalase
   Glutathione peroxidase/glutathione reductase
Preventative antioxidants
   Caeruloplasmin
   Transferrin
   Albumin
Chain-breaking antioxidants
   Water soluble
      Urate
      Ascorbate
      Thiols
      Bilirubin
      Flavonoids
   Lipid soluble
      Tocopherols
      Ubiquinol-10
      Beta-carotene

Source: From Simon R. J., Maxwell, G., and Lip, Y.H., *Clin. Pharmacol.*, 44, 307, 1997. With permission.)

3.5 ANTIOXIDANT SUPPLEMENTATION AND OXIDATIVE STRESS

As mentioned earlier, ETS not only produces free radicals and prooxidants, but also induces oxidative metabolism of inflammatory phagocytes.[26,30–33] Fortunately, a variety of antioxidant mechanisms have evolved to combat the potential threat of damage to vital biological structures from the reactive oxygen species sources (Table 3.2) and to protect the cells from the toxic reactions of these oxygen species.[46–49]

A cell may defend itself against oxidative stress through the use of antioxidants. These antioxidants are consumed in the process of scavenging free radicals before more important structures are damaged.[1] Oxygen free radicals reduce the level of vitamin E in vascular endothelial cells, which has been shown to be crucial for cellular integrity.[50] Cigarette smoke also has been reported to result in a depletion of antioxidant vitamins in the body, particularly ascorbic acid and beta-carotene.[51,52]

As smokers are being subjected to oxidative stress resulting from oxidants and free radicals present in smoke, as well as reactive oxygen species generated by increased activated phagocytes, their antioxidant status is likely to be adversely affected. Indeed, decreased plasma and leukocyte concentrations of vitamin E and vitamin C in smokers are associated with increased numbers and activity of neutrophils. This suggests an increased consumption of these micronutrients during neutralization of phagocyte-derived extracellular oxidants.[53–55]

Antioxidants can be conveniently divided into those that are water soluble and those that are lipid soluble and exist in environments such as lipoproteins and cell membranes to prevent the propagation phase of lipid peroxidation (chain-breaking antioxidants). Of the aqueous molecules, the best known is vitamin C (ascorbate), which is the most powerful electron donor and is the first plasma antioxidant to be sacrificed upon exposure to oxidative stress.[49] Ascorbate contributes up to 24% of the total peroxyl radical-trapping antioxidant capacity in human plasma.[56–58] There is evidence that ascorbic acid inhibits both phagocyte-induced, endothelial cell-induced, and lipid peroxidation processes.[59]

Furthermore, there are reports that antioxidant protection from vitamin C may reduce the risk or slow development of certain diseases, including cancer and coronary heart disease.[60,61] Elevated lipid peroxidation products formed by cigarette smoke exposure may generate a tissue antioxidant/oxidant imbalance that could represent a crucial link between smoke and atherosclerosis. Simultaneous administration of a megadose of vitamin C somehow rectifies the oxidant/antioxidant imbalance induced by elevated lipid peroxidation products in cigarette smoke.[59]

Vitamin E is a lipid-soluble, powerful chain-breaking antioxidant.[50] It has a protective role against the damaging effects of smoke, including in the immune cells that produce large amounts of oxidants.[62–64] Vitamin E is incorporated into the hydrophobic core of lipoprotein, where it plays a role in terminating the free radical-mediated reaction.[65] It has been extensively studied in animals as an effective prophylactic agent that potentially can protect humans from oxidant activity.[58,66]

Studies in humans have also shown protective effects of vitamin E against oxidative stress caused by cigarette smoke.[67] A study of young adult smokers showed that the lower respiratory tract fluid of smokers was deficient in vitamin E in comparison to nonsmokers. Vitamin E supplementation (2400 IU/day for 3 weeks) resulted in increased vitamin E concentrations in the lower respiratory tract fluid, but levels remained much lower than baseline levels of nonsmokers. Vitamin E may be an important antioxidant in the lung's defense against free radical damage due to cigarette smoke.[67] The red blood cells of smokers showed increased peroxidation when incubated with hydrogen peroxide compared to nonsmokers; this effect was inhibited in smokers supplemented with vitamin E (1000 IU/day for 2 weeks).[68]

Consumption of antioxidant vitamins has been associated with a reduced risk of cardiovascular disease. Antioxidant vitamins, vitamins C and E, inhibit LDL oxidation in smokers.[69] One study provided strong evidence that the increased oxidative damage that is caused by exposure to ETS in the workplace can be largely prevented through antioxidant supplementation.[1] Several studies showed that antioxidant diets improve endothelium-dependent vasomotion in hypercholesterolemic rabbits and in individuals who smoke, suggesting a direct protective effect of antioxidants on endothelial cell function.[70]

3.6 ANTIOXIDANT SUPPLEMENTATION AND IMMUNE RESPONSE

Antioxidant nutrients have a major potential for modulating inflammatory aspects of immune function by regulating aspects of cytokine and reactive oxygen species

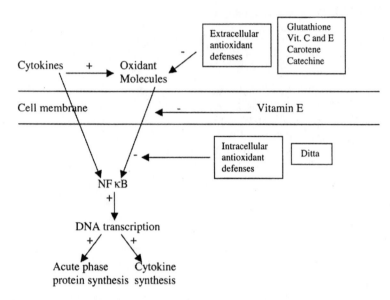

FIGURE 3.2 The interaction between proinflammatory cytokines, oxidants, and antioxidants in modulating the inflammatory response. (From Grimble, R.F., *Nutr. Res.*, 18, 1297, 1998. With permission.)

biology.[71,72] The proinflammatory cytokines interleukin 1 (IL-1), interleukin 6 (IL-6), and tumor necrosis factor-alpha (TNF-α), as well as reactive oxygen species, play a major role in inflammatory aspects of immune function.[71] Proinflammatory cytokines and oxidants have the ability to stimulate production of each other. The stimulation of cytokine production by oxidants is due to activation of nuclear factor kappa B (NF κB) by oxidants. NF κB is a multicomponent protein present in the cytoplasm of a wide range of cells, including macrophages, lymphocytes, and hepatocytes. In unstimulated cells, it is present as an inactive form due to the presence of the inhibitory subunit. Cytokines and oxidants result in the detachment of the inhibitory component from the NF κB complex. The activated transcription factor migrates to the nucleus and results in transcription of genes for the synthesis of a number of molecules associated with immune function, including acute phase proteins and proinflammatory cytokines. The interaction of cytokines and oxidants with the transcription factor leads to enhancement of cytokine production by oxidants. These effects are opposed by antioxidants (Figure 3.2). A number of studies have shown that a decrease in antioxidant defense, or an enhanced oxidant concentration in tissues, increases a number of aspects of inflammatory response.[71]

Vitamin E exerts modulatory effects on both inflammatory and immune components of immune function. In general, vitamin E deficiency and low tissue vitamin E content enhance components of the inflammatory response and suppress components of the immune response. Dietary vitamin E supplementation brings about the opposite effect. Studies in animals have demonstrated that vitamin E deficiency impairs cellular and humoral immunity.[71,73,74] In smokers, the incidence of inflammation inversely relates to the intakes of vitamins C and E.[75,76]

3.7 SUMMARY

Environmental smoke contains a large variety of compounds, including oxidants and free radicals, that are capable of initiating or promoting oxidative damage. Also, oxidative damage may result from reactive oxygen species generated by the increased and actived neutrophils and macrophages following TS.[77] Oxidative stress is a major factor in the etiology of TS-related diseases. Antioxidant vitamins could benefit the prevention of oxidative stress induced by ETS. However, several factors affect environmental smoke exposure;[76] the degree of oxidative damage is variant, and the dose of antioxidant supplementation is undetermined. More studies in animal models and humans are needed to better understand the role of antioxidant vitamins in the prevention of the diseases related to ETS in the future.

ACKNOWLEDGMENT

Preparation of this chapter was supported in part by a grant to Dr. Mark L. Witten from the Arizona Diseases Control Research Commission.

REFERENCES

1. Howard, D.J., Ota, R.B., Briggs, L.A., Hampton, M., and Pritsos, C.A., Oxidative stress induced by environmental tobacco smoke in workplace is mitigated by anti-oxidant supplementation, *Cancer Epidemiol. Biomark. Prev.*, 7, 981–988, 1998.
2. Sies, H., *Oxidative Stress: Oxidants and Antioxidants*, Academic Press, San Diego, CA, 1985, pp. XV–XXII.
3. Stedman, R.L., The chemical composition of tobacco and tobacco smoke, *Chem. Rev.*, 68, 153–207, 1968.
4. Schmeltz, I. and Hoffmann, D., Nitrogen-containing compounds in tobacco and tobacco smoke, *Chem. Rev.*, 77, 295–311, 1977.
5. Gairola, C., Genetic effects of fresh cigarette smoke in *Saccharomyces cerevisiae*, *Mutat. Res.*, 102, 123–136, 1982.
6. Hoffmann, D., Rivenson, A., Hecht, S.S., Hilfrich, I., Kobayashi, N., and Wynder, E.L., Model studies in tobacco carcinogenesis with Syrian golden hamster, *Prog. Exp. Tumor Res.*, 24, 370–390, 1979.
7. Pryor, W.A. and Stone, K., Oxidants in cigarette smoke. Radicals, hydrogen peroxide, peroxynitrate, and peroxynitrite, *Ann. N.Y. Acad. Sci.*, 686, 12–28, 1993,
8. Vayssier, M., Banzet, N., Francois, D., Bellmann, K., and Polla, B.S., Tobacco smoke induces both apoptosis and necrosis in mammalian cells: differential effects of HSP70, *Am. J. Physiol.*, 275 (*Lung Cell Mol. Physiol.*, 19), L771–L779, 1998.
9. Diana, J.N., Tobacco smoking and nutrition, *Ann. N.Y. Acad. Sci.*, 686, 1–11, 1993.
10. Duthie, G.G., Arthur, J.R., Beattie, J.A.G., Brown, K.M., Morrice, P.C., Robertson, J.D., Shortt, C.T., Walker, K.A., and James, W.P.T., Cigarette smoking, antioxidants, lipid peroxidation, and coronary heart disease, *Ann. N.Y. Acad. Sci.*, 686, 120–129, 1993.
11. U.S. Department of Health and Human Services, The Health Consequences of Involuntary Smoking, U.S. Government Printing Office, Washington, D.C., 1986.
12. Sterling, T.D. and Kobayyashi, D., Indoor byproduct level of tobacco smoke: a critical review of literature, *J. Air Pollut. Control Assoc.*, 32, 250–259, 1982.

13. Natural Research Council Committee on Passive Smoking, *Environmental Tobacco Smoke: Measuring Exposures and Assessing Health Effects*, National Academy Press, Washington, D.C., 1986.

14. Samet, J.M., Environmental tobacco smoke, in *Environmental Toxicants: Human Exposures and Their Health Effects*, Lippmann, M., Ed., Van Nostrand Reinhold, New York, 1992, 231–265.

15. Hoffmann, D. and Wynder, E.L., Chemical constituents and bioactivity of tobacco smoke, *IARC Sci. Publ.*, 74, 145–165, 1986.

16. Centers for Disease Control, State-specific prevalence of cigarette smoking — United States, 1995, *J. Am. Med. Assoc.*, 276, 1713, 1996.

17. California Environmental Protection Agency, Health Effects of Exposure to Environmental Tobacco Smoke — Final Draft, 1997.

18. California Environmental Protection Agency, *The Health Consequences of Involuntary Smoking: A Report of the Surgeon General*, Publication No. DHHS (CDC) 87-(8398), U.S. Government Printing Office, Washington, D.C., 1986.

19. Hennig, B., McClain, C.J., and Diana, J.N., Function of vitamin E and zinc in maintaining endothelial integrity: implications in atherosclerosis, *Ann. N.Y. Acad. Sci.*, 686, 99–109, 1993.

20. Harats, D., Ben-Naim, M., Dabach, Y., Hollander, G., Stein, O., and Stein, Y., Cigarette smoking renders LDL susceptible to peroxidative modification and enhanced metablism by macrophages, *Atherosclerosis*, 79, 245–252, 1989.

21. Howard, D.J., Briggs, L.A., and Pritsos, C.A., Oxidative DNA damage in mouse heart, liver, and lung tissue due to acute side-stream tobacco smoke exposure, *Arch. Biochem. Biophys.*, 352, 293–297, 1998.

22. Priddy, S. and Pritsos, C.A., Superoxide dismutase and lipid peroxidation effects of sidestream cigarette exposure on heart, liver and lung tissues, *FASB J.*, 8(4), A405, 1994.

23. Halliwell, B. and Cross, C.E., Oxygen-derived species: their relation to human disease and environmental stress, *Environ. Health Perspect.*, 102(Suppl. 10), 5–12, 1994.

24. Ames, B.N., Dietary carcinogens and anticarcinogens: oxygen radicals and degenerative diseases, *Science*, 221, 1256–1264, 1983.

25. Law, M.R. and Hackshaw, A.K., Environmental tobacco smoke [Review], *Br. Med. Bull.*, 52, 22–34, 1996.

26. Church, D.F. and Pryor, W.A., Free-radical chemistry of cigarette smoke and its toxicological implications, *Environ. Health Perspect.*, 64, 111–126, 1985.

27. Schunemann, H.J., Muti, P., Freudenheim, J.L., Armstrong, D., Browne, R., Klocke, R.A., and Trevisan, M., Oxidative stress and lung function, *Am. Epidemiol.*, 146, 939–948, 1997.

28. U.S. Department of Health and Human Services, *The Health Consequences of Smoking. Chronic Obstructive Lung Disease. A Report of Surgeon General*, DHHS Publication No. 84-50205, U.S. Government Printing Office, Washington, D.C., 1984.

29. Blomberg, A., Krishna, M.T., Helleday, R., Soderberg, M., Ledin, M.C., Kelly, F.J., Frew, A.J., Holgate, S.T., and Sandstrom, T., Persistent airway inflammation but accommodated antioxidant and lung function responses after repeated daily exposure to nitrogen dioxide, *Am. Respir. Crit. Care Med.*, 159, 536–543, 1999.

30. Bridges, R.B., Chow, C.K., and Rehm, S.R., Micronutrients and immune function in smokers, *Ann. N.Y. Acad. Sci.*, 587, 218–231, 1990.

31. Hunninghake, Chow, C.K., and Crystal, R.G., Cigarette smoking and lung destruction: accumulation of neutrophils in the lungs of cigarette smokers, *Am. Rev. Respir. Dis.*, 128, 833–838, 1983.

32. Hoidal, J.R., Fox, R.B., LeMarbre, P.A., Perri, R., and Repine, J.E., Altered oxidative metabolic responses in vitro of alveolar macrophages from asymptomatic cigarette smokers, *Am. Rev. Respir. Dis.*, 123, 85–89, 1981.

33. Hoidal, J.R. and Niewoehner, D.E., Cigarette-smoke induced phagocyte recruitment and metabolic alterations in human and hamsters, *Am. Rev. Respir. Dis.*, 126, 548–552, 1982.

34. Bridges, R.B., Fu, M.C., and Rehm, S.R., Increased neutrophil myeloperoxidase activity associated with cigarette smoking, *Eur. J. Respir. Dis.*, 67, 84–93, 1985.

35. Glantz, S.A. and Parmley, W.W., Passive smoking and heart disease: mechanisms and risk, *J. Am. Med. Assoc.*, 273, 1047–1053, 1995.

36. Kritz, H., Schmid, P., and Sinzinger, H., Passive smoking and cardiovascular risk, *Arch. Intern. Med.*, 155, 1942–1948, 1995.

37. Zhu, B.Q. and Parmley, W.W., Hemodynamic and vascular effects of active and passive smoking, *Am. Heart J.*, 130, 1270–1275, 1995.

38. Zeiher, A., Schaechinger, V., and Minners, J., Long-term cigarette smoking impairs endothelium-dependent coronary arterial vasodilator function, *Circulation*, 92, 1094–1100, 1995.

39. Radi, R., Beckman, J.W., Bush, K.M., and Freeman, B.A., Peroxynitrite induced membrane lipid peroxidation: the cytotoxic potential of superoxide and nitric oxide, *Arch. Biochem. Biophys.*, 288, 481–487, 1991.

40. Celermajer, D.S., Adams, M.R., Clarkson, P., Robinson, J., McCredie, R., Donald, A., and Deanfield, J.E., Passive smoking and impaired endothelium-dependent arterial dilatation in healthy young adults, *N. Engl. J. Med.*, 334, 150–154, 1996.

41. Diez Roux, A.V., Nieto, F.J., Comstorck, G.W., Howard, G., and Szklo, M., The relationship of active and passive smoking to carotid atherosclerosis 12-14 years later, *Prev. Med.*, 24, 48–55, 1995.

42. Penn, A., Chen, L.C., and Snyder, C.A., Inhalation of steady-state side-stream smoke from one cigarette promotes arteriosclerotic plaque development, *Circulation*, 90, 1363–1367, 1994.

43. Brown, M.S. and Goldstein, J.L., Lipoprotein metabolism in the macrophage: implications for cholesterol deposition in atherosclerosis, *Annu. Rev. Biochem.*, 52, 223–261, 1983.

44. Steinberg, D., Parthasarathy, S., Carew, T.E., Khoo, J.C., and Witztum, J.L., Beyond cholesterol. Modifications of low-density lipoprotein that increase its atherogenicity, *N. Engl. J. Med.*, 320, 915–924, 1989.

45. Hoeschen, R.J., Oxidative stress and cardiovascular disease, *Can. Cardiol.*, 13, 1021–1025, 1997.

46. Halliwell, B., How to characterize a biological antioxidant, *Free Rad. Res. Commun.*, 9, 1–32, 1990.

47. Halliwell, B. and Gutteridge, J.M.C., The antioxidants of human extracellular fluids, *Arch. Biochem. Biophys.*, 280, 1–8, 1990.

48. Stocker, R. and Frei, B., Endogenous antioxidant defenses in human blood plasma, in *Oxidative Stress: Oxidants and Antioxidants*, Academic Press, San Diego, CA, 1991, 213–242.

49. Maxwell, S.R. and Lip, G.Y., Free radical and antioxidants in cardiovascular disease, *Clin. Pharmacol.*, 44, 307–317, 1997.

50. Lucy, J.A., Functional and structure aspects of biomembranes: a suggested structure role of vitamin E in the control of membrane permeability and stability, *Ann. N.Y. Acad. Sci.*, 230, 4–16, 1972.

51. Smith, J.L. and Hodges, R.E., Serum levels of vitamin C in relation to dietary and supplemental intake of vitamin C in smokers and nonsmokers, *Ann. N.Y. Acad. Sci.*, 489, 144–152, 1987.

52. Stryker, W.S., Kaplan, L.A., Stein, E.A., Stampfer, M.J., Sober, A., and Willett, W.C., The relation of diet, cigarette smoking, and alcohol consumption to plasma beta-carotene and alpha-tocopherol levels, *Am. J. Epidemiol.*, 127, 283–296, 1988.

53. Theron, A.J., Richards, G.A., Van Rensburg, A.J., Van der Merwe, A.A., and Anderson, R., Investigation of the role of phagocytes and antioxidant nutrients in oxidant stress mediated by cigarette smoke, *Int. J. Vitam. Nutr. Res.*, 60, 261–266, 1990.

54. Barton, G.M. and Roath, O.S., Leukocyte ascorbic acid in abnormal leukocyte states, *Int. J. Vitam. Nutr. Res.*, 46, 271–274, 1976.

55. Hemila, H., Roberts, P., and Wikstrom, M., Activated polymorphonuclear leukocytes consume vitamin C, *FEBS Lett.*, 178, 25–30, 1984.

56. Bendich, A., Machlin, L.J., Scandurra, O., Burton, G.W., and Wayner, D.D.M., The antioxidant role of vitamin C, *Adv. Free Rad. Biol. Med.*, 2, 419–444, 1986.

57. Frei, B., England, L., and Ames, B.N., Ascorbate is an outstanding antioxidant in human blood plasma, *Proc. Natl. Acad. Sci. U.S.A.*, 86, 6377–6381, 1989.

58. Chakraborty, S., Nandi, A., Mukhopadhyay, C.K., and Chatterjee, I.B., Protective role of ascorbic acid against lipid peroxidation and myocardial injury, *Mol. Cell. Biochem.*, 111, 41–47, 1992.

59. Helen, A. and Vijayammal, P.L., Vitamin C supplementation on hepatic oxidative stress induced by cigarette smoke, *J. Appl. Toxicol.*, 17, 289–295, 1997.

60. Jacob, R.A. and Burri, B.J., Oxidative damage and defense, *Am. J. Clin. Nutr.*, 63, 985S–990S, 1996.

61. Jialal, I., Vega, G.L., and Grundy, S.M., Physiologic levels ascorbate inhibit the oxidative modification of low density lipoprotein, *Atherosclerosis*, 82, 185–190, 1990.

62. U.S. Environmental Protection Agency, Air Quality Criteria for Ozone and Other Photochemical Oxidants, Vol. I, EPA-600/8-84-020 aF, Research Triangle Park, NC, 1986.

63. Elsayed, N.M., Kass, R., Mustafa, M.G., Hacker, A.D., Ospital, J.J., Chow, C.K., and Cross, C.E., Effect of dietary vitamin E level on the biochemical response of rat lung to ozone inhalation, *Drug-Nutr. Interact.*, 5, 373–386, 1988.

64. Sevanian, A., Hacker, A.D., and Elsayed, N., Influence of vitamin E and nitrogen dioxide on lipid peroxidation in rat lung and liver microsomes, *Lipids*, 17, 269–277, 1982.

65. Kayden, H.J. and Traber, M.G., Absorption, lipoprotein transport, and regulation of plasma concentrations of vitamin E in humans, *J. Lipid Res.*, 34, 343–358, 1993.

66. Chakraborty, S., Nandi, A., Mukhopadhyay, M., and Mukhopadhyay, C.K., Ascorbate protects guinea pig tissues against lipid peroxidation, *Free Rad. Biol. Med.*, 16, 417–426, 1994.

67. Pacht, E.R., Kaseki, H., Mohammed, J.R., Cornwell, D.G., and Davis, W.B., Vitamin E in the alveolar fluid of cigarette smokers, *J. Clin. Invest.*, 77, 789–796, 1986.

68. Duthie, G.G., Arthur, J.R., James, W.P.T., and Vint, H.M., Antioxidant status of smokers and nonsmokers — effects of vitamin E supplementation, *Ann. N.Y. Acad. Sci.*, 393, 369–375, 1982.

69. Steinberg, F.M. and Chait, A., Antioxidants vitamin supplementation and lipid peroxidation in smokers, *Am. Soc. Clin. Nutr.*, 68, 319–327, 1998.
70. Shwarzaher, S.P., Hutchison, S., and Chou, T.M., Antioxidant diet preserves endothelium-dependent vasodilatation in resistance arteries of hypercholesteralemic rabbits exposed to environmental tobacco smoke, *J. Cardiol. Pharmacol.*, 31, 649–653, 1998.
71. Grimble, R.F., Modification of inflammation aspects of immune function by nutrients, *Nutr. Res.*, 18, 1297–1317, 1998.
72. Cross, C.E., van der Vliet, A., and Eiserich, J.P., Cigarette smokers and oxidant stress: a continuing mystery, *Am. J. Clin. Nutr.*, 67, 184–185, 1998.
73. Tengerdy, R.P., Heinzerling, R.H., and Mathias, M.M., Effect of vitamin E on disease resistance and immune responses, in *Tocopherol, Oxygen, and Biomembranes*, de Duve, C. and Hayaishi, O., Eds., Elsevier/North Holland Biomedical Press, Amsterdam, 1978, 191–200.
74. Colnago, G.L., Jensen, L.S., and Long, P.L., Effect of selenium and vitamin E on the development of immunity to coccidiosis in chickens, *Poult. Sci.*, 63, 1136–1143, 1984.
75. Tappia, P.S., Troughton, S.C., Langley-Evans, S.C., and Grimble, R.F., Smoking influences cytokine production and antioxidant defenses, *Clin. Sci.*, 88, 485–489, 1995.
76. Pringle, K.L., Modulation of the Inflammatory Response by Antioxidants, Ph.D. thesis, University of Southampton, Highfield, Southampton, U.K., 1994.
77. Chow, C.K., Cigarette smoking and oxidative damage in the lung, *Ann. N.Y. Acad. Sci.*, 686, 289–298, 1993.

4 Effects of Antioxidants on Tobacco Smoke-Associated Cardiovascular Damage

Ronald R. Watson, Julian Rodriquez, and Yinhong Chen

CONTENTS

Tobacco smoke exposure contributes significantly to cardiovascular damage through a number of pathways, including direct toxicity to the endothelium by free radicals, indirect effect of platelet aggregation, as well as increased catecholamine, homocysteine, and chronic oxidative changes occurring in LDL cholesterol. All ultimately disrupt myocardial structure and function for increased coronary heart disease.

4.1 ANTIOXIDANTS AND HEART DISEASE PREVENTION

Tobacco smoke produces excessive harmful oxidant-free radicals. A number of studies have demonstrated that free radical reactions mediate in the development of coronary heart disease and antioxidants play a protective role. Oxidants are substances and molecules with excess energy that have the ability to react with lipids, protein, and DNA, thereby modifying and sometimes destroying unprotected molecules. The body's defense against harmful oxidants is antioxidants. Antioxidants receive the excess energy and, consequently, become damaged; but in comparison to other molecules like lipids or DNA, antioxidants are easily repairable. Antioxidants, therefore, are defined as substances that readily receive excess energy in the form of electrons and are easily repairable. Common antioxidants are alpha-tocopherol (vitamin E), asorbic acid (vitamin C), beta-carotene, urate, albumin, and various other proteins and enzymes.

0-8493-0311-7/00/$0.00+$.50
© 2001 by CRC Press LLC

Powerful antioxidant nutrients such as vitamins E and C and beta-carotene are inversely related to the incidence of coronary heart disease caused by smoking.[1-6] Vitamin E especially shows great promise in preventing LDL oxidation, offering a tremendous impact on cardiovascular disease. Decreased vitamin A and E concentrations are associated with increased risk of coronary artery disease (CAD), even after adjusting for CAD risk factors.[7] The B complex can prevent hyperhomocysteinemia.[8-14] Nonesterified polyunsaturated fatty acids (PUFA), a major component in fish oil, not only decrease in LPL activity and inhibit the synthesis of apolipoprotein B, but also defect platelet aggregation.[15-17] Thus, the combination of a healthy diet with PUFA will be useful to retard the procedure of tobacco-associated cardiovascular damage.

4.2 CIGARETTE SMOKE AND STIMULATION OF FREE RADICALS

Oxidants are tangible energy made up of free radicals and reactive substances that come in contact with oxygen. Oxidants can harm and mutate molecules, rendering them useless in their function for the cell. Smoking induces the body to produce increased amounts of free radicals. Tobacco smoke has two phases: the tar phase containing 10^{29} free radicals and the gas phase containing 10^{30} free radicals. Thus, antioxidant levels will be lower because the body requirement is enough to repair its own oxidant processes. Cells exposed to constant oxidative stress produced by normal body processes will consume much vitamin C, as well as other antioxidants. Of the oxidants produced by the body remaining after vitamin C consumption, 64% is captured by other antioxidants, leaving approximately 36% to do further damage. This excess of 36% is joined by many oxidants during smoking.

Free radical-induced oxidative damage is thought to be involved in the pathogenesis of disease associated with cigarette smoking. Free radicals cause lipid peroxidation: the conversion of the lipid bilayer membrane around cells into fatty, radical-containing acids. Lipids and hydroxyl radicals react together to start a chain reaction, creating more radicals in the presence of oxygen. This chain reaction renders the lipid useless to serve its cellular function. Peroxidation can be stopped and prevented by antioxidants. Vitamin E, found mostly in the lipid bilayer membrane, is a chain-breaking antioxidant that traps peroxyl and hydroxyl radicals. Vitamin E either transfers a hydrogen atom to the radical–oxygen complex, resulting in a fatty acid and another unreactive radical, or transforms both the radical–oxygen complex and the resulting radical to nonradical products. Vitamin C is concentrated in the cytoplasm of the cell. It works with the vitamin E by repairing it. After vitamin E transfers a hydrogen atom, vitamin C replenishes it where both vitamins come in contact. Smokers are generally vitamins E and C deficient, caused by an imbalance in antioxidant nutrient intake and status,[4] whereas vitamin C and other preventive antioxidants inside the cytosol increase.[2]

Tobacco smoke contains many oxidizing compounds. Indeed, smokers receive a high and sustained free radical load. Free radicals may play a fundamental role in life processes such as mitochondrial respiration, platelet activation, prostaglandin synthesis, metabolism of lipid membrane organelles, protein synthesis, and even DNA itself.[5,18] The heart is very susceptible to free radical oxidative stress. The most potent destructive free radical, the hydroxyl radical, initiates the breakdown of phospholipid in cellular and inner mitochondrion membranes, and the viability of the cell is threatened.[19] Lipid peroxides can form lipid alkoxyl or peroxl radicals that are capable of abstracting hydrogen from adjacent polyunsaturated lipid molecules to propagate the lipid peroxidation reaction.[20] The alkoxyl and peroxyl radicals can also damage proteins by affecting all amino acid side chains and attacking the a-carbon atom of the peptide bond. Oxidative modification of the proteins makes them more susceptible to protease.[21] There are several potential sources of oxygen-derived free radicals in ischemic areas. The endothelial-derived xanthine oxidase, a dehydrogenase, cannot reduce molecular O_2 in tissue. However, in ischemia, xanthine dehydrogenase in the endothelial cell is converted to xanthine oxidase. The hypoxanthine, a degradation product of ATP, in the presence of xanthine oxidase will combine with oxygen during postischemic reperfusion. This results in the generation of superoxide anions and other free radicals that cause widespread lipid peroxidation and damage to cellular membranes.[20,22,23] Free radicals will excessively stimulate intracellular metabolism of myocardiovascular cells, promote myocyte calcium overload, induce autooxidation of catecholamine to toxic adrenochromes, and direct toxicity to the endothelium by hyperhomocysteine and the chronic oxidative LDL.[8–13,16,24–27] Thus, the myocardial structure and function are damaged.

4.3   CIGARETTE SMOKE AND PLATELET AGGREGATION IN CARDIOVASCULAR DISEASE

When platelets aggregate inappropriately and form a thrombus in the coronary circulation, they can precipitate a myocardial infarction. The endocardium is most vulnerable to injury by tobacco smoking. Tobacco smoking-induced free radicals oxidize LDL that is cytoxic to endothelial cells. Oxidized LDL can directly damage the endothelial cells of cardiac vessels, distort vascular integrity, and initiate the coagulation process.[28,29] Platelets adhere to injured vessels and then aggregate at the injury site. Eventually, they form clots and, thus, increase the likelihood of thrombus formation and myocardial infarction. Furthermore, accelerated atherosclerotic plaque development has been found in animals subject to sidestream cigarette smoke.[30] Aggregated platelets release arachidonic acid and its metabolite thromboxane A2 (TXA2), a strong platelet aggregator which enlarges myocardial ischemia. There is a significant elevation of arachidonic acid in the ischemic area occurring within 20 to 45 min of ischemia.[31] Long-term exposure to tobacco smoke can reduce platelet sensitivity to the antiaggregating substances, prostaglandins I2 (PGI2) and E1; however, short periods of exposure to smoke did not significantly change their levels.[31,32] In contrast, platelets were more sensitive to PGI2 in nonsmokers.[31] The results suggest that the platelets of smokers are already desensitized to the antiaggregatory substance PGI2 so that less further decrease in platelet aggregation occurs upon smoke exposure. Prostaglandins and leukotrienes derived from arachidonic acid may influence arrhythmogenesis and the electrophysiological properties of ischemic tissue. Arrhythmia can occur when calcium channels in the myocyte become hyper-active and fatty acids are well positioned to act as messenger molecules that regulate ion channels.[16,24]

4.4   ANTIOXIDANTS AND SMOKERS' RISK OF HEART DISEASE

Coronary heart disease (CHD) risk increases markedly with cigarette smoking, age, gender, and so forth. The major risk for CHD is free radicals produced by cigarette smoking. There are a variety of mechanisms providing defenses against free radical damage. Vitamins E and C and beta-carotene have powerful antioxidant effects. They can quench oxygen radicals, block the chain reaction of lipid peroxidation by scav-enging intermediate peroxyl radicals, and protect tissues against damage by trapping free radicals.[1] The lag phase of conjugated diene formation lengthened and the propagation rate decreased, indicating a decreased susceptibility of LDL to oxidative modification. Increased concentrations of plasma vitamin C, beta-carotene, and vitamin E were found to be significantly correlated with the conjugated diene lag phase and rate of formation.[1,33] Antioxidant supplementation in smokers also reduces the capacity of platelets to aggregate and to produce TXA2, and *in vivo* platelet activation will contribute to further progress of atherosclerosis, leading eventually to complex atherosclerotic lesions.[29,31,34,35] However, vitamin E supplementation in a healthy population has not shown any effect on platelet aggregation.[36] Vitamin E has a stabilizing effect on heart phospholipids by preventing changes in their fatty acid composition and peroxidative deterioration. Vitamin E appears especially prom-ising because it is carried directly in LDL particles. Several studies have shown long-term supplementation where large doses of vitamin E (minimum dose 400 IU) alone increased LDL resistance to oxidative damage.[33,37] Several studies have shown that vitamin E supplement users have an approximately 40% lower rate of CHD.[38] Some epidemiological studies showed that fatality from heart disease was strongly correlated with lowest levels of plasma vitamin E.[38,39] The reduced alpha-tocopherol levels in red blood cells from smokers may be due to impairment of alpha-tocopherol uptake activity. However, at present, it is not known whether supplementation of smokers with vitamin E would normalize the alpha-tocopherol uptake activity of red blood cells.[2] Vitamin C not only regenerates tocopherol, but also enhances the body's total antioxidant system by raising levels of glutathione, a polypeptide amino acid and potent free radical scavenger.

Vitamin C is a highly potent aqueous-phase antioxidant, which has been shown *in vitro* to retard LDL oxidation. Cigarette smokers have reduced concentrations of ascorbate in their plasma, and their LDL may be more prone to oxidation. It has been shown that the ascorbate-supplemented group has a significant reduction in LDL oxi-dative susceptibility, as measured by thiobarbituric acid-reactive substances (TBARS) and the formation of conjugated dienes in smokers. The ascorbate-supplemented group demonstrated significantly an increased lag phase and a decreased oxidation rate at 4 weeks compared to 0 weeks. Supplementation of otherwise healthy smokers for 4 weeks with 1000 mg of ascorbate per day resulted in increased plasma ascorbate and reduced LDL oxidative susceptibility.[3] 8-Epi-PGF2 alpha, not only a platelet aggregator but also a stable product of lipid peroxidation *in vivo*, will increase in smokers. Vitamin C and a combination of vitamins C and E suppressed urinary 8-epi-PGF2 alpha, whereas vitamin E alone had no effect.[40] However, some research indicates that adequate levels of vitamins E and C may protect the plasma from oxidative damage elicited by smoking-mediated reactive oxygen species (ROS) and free radicals in young smokers. The antioxidant activities of vitamins E and C may be overwhelmed by the long-standing oxidative stress elicited by cigarette smoking in elderly subjects.[41] Beta-carotene not only modifies LDL, but may also have a favorable effect on prostaglandins. The B complex, particularly folic acid, B12, and B6, are also essential in the prevention of hyperhomocysteinemia, another major risk factor for the circulatory system.[9,10,13,14] PUFA may provide the best protection for lowering cholesterol and reducing susceptibility to lipid peroxidation at the same time.[15-17]

According to the American Heart Association, passive smoking is an important risk for CHD. Exposure of a nonsmoking subject to secondhand smoke breaks down the serum antioxidation defense, leading to accelerated lipid peroxidation, LDL modification, and accumulation of LDL cholesterol in human macrophages. Many studies suggest that a link exists between fruit and vegetable in the diet or the amount of plasma antioxidant vitamins and risk of death from CHD.[39,42,43] Observational studies have found that persons who consume large amounts of fruit and vegetables have lower rates of CDH.[43,44] In postmenopausal women, the intake of vitamin E from food is inversely associated with the risk of death from CHD, and such women can lower their risk without using vitamin supplements. However, the intake of vitamins A and C have not shown lower risks of dying from CHD.[42] Plasma vitamin C concentrations were also not associated with significant CAD risk after adjusting for CHD risk factors.[7] High doses of vitamin E (400 and 800 IU/day) strongly reduce the risk of CHD in nonfatal myocardial infarction, but short-term intake and doses of less than 100 IU/day have no significant effect.[38] Some studies[45,46] suggest that antioxidant vitamin intake again protects CHD only for men who smoke. No significant interaction is found in women who smoke.

4.5 POLYUNSATURATED FATTY ACID DEFENSE

PUFA are predominantly present in fish oils. Studies demonstrate that ischemia-induced cardiac arrhythmia can be prevented by infusion of a fish oil emulsion just 50 to 60 min prior to the test. Fish oil decrease in LPL activity was accompanied by accelerated clearance of VLDL apolipoprotein B-100 in swine;[17] whereas in humans, fish oil inhibited the synthesis of apolipoprotein B. PUFA may exhibit their beneficial effects via incorporation into sarcolemmal phospholipids, changing membrane properties and altering the eicosanoid metabolism.[15] Fish oil supplementation can reverse the accumulation of arrhymogenic arachidonic acid, inhibit platelet aggregation, prevent lethal ventricular arrhythmias after myocardial ischemia, and enhance the effect of antioxidant against free radicals. It is also shown that fish oil can modulate the calcium current through l-type calcium channels in sarcolemma of the rat myocyte.[16] Pepe et al. found[47,48] that the activation of Ca^{2+}-dependent pyruvate dehydrogenase can be augmented when the n-3 to n-6 PUFA ratio is low or attenuated when this ratio is high. One of the sequences of PUFA dietary-induced manipulation of membrane phospholipids may be changed and, thus, PUFA may alter the flux of Ca^{2+} across the mitochondria membrane and intramitochondrial Ca^{2+}-dependent processes. Clinical studies show that PUFA diet directly affects myocardial properties which may contribute to cardiac mortality associated with fish oil consumption.

4.6 SUMMARY

Clinical and epidemiological studies have shown approximately 35,000 to 40,000 annual deaths in a total of 150,000 smoking-related annual deaths in the U.S. and have revealed a 20% higher death rate with tobacco smoke-induced CHD than unexposed nonsmokers. One fifth of all heart disease-related deaths are due to cigarette smoking. Smoking alone doubles the risk of heart failure. African Americans suffer the highest death rates from smoking among all U.S. population groups.[49] Epidemiological studies[50] have shown that the mortality from CHD began to decline in many countries after rising for many years in the mid to late 1930s. Even though its impact on the differing trends in various countries is not clear, the benefit of a healthy diet and smoking cessation is a clear one. Furthermore, it is imperative to find an effective regimen to prevent active and passive smoke-induced heart disease and to decrease the death rate. A lot of studies have shown that antioxidants and other strategies can limit cardiovascular damage caused by exposure to tobacco smoke, but this effect was not uniformly observed.[38,42,52-55] Therefore, it is most important that the whole world plans to start a campaign against smoking. The risk of heart disease caused by smoking can be reduced by half after just 1 year of quitting smoking.[49] We also recommend a healthy diet rich in fruits and vegetables to reduce risk factors of CHD. With proper nutrition and cessation of smoking, the risk factor associated with heart disease could be greatly reduced.

ACKNOWLEDGMENTS

Julian Rodriquez was supported by a summer research grant given to Marly Witte, NIH R25.RR10163; the preparation part of NIH grants given to Ronald Watson were HL 63667 and HL 59794.

REFERENCES

1. Steinberg, F.M. and Chait, A., Antioxidant vitamin supplementation and lipid peroxidation in smokers, *Am. J. Clin. Nutr.*, 68(2), 319–27, 1998.
2. Bellizzi, M.C., Dutta-Roy, A.K., Duthie, G.G., and James, W.P., Alpha-tocopherol binding activity of red blood cells in smokers, *Free Rad. Res.*, 27(1), 105–112, 1997.
3. Fuller, C.J., Grundy, S.M., Norkus, E.P., and Jialal, I., Effect of ascorbate supplementation on low density lipoprotein oxidation in smokers, *Atherosclerosis*, 119(2), 139–50, 1996.
4. Faruqe, M.O., Khan, M.R., Rahman, M.M., and Ahmed, F., Relationship between smoking and antioxidant nutrient status, *Br. J. Nutr.*, 73(4), 625–632, 1995.

5. Van Poppel, G., Verhagan, H., van't Veer, P., and van Bladeren, P.J., Markers for cytogenetic damage in smokers: associations with plasma antioxidants and glutathione S-transferase mu, *Cancer Epidemiol. Biomark. Prev.*, 2(5), 441–447, 1993.

6. Salonen, J.T., Salonen, R., Seppanen, K., Rinta-Kiikka, S., Kuukka, M., Korpela, H., Alfthan, G., Kantola, M., and Schalch, W., Effects of antioxidant supplementation on platelet function: randomized pair-matched, placebo-controlled, double-blind trial in men with low antioxidant status, *Comm. Am. J. Clin. Nutr.*, 53(5), 1222–1229, 1991.

7. Delport, R., Ubbink, J.B., Human, J.A., Becker, P.J., Myburgh, D.P., and Vermaak, W.J., Antioxidant vitamins and coronary artery disease risk in south African males, *Clin. Chim. Acta*, 278(1), 55–66, 1998.

8. Sinatra, S.T. and Demarco, A., Free radicals, oxidative stress, oxidized low density lipoprotein, and the heart: antioxidants and other strategies to limit cardiovascular damage, *Conn. Med.*, 59, 579–587, 1995.

9. Stampfer, M.J., Malinow, M.R., Willett, W.C., et al., A prospective study of plasma homocysteine and risk of myocardial infarction in U.S. physicians, *J. Am. Med. Assoc.*, 268, 87–91, 1992.

10. Ubbink, J.B., Vermaak, W.J.H., van der Merwe, A., et al., Vitamin B-12, vitamin B-6 and folate nutritional status in men with hyperhomocysteinemia, *Am. J. Clin. Nutr.*, 57, 47–53, 1993.

11. Clark, R., Daly, L., Robinson, K., et al., Hyperhomocysteinemia: an independent risk factor for vascular disease, *N. Engl. J. Med.*, 17, 1149–1155, 1991.

12. Joosten, E., Metabolic evidence that deficiencies of vitamin B-12, folate and vitamin B-6 occur commonly in elderly people, *Am. J. Clin. Nutr.*, 58, 468–476, 1993.

13. Selhub, J., Jacques, P.F., Bostom, A.G., et al., Association between plasma homocyteine concentrations and extracranial carotidartery stenosis, *N. Engl. J. Med.*, 32, 286–289, 1995.

14. Wu, L.L., Wu, I., Hunt, S.C., et al., Plasma homocysteine as a risk factor for early familial coronary artery disease, *Clin. Chem.*, 40, 552–561, 1994.

15. Billman, G.E., Kang, J.X., and Leaf, A., Prevention of ischemia-induced cardiac sudden death by n-3 polyunsaturated fatty acids in dogs, *Lipids*, 32, 1161–1168, 1997.

16. Pepe, S., Bogdanov, K., Hallaq, H., Spurgeon, H., Leaf, A., and Lakatta, E., Omeg polyunsaturated fatty acid modulates dihydropyridine effects on L-type Ca^{2+} channels, cytosolic Ca^{2+}, and contraction in adult rat cardiac myocytes, *Proc. Natl. Acad. Sci. U.S.A.*, 91, 8832–8836, 1994.

17. Nair, S.S.D., Leitch, J., Falconer, J., and Garg, M.L., Cardiac(n-3) non-esterified fatty acid are selectively increased in fish oil-fed pigs following myocardial ischemia, *Nutrition*, 129, 1518–1523, 1999.

18. Randerath, E., Mittal, D., and Randerath, K., Tissue distribution of covalent DNA damage in mice treated dermally with cigarette 'tar': preference for lung and heart DNA, *Carcinogenesis*, 9, 75–80, 1988.

19. Farber, J.L., Kyle, M.E., and Coleman, J.B., Biology of disease: mechanisms of cell injury by activated oxygen species, *Lab. Invest.*, 62, 670–678, 1990.

20. Mehta, J., Yang, B., and Nichols, W., Free radicals, antioxidants and coronary heart disease, *Myocard. Ischemia*, 5, 31–41, 1993.

21. Davies, K.J.A., Intracellular proteolytic systems may function as secondary antioxidant defences: an hypothesis, *Free Rad. Biol. Med.*, 2, 155–173, 1986.

22. Coghlan, J.G., Madden, B., Norell, M.N., et al., Lipid peroxidation and changes in vitamin E levels during coronary artery bypass grafting, *J. Thorac. Cardiovasc. Surg.*, 106, 268–274, 1993.

23. Goldhaber, J.L. and Weiss, J.N., Oxygen free radicals and cardiac reperfusion abnormalities, *Hypertension*, 20, 118–127, 1992.
24. Kaneko, M., Matsumoto, Y., Hayashi, H., Kobayashi, A., and Yamazaki, N., Oxygen free radical and calcium homeostasis in the heart, *Mol. Cell. Biochem.*, 139, 91–100, 1994.
25. Esterbauer, H., Dieber-Rotheneder, M., Striegl, G., and Waeg, G., Role of vitamin E in preventing the oxidation of low-density lipoprotein, *Am. J. Clin. Nutr.*, 53(Suppl.), 314S–321S, 1991.
26. Galle, J., Mulsch, A., Busse, R., and Bassenge, E., Effects of native and oxidized low-density lipoproteins on formation and inactivation of endothelium-derived relaxing factor, *Arterioscler. Thromb.*, 11, 198–203, 1991.
27. Parthasarathy, S. and Rankin, S.M., Role of oxidized low density lipoprotein in atherogenesis, *Prog. Lipid Res.*, 31, 127–143, 1992.
28. Davis, J., Shelton, L., Watanabe, I., and Arnold, J., Passive smoking affects endothelium and platelets, *Arch. Intern. Med.*, 149, 386–389, 1989.
29. Chait, A. and Heinecke, J.W., Lipoprotein modification cellular mechanisms, *Curr. Opin. Lipidol.*, 5, 365–370, 1994.
30. Ross, R., The pathology of atherosclerosis — an update, *N. Engl. J. Med.*, 314, 488–500, 1986.
31. Glantz, S.A. and Parmley, W.W., Passive smoking and heart disease. Epidemiology, physiology, and biochemistry, *Circulation*, 83(1), 1–12, 1991.
32. Sinzinger, H. and Kefalides, A., Passive smoking severely decreases platelet sensitiving to antiaggregatory prostaglandins, *Lancet*, 2, 392–393, 1982.
33. Duthie, G.G., Arthur, J.R., and James, W.P., Effects of smoking and vitamin E on blood antioxidant status, *Am. J. Clin. Nutr.*, 53, 1061–1063, 1991.
34. Witztum, J.L. and Steinberg, D., Role of oxidized low density lipoprotein in atherogenesis, *J. Clin. Invest.*, 88, 1785–1792, 1991.
35. Berliner, J.A. and Haberland, M.E., The role of oxidized low-density lipoprotein in atherogensis, *Curr. Opin. Lipidol.*, 4, 373–381, 1993.
36. Salonen, J.T., Antioxidants and platelets, *Ann. Med.*, 21, 59–62, 1989.
37. Duthie, G.G., Arthur, J.R., Beattie, J.A., Brown, K.M., Morrice, P.C., Robertson, J.D., Shortt, C.T., Walker, K.A., and James, W.P., Cigarette smoking, antioxidants, lipid peroxidation, and coronary heart disease, *Ann. N.Y. Acad. Sci.*, 686, 120–129, 1993.
38. Rimm, E.B. and Stampter, M.J., The role of antioxidants in preventive cardiology, *Curr. Opin. Cardiol.*, 12(2), 188–194, 1997.
39. Knekt, P. and Reunanen, A., Antioxidant vitamin intake and coronary mortality in a longitudinal population study, *Am. J. Epidemiol.*, 39, 1180–1189, 1994.
40. Reilly, M., Delanty, N., Lawson, J.A., and Fitzgerald, G.A., Modulation of oxidant stress in vivo in chronic cigarette smokers, *Circulation*, 94(1), 19–25, 1996.
41. Liu, C.S., Chen, H.W., Lii, C.C.K., Chen, S.C., and Wei, Y.H., Alterations of small-molecular-weight antioxidants in the blood of smokers, *Chem. Biol. Interact.*, 116(1–2), 143–154, 1998.
42. Kushi, L.H., Folsom, A.R., Prineas, R.J., Mink, P.J., Wu, Y., and Bostick, R.M., Dietary antioxidant vitamins and death from coronary heart disease in postmenopausal women, *N. Engl. J. Med.*, 334(18), 1156–1162, 1996.
43. Ghiselli, A., Dámicis, A., and Giacosa, A., The antioxidant potential of the mediterranean diet, *Eur. J. Cancer Prev.*, 6(Suppl. 1), S15–S19, 1997.
44. Rexrode, K.M. and Manson, J.E., Antioxidants and coronary heart disease: observational studies, *J. Cardiovasc. Risk*, 3(4), 363–367, 1996.

45. Toddo, S., Woodward, M., Bolton-Smith, C., and Tunstall-Pedie, H., An investigation of the relationship between antioxidant vitamin intake and coronary heart disease in men and women using discriminant analysis, *J. Clin. Epidemiol.*, 48(2), 297–305, 1995.

46. Todd, S., Woodward, M., and Bolton-Smith, C., An investigation of the relationship between antioxidant vitamin intake and coronary heart disease in men and women using logistic regression analysis, *J. Clin. Epidemiol.*, 48(2), 307–316, 1995.

47. Pepe, S. and McLennan, P.L., Dietary fish oil confers direct antiarrhythmic properties on the myocardium of rats, *J. Nutr.*, 126(1), 34–42, 1996.

48. Pepes, S., Tsuchiya, N., Lakatta, E.G., and Hansford, R.G., PUFA and aging modulate cardiac mitochondrial membrane lipid composition and Ca^{2+} activation of PDH, *Am. J. Physiol.*, 276(1pt2), H149–158, 1999.

49. Manley, A.F., Cardovascular implication of smoking: the surgeon general's point of view, *J. Health Care Poor and Underserved*, 8(3), 303–310, 1997.

50. Lioyd, B.L., Delining cardiovascular disease incidence and environmental components, *Aust. N.Z. J. Med.*, 24(1), 124–132, 1994.

51. Klipstein-Grobusch, K., Geleijnse, J.M., den breeijen, J.H., Boeing, H., Hofman, A., Grobbee, D.E., and Wittean, J.C., Dietary antioxidants and risk of myocardial infarction in the elderly: the Rotterdam study, *Am. J. Clin. Nutr.*, 69(2), 261–266, 1999.

52. Van de Vijver, L.P.L., Kardinaal, A.F.M., Grobbee, D.E., Princen, H.M.G., and van Poppel, G., Lipoprotein oxidation, antioxidants and cardiovascular risk: epidemiologic evidence, *Prostagl. Leukotr. Essent. Fatty Acid*, 57(4–5), 479–487, 1997.

53. Simon, J.A., Vitamin C and cardiovascular disease: a review, *J. Am. Coll. Nutr.*, 11, 107–125, 1992.

54. Keaney, J.F., Gaziano, J.M., Xu, A., et al., Low-dose a-tocopherol improves and high-dose a-tocopherol worsens endothelial vasodilator function in cholesterol-fed rabbits, *J. Clin. Invest.*, 93, 844–851, 1994.

55. Keaney, J.F., Gaziano, J.M., Xu, A., et al., Dietary antioxidants preserve endothelium-dependent vessel relaxation in cholesterol-fed rabbits, *Proc. Natl. Acad. Sci. U.S.A.*, 90, 11880–11884, 1993.

5 Tobacco Allergy

Nancy Ortega, Carlos Blanco, Rodolfo Castillo, and Teresa Carrillo

CONTENTS

5.1 INTRODUCTION

Allergy is a condition that involves genetic and environmental factors. The epidemiology of these conditions is the science that deals with their etiology, the relationship between various factors that determine its frequency, distribution, and control.

Until the turn of the 20th century, the immune system and the presence of humoral antibodies were only thought of in terms of body defense systems, such as the response to infection. In 1923, Coca and Cooke[1] proposed the term atopy to identify a subgroup of clinical allergies that involved reaginic or skin-sensitizing antibodies. Later, the discovery that human reaginic or skin-sensitizing antibodies reside in a unique immunoglobulin (Ig) isotype, IgE, was a major advance in the field of allergy.[2] In this chapter, atopy is defined as adverse immune reactions involving IgE antibodies.

It is difficult to investigate the frequency, incidence, and prevalence of allergy and specific allergy-related conditions such as asthma, rhinitis, urticaria/angioedema, anaphylaxis, and atopic dermatitis, as well as components involved in these conditions, such as total serum IgE and specific IgE as measures of bronchial hyperactivity. This difficulty arises due to problems of definition and the influence of multiple variables on the ultimate clinical picture or phenotype, such as age, sex, and environmental factors (such as the amount of the allergen exposure and

0-8493-0311-7/00/$0.00+$.50
© 2001 by CRC Press LLC

the presence of non-specific adjuvant factors, including infections, air pollution, and smoke).

It has been mentioned that allergies occur in approximately one out of every six Americans. Of these, 41% are due to hay fever, 25% to asthma, and the remainder to other allergies.[3]

5.2   RELATIONSHIPS BETWEEN CIGARETTE SMOKING AND SERUM IGE

Total serum IgE levels are measured by immunoassay. IgE antibody production is normally highly regulated, resulting in minimal concentration in body fluids. Although the human fetus can synthesize IgE as early as the 11th week of gestation,[4] cord serum contains very little IgE protein. The concentration of IgE in maternal serum and in cord serum is not correlated, indicating that there is no appreciable placental transfer of maternal IgE.[5]

Most infants, at least in developed countries, are not stimulated to produce IgE *in utero* because maternal IgE does not normally cross the placenta. The total IgE concentration in cord serum should provide an estimate of the infant's basal genetic potential for IgE production and, hence, an estimate of the infant's risk of developing allergic disease. Racial[6] and genetic[7] factors seem to be important in controlling IgE levels, although it is often difficult to separate racial and genetic factors from environmental effects.

Serum IgE levels vary two- to fourfold annually in sensitive patients due to pollen seasons[8] and the fact that total serum IgE concentrations tend to be higher in allergic adults and children compared to nonallergic individuals.[9]

Multiple studies have evaluated the relationship between cigarette smoking and serum IgE levels. It has been reported that maternal cigarette smoking increases the concentration of cord blood IgE without reducing the predictive value of the cord blood IgE measurements.[10]

The effect of smoking on serum IgE levels is usually greater in men than in women,[11] and there is increased serum IgE concentrations in smokers compared to non-smokers and an intermediate concentration in ex-smokers.[8]

5.2.1   INFLUENCE OF SMOKING ON IGE AND ATOPY

Patients with skin test reactivity had statistically significant higher levels of total serum IgE than those without skin test reactivity, independent of their smoking status. However, IgE levels of current smokers did not decrease with age at the same rate as non-smokers, and so they have elevated levels throughout adulthood. The mechanism for the elevation of IgE in smokers is not known, but the significant dose-response relationship is suggestive of a causal association. A further study of 137 patients, 12 to 16 year olds in the U.S., found a significant association between maternal smoking and a positive skin prick test independent of the history of parental asthma.[12]

There is an association between environmental tobacco smoke exposure and total serum IgE levels in children.[13] These data suggest that environmental pollutants may act synergistically to increase airway reactivity and allergy, especially in children with a family history of asthma.

5.3   ATOPY: GENETIC VS. ENVIRONMENTAL FACTORS

It is clear that genetic factors play a central role in predisposition to atopy, but it is equally clear that environmental factors provide the obligatory final "triggers" required for ultimate expression of the allergic phenotype.

A number of additional environmental "risk" factors for primary sensitization have been identified apart from intensity of aeroallergen exposure. The most noteworthy of these is respiratory infection,[14,15] which is likely to increase the "permeability" of the airway epithelium to inhaled allergens, as well as non-specifically suppressing regulatory T-cell functions.[16]

Exposure during early childhood to airborne chemical irritants appears to constitute a further series of risk factors. These include "urban" pollution,[17,18] much of which may be attributable to motor-vehicle exhaust;[19] industrial pollution;[20] and domestic pollution generated via cooking and heating,[21] inadequate ventilation,[20] and, in particular, maternal smoking.[22]

Tobacco smoke is generally recognized as a respiratory irritant. Furthermore, there is increasing evidence that passive exposure to smoke can significantly increase asthma morbidity, particularly among children with smokers in the household. In addition to the aggravating effect on pre-existing asthma, exposure to smoke may increase the prevalence of allergic sensitization, as measured by IgE levels,[23] and has been found to increase the risk of developing asthma in children with atopic dermatitis.[24]

There is some evidence that an association between exposure to environmental tobacco smoke (ETS) increases the risk of being sensitized to common allergens in later childhood. One study conducted in Italy on patients of up to 12 years of age found that exposure to ETS increases the risk of being sensitized to common allergens or raised serum IgE and that this effect is greater in boys.[25] Martinez et al. found that boys who had been exposed to ETS had an increased risk of positive skin prick test reactions (p <0.01), but not girls.[26] Differential effects between boys and girls are plausible because there is evidence that boys develop atopy and airway hyperresponsiveness (AHR) at an earlier age than girls.[27] The mechanism(s) of action of these factors remains to be established, but it appears likely that (as in animal models) they potentially influence local immunologic homeostasis in the lung and airways at a number of levels, including epithelial and T-cell functions.[28]

Adults with asthma differ significantly in their airway reactivity to cigarette smoke. In one study, only 7 of 21 subjects showed a significant reduction in ventilatory function following smoke inhalation challenge. There was no evidence for IgE-mediated tobacco leaf allergy in these reactive subjects.[29]

Why the prevalence of atopy is increasing is unknown. An inverse relationship between infection and allergy could explain the apparent rise. Another possibility is that air pollution enhances allergic sensitization to aeroallergens, thereby accounting for both temporal and urban–rural differences in the prevalence of allergic disease. In Japan, cedar pollen sensitivity was found to be more common among residents of an area heavily polluted with vehicle exhaust.[30]

5.4   TOBACCO AND PREGNANCY

Male children who have a low birthweight or who are not breastfed appear to be at greater risk from the effects of ETS. Boys are more likely to become atopic at an earlier age than girls,[31] and smoking during pregnancy may lead to a lower birthweight[32] or to smaller stature,[33] low birthweight babies who have smaller caliber or increased rates of morbidity.[34] McConnochie and Roghmann demonstrated a significant relationship between an increase in maternal smoking with an increase in wheezing,[35] suggesting that there is a dose-response relationship. The risk of having a respiratory infection is higher in children who attend day care, but the risk of both wheezing and non-wheezing respiratory infections attributable to ETS appears to be lower, perhaps because of reduced exposure to ETS.

Prevention of the apparent *in utero* effect of exposure to ETS that appears to lead to a narrower airway caliber and that may result in damage to the epithelium and a susceptibility to sensitization would also be expected to result in a decrease in the prevalence of wheeze and asthma.

The issue of whether prenatal or postnatal smoking poses a greater risk has not been resolved, largely because few mothers change their smoking habits at the birth of their children.[36] It is thought that children whose mothers smoke both during and after pregnancy are at greater risk of having wheezy bronchitis than those whose mothers smoked in either period alone.[37] It is also likely that prenatal exposure causes this as well,[36] but no study to date has had sufficient scope to test this.

A study in Sweden found that children whose mothers smoked had a cord IgE level at birth that was approximately double the expected value ($p > 0.05$) and that "infant allergy" was increased with an odds ratio of 2.2 (1.1, 7.7),[10] but a study of Australian children failed to confirm this association.[38] Some authors think that although exposure to ETS may increase cord IgE levels, there is no evidence that neonates with high cord IgE levels are at increased risk for the later development of atopy.[39]

The dominant effect of maternal smoking may simply reflect greater exposure because of longer contact hours or may be a cumulative effect after *in utero* exposure. Exposure during early childhood may be increased by maternal smoking because cotinine can be transferred to the infant via breast milk.[32,40] Because children do not smoke,[41] the effects of active smoking may need to be considered in children older than 12 years of age.

5.5   EVIDENCE OF A CAUSAL ASSOCIATION BETWEEN EXPOSURE TO ETS ASTHMA AND ATOPIA

The stimuli most likely to increase responsiveness are viral respiratory infections, allergen exposure, and air pollutants.

Viral respiratory infections are recognized as one of the commonest clinical reasons for loss of asthma control.

Clinical investigation in adults with asthma have documented that allergen exposure leads to an inflammatory reaction within the airways, which is associated with obstruction and increased responsiveness.[42]

ETS has both a secondary (indirect) role and primary (direct) role in the etiology of childhood asthma. Cigarette smoke should be considered an environmental air pollutant. Several studies have indicated that this is an especially important environmental factor for infants and children in households in which there is maternal smoking. In this respect, maternal smoking has been associated with an increased incidence of lower respiratory illness and diminished pulmonary function, as well as higher rates of asthma, an increased likelihood of using asthma medications, and an earlier onset of disease.[43] Exposure to passive smoke is also a risk factor for the first episode of bronchiolitis,[35] as well as for recurrent wheezing later in childhood and adolescence.[44] A remarkably clear dose-response relationship between hospitalization for a serious respiratory infection in early childhood and exposure to ETS was demonstrated by Chen et al.[45] Subjects exposed to cigarette smoke through active[46] or passive routes[26] may demonstrate an increase in airway responsiveness. In a recent study, enhanced airway responsiveness to methacholine was demonstrated as long as 3 weeks after exposure to cigarette smoke in a static test chamber, suggesting that prolonged subclinical airway inflammation can occur, in the absence of a demonstrable change in airway caliber, on exposure to ETS.[47] Some authors suggest that parental smoking may contribute to elevated levels of airway responsiveness in their children as early as the first 2 to 10 weeks of age.[38] Studies that have examined an association have enrolled children over 6 years of age from whom lung function measurements are easier to obtain. In this age group, the response to carbacol was found to be significantly higher in male children exposed to ETS, but not in girls, the possible explanation being that boys develop atopy at an earlier age than girls.[31] It also has been shown that changes in the airway caliber (FEF_{25-75}) after use of a bronchodilator were 4% higher in children exposed to ETS, which suggests that the children exposed to ETS develop more reactive airways.[48]

Exposure to ETS has been reported to increase both the incidence and the prevalence of asthma in school-age children and also to trigger acute exacerbation of asthma. The frequency of both wheeze and asthma may reflect many factors, including a genuine difference in the prevalence of wheeze between countries, cultural differences, or an increased awareness in some regions.

The mechanism by which childhood exposure to ETS exerts its effects is unknown, but may include effects on the IgE immune system, which can be elicited both *in utero* and postnatally. However, airway inflammation facilitates allergens crossing the epithelium of the airways, where they can promote sensitization and thus facilitate the development of AHR and asthma.

5.5.1 Occupational Asthma

Asthma is now the most common occupational respiratory ailment in developed countries. Patients with pre-existing asthma, whatever the cause, are particularly susceptible to primary respiratory irritants of any sort, occupational or otherwise, including dust, tobacco and other smoke, strong odors, cold air, and various general atmospheric pollutants such as sulfur dioxide.[49,50]

The difficulties in getting a true estimate on the prevalence of asthma due to occupational exposure are many because occupational asthma is still poorly recognized and because the method of confirming the relationship between asthma and working conditions is still rather crude; affected workers are scattered through many small workplaces often employing few workers, and there is no good reporting system in many countries for occupational asthma.

The provoking agents fall into three categories: protein allergens, low-molecular-weight chemicals, and non-specific primary irritants.

Three separate factors have been reported to contribute to the development of occupational asthma in populations exposed to its causes: intensity of exposure, atopy, and tobacco smoking.

> *Exposure:* The development of inhibition immunoassays has now allowed measurements, as several recently reported studies have found evidence for a relationship between measured intensity of exposure the prevalence of sensitization and asthma. Burge et al.[52] found a gradient of work-related respiratory symptoms related to measured concentrations of airborne colophony. The study reported by Cullinan and co-workers[53] found an exposure-response relationship between airborne rat urine protein concentration and the prevalence of both skin test reactions to rat urine protein and respiratory symptoms.

> *Atopy:* Atopy is defined in immunological terms as that which readily produces IgE antibodies on contact with environmental allergens encountered everyday. It is commonly identified by the presence of one or more immediate skin prick test responses to common inhalant allergens (which would include grass pollen, *Dermatophagoides pteronyssinus*, and cat fur). Asthma and IgE antibodies induced by several causes of occupational asthma have been reported to occur more commonly among atopic individuals. Atopy has been shown to be strongly associated with sensitization to high-molecular-weight agents.

> *Tobacco smoking:* Tobacco smoking has been reported to increase the risk of developing specific IgE antibodies that may be responsible for several different causes of occupational asthma. This adjuvant effect of tobacco smoking is unknown, but may be related to some injury or to the respiratory mucous concurrent with inhalation of new antigens. Inhaled tobacco smoke potentiates the IgE response to inhaled, but not subcutaneous, ovalbumin.[54] Other respiratory irritants can exert a similar effect.

The effect of smoking on occupational asthma is not clear and appears to be dependent on the type of occupational agent. When the agent induces asthma by producing specific IgE antibodies, cigarette smoking may enhance sensitization. Venables et al.[55] found a relationship between smoking and atopy in workers exposed to tetrachlorophthalic anhydride (TCPA); atopic smokers had the highest prevalence of sensitization, with nonatopic non-smokers having the lowest prevalence of sensitization. A specific IgE antibody or an immediate skin test response has been found some four to five times more frequently in smokers than in non-smokers exposed to green coffee bean and ispaghula[56] and ammonium hexachloroplatinate;[57] the risk of developing asthma in this case is also increased, although less than for specific IgE.

When the agent induces asthma independent of IgE antibodies, non-smokers may be more frequently affected than smokers, as in isocyanate-induced asthma[58] and red cedar asthma.[59] However, the development of asthma is determined by a number of endogenous modifiers, such as age, gender, bronchial hyperresponsiveness, atopy, and smoking, and a number of exogenous modifiers, such as smoking and indoor environment.

5.5.2 EXTRINSIC ALLERGIC ALVEOLITIS

Extrinsic allergic alveolitis, also know as hypersensitivity pneumonitis, constitutes a group of pulmonary diseases in which diffuse inflammation of the lung parenchyma occurs because of recurrent exposure to inhaled dust. Carrillo et al.[63] studied one of the most frequent clinical forms, called pigeon breeder's disease, and found that non-smokers had significantly higher levels of sensitization to pigeon serum. They concluded that this disease is less prevalent in smokers, probably due to the changes induced by tobacco smoke in the immune system, as it occurs in another lung disease called sarcoidosis.[61]

5.6 RHINITIS RISK FACTOR

Habitual cigarette smokers reported less hay fever than non-smokers, independent of other factors. It has been shown that, although cigarette smoking is associated with increased serum IgE level, a direct protective effect of tobacco smoke is unlikely, as allergic sensitization to occupational allergens appears to be enhanced by cigarette smoking.[55] The discrepancy has instead been attributed to fewer smokers and a higher rate of ex-smokers among "allergic" individuals.[62]

There was a similar proportion of hay fever sufferers among non-smokers and ex-smokers, suggesting that affected individuals are not more likely to give up smoking once they have started.

Although genetic predisposition and allergen exposure are prerequisites for development of allergic rhinitis, the epidemiologic evidence suggests that there may be powerful influences relating to environment in early life. The immaturity of the immune system in the first 3 months of life and the association of transient immunologic abnormalities with subsequent infant eczema[63] support the notion that this may be a period of particular vulnerability in the development of allergic sensitization.

The reduced prevalence of cigarette smoking among adults with rhinitis is intriguing. Nevertheless, as there clearly are effects of smoking on IgE-mediated immunity, this is a potentially fruitful area for further research into mechanisms.

5.7 HYPERSENSITIVITY TO TOBACCO

Tobacco and its products can interact with the immune system in at least two ways: either as antigens or as irritant/toxic agents or both. Cigarette smoke has been reported to produce several stimulatory and inhibitory effects on the immune system.

The tobacco plant belongs to the botanical family Solanaceae. Some foods such as potatoes, tomatoes, eggplants, and green peppers are also members of this botanical family. The tobacco leaf is a heterogeneous mixture of proteins, insecticide residues, fungal contaminants, and other additives.[64]

Studies have shown that the tobacco leaf has at least 37 different antigens, and these antigens show the capacity to react specifically with human IgE antibodies.[65] Occupational allergy to tobacco has been reported a few times; one of them was a patient who worked in a cigarette factory.[66] The subject experienced rhinitis and asthma while working. The patient was evaluated for sensitivity to tobacco leaf. An *in vivo* test was done, including a skin prick test, nasal provocation, and bronchial provocation. An *in vitro* test was also done, including IgE antibodies with a radioallergosorbent test (RAST) and a test of the ability of the patient leukocytes to release histamine upon exposure to green tobacco leaf. This demonstrated that the patient was sensitive to allergens in tobacco leaf, but sensitivity to tobacco smoke was not found.

We did a study on 50 atopic patients. All subjects had a clinical history of rhinoconjunctivitis, asthma, urticaria, and anaphylaxis, shown by skin test inhalant or food allergen sensitivity. Differences can be seen in both smoker and non-smokers in the levels of total IgE. The smoker had increased levels of total serum IgE compared to non-smokers or ex-smokers. In all patients, levels of specific IgE to tobacco leaf was measured. Only one individual studied had a positive response. This subject was a non-smoker who had a clinical history of anaphylaxis with peanut and mustard and who tolerated all members of the botanical family Solanaceae, including potato, tomato, eggplant, and tobacco. CAP and skin test responses to tobacco leaf did not correlate with either the subject's reported symptoms of tobacco smoke "sensitivity" or his smoking history. Neither the intensity of the reactions nor the incidence of reactivity with any of these tobacco antigens correlated with smoking or clinical smoke sensitivity. However, the results did correlate with the subject's atopic status. Some surveys of molecular biology have found a relationship between the main allergen of birch and apple. Moreover, it has been possible to clone the second allergen of birch, which represents the profilin of birch. Profilins are ubiquitous proteins present in plant, pollen, fruit, leaves, and roots. Any of these circumstances may be a cause of the cross-reactivity in different vegetables, as in the patient of this study. Many previous studies of tobacco smoke hypersensitivity have been flawed because antigens of questionable relevancy to tobacco were employed.

Recently, a case report of occupational sensitivity to tobacco has been described by us.[67] A lot of plants can be fumigated with an aqueous solution of cut tobacco. Nicotine is an alkaloid of tobacco plants and is used for destroying insect eggs. The patient had a clinical history of rhinoconjunctivitis, shown by skin test inhalant allergen sensitivity to mugwort, *Artemisia vulgaris*. The patient worked in a garden for 3 years. After that, he experienced rhinoconjunctivitis and generalized urticaria while fumigating with an aqueous solution of cut tobacco. These symptoms disappeared within 2 h after work or when he stopped fumigating. Since the onset of this illness, he has reacted to cigarette smoke, particularly in crowded rooms, and induced rhinoconjunctivitis. Skin prick test and specific serum IgE were positive to tobacco leaf, *A. vulgaris*, tomato, and potato. A CAP inhibition was carried out on tobacco. Mugwort pollen and tomato extracts were found to inhibit the binding of the patient serum to tobacco leaf. The presence of reactions to tobacco was followed by *in vivo* and *in vitro* tests positive to tomato and potato allergens. This correlated on the basis of botanically related vegetables, and all these vegetables belong to the Solanaceae family. The patient did not experience symptoms after eating a tomato or potato, perhaps because the tobacco leaf allergy precedes a possible appearance of food allergy to tomato or potato.

An association of vegetables with *Artemisia* has been suggested by other authors.[68] Our CAP inhibition results suggest that tobacco leaf and mugwort pollen probably share common antigenic determinants, at least for the allergens to which our patient is sensitized. However, the existence of vegetable panallergens such as profilin or other common antigens could explain these reactions. The patient was a dedicated non-smoker. We do not know if tobacco smoke only acted as an irritant or if the patient was sensitized to cigarette smoke. Allergy to tobacco can be an occupational hazard. In a review of the literature on this subject and in view of the lack of information relating to tobacco hypersensitivity in the literature review, it is necessary to conduct epidemiological studies to discover the importance of the tobacco allergen in the population of tobacco-growing areas and in the general population. Furthermore, we can recommend specific diagnostic or therapeutic strategies.

REFERENCES

1. Coca, A.F. and Cooke, R.A., On the classification of the phenomena of hypersensitiveness, *J. Immunol.*, 8, 163, 1923.
2. Tada, T. and Ishizaka, K., Distribution of γE-forming cells in lymphoid tissues of the human and monkey, *J. Immunol.*, 104, 377, 1970.
3. Burr, M., Epidemiology of asthma, in Burr, M., Ed., *Epidemiology of Clinical Allergy, Monographs in Allergy 31*, S. Karger, Basel, 1993, 80–102.
4. Miller, D.L., Hirvonen, T., and Gitlin, D., Synthesis of IgE by the human conceptus, *J. Allergy Clin. Immunol.*, 52, 182, 1973.
5. Michael, F.B., Bousquet, J., Greillier, P., et al., Comparison of cord blood immunoglobulin E concentration and maternal allergy for the prediction of atopic diseases in infancy, *J. Allergy Clin. Immunol.*, 65, 422, 1980.
6. Grundbacher, F.J. and Massie, F.S., Levels of immunoglobulin G, M, A, and E at various ages in allergic and nonallergic black and white individuals, *J. Allergy Clin. Immunol.*, 74, 651, 1985.
7. Bazaral, M., Orgel, H.A., and Hamburger, R.N., Genetics of IgE and allergy: serum IgE levels in twins, *J. Allergy Clin. Immunol.*, 54, 288, 1974.
8. Omenaas, E., Bakke, P., Elsayed, S., et al., Total and specific serum IgE levels in adults: relation ship to sex, age and environmental factors, *Clin. Exp. Allergy*, 24, 530, 1994.
9. Klink, M., Cline, M.G., Halonen, M., et al., Problems in defining normal limits for serum IgE, *J. Allergy Clin. Immunol.*, 85, 440, 1990.
10. Magnunsson, C.G.M., Maternal smoking influences cord serum IgE and IgD levels and increases the risk for subsequent infant allergy, *J. Allergy Clin. Immunol.*, 78, 898, 1986.
11. Jensen, E.J., Pedersen, B., Schmidt, E., et al., Serum IgE in nonatopic smokers, nonsmokers, and recent ex-smoker: relation to lung function, airway symptoms and atopic predisposition, *J. Allergy Clin. Immunol.*, 90, 224, 1992.

12. Weiss, S.T., Tager, I.B., Muñoz, A., and Speizer, F.E., The relationship of respiratory infections in early childhood to the occurrence of increased levels of bronchial responsiveness and atopy, *Am. Rev. Respir. Dis.*, 131, 573, 1985.
13. Weiss, S.T., Sparrow, D., and O'Connor, G.T., The interrelationship among allergy, airways responsiveness, and asthma, *J. Asthma*, 30, 329, 1993.
14. Frick, O.L., German, D.C., and Mills, J., Development of allergy in children. I. Association with virus infections, *J. Allergy Clin. Immunol.*, 63, 228, 1979.
15. Busse, W.W., Relationship between viral infections and onset of allergic diseases and asthma, *Clin. Exp. Allergy*, 19, 1, 1989.
16. Mims, C.A., Interactions of viruses with the immune system, *Clin. Exp. Immunol.*, 66, 1, 1983.
17. Brabäck, L. and Kälvestein, L., Urban living as a risk factor for atopic sensitization in Swedish schoolchildren, *Pediatr. Allergy Immunol.*, 2, 14, 1991.
18. Linna, O., Environmental and social influence on skin test results in children, *Allergy*, 38, 513, 1983.
19. Kaneko, S., Shimada, K., Horikuchi, H., et al., Nasal allerfy and air pollution, *Oto-Rhino-Laryngology Tokyo*, 23, 270, 1980.
20. Andrae, S., Axelson, O., Björksten, B., Fredriksson, M., and Kjellman, N.I.M., Symptoms of bronchial hyperactivity and asthma in relation to environmental factors, *Arch. Dis. Child.*, 63, 473, 1988.
21. Reidel, F., Influence of adjuvant factors on development of allergy: evidence from animal experiments, *Pediatr. Allergy Immunol.*, 2, 1, 1991.
22. Kjellman, N.-I.M., Effect of parental smoking on IgE levels in children, *Lancet*, 1, 993, 1981.
23. Tager, I.B., Passive smoking-bronchial responsiveness and atopy, *Am. Rev. Respir. Dis.*, 138, 507, 1988.
24. Murray, A.B. and Morrison, B.J., It is children with atopic dermatitis who develop asthma more frequently if the mother smokes, *J. Allergy Clin. Immunol.*, 86, 732, 1990.
25. Palmieri, M., Longobardi, G., Napolitano, G., and Simonetti, D.M.L., Parental smoking and asthma in childhood, *Eur. J. Pediatr.*, 149, 738, 1990.
26. Martinez, F.D., Antognoni, G., Macri, F., Bonci, E., Midulla, F., De Castro, G., and Ronchetti, R., Parental smoking enhances bronchial responsiveness in nine-year-old children, *Am. Rev. Respir. Dis.*, 138, 518, 1988.
27. Peat, J.K., Salome, C.M., and Woolcock, A.J., Factors associated with bronchial hyperresponsiveness in Australian adults and children, *Eur. Respir. J.*, 5, 921, 1992.
28. Holt, P.G. and Turner, K.J., Respiratory symptoms in children of smokers: an overview, *Eur. J. Respir. Dis.*, 65(Suppl. 133), 109, 1984.
29. Stankus, R.P. et al., Cigarette smoke-sensitive asthma: challenge studies, *J. Allergy Clin. Immunol.*, 82, 331, 1988.
30. Ishizaka, T., Koizumi, K., Ikemori, R., Ishiyama, Y., and Kushibiki, E., Studies of prevalence of Japanese cedar pollinosis among the residents in a densely cultivated area, *Ann. Allergy*, 58, 265, 1987.
31. Sears, M.R., Burrows, B., Flannery, E.M., Herbison, G.P., and Holdaway, M.D., Atopy in childhood. I. Gender and allergen related risk for development of hay fever and asthma, *Clin. Exp. Allergy*, 23, 941, 1993.
32. Shulte-Hobein, B., Schwartz-Bickenbach, D., Abt, S., Plum, C., and Nau, H., Cigarette smoke exposure and development of infants throughout the first year of life: influence of passive smoking and nursing on cotinine levels in breast milk and infant's urine, *Acta Paediatr.*, 81, 550, 1992.

33. Rantakallio, P., A follow-up study up to the age of 14 of children whose mothers smoked during pregnancy, *Acta Pediatr. Scand.*, 72, 747, 1982.

34. Anderson, L.J., Parker, R.A., Farrar, J.A., Gangarosa, E.J., Keyserling, H.L., and Sikes, R.K., Day-care center attendance and hospitalization for lower respiratory tract illness, *Pediatrics*, 82, 300, 1988.

35. McConnochie, K.M. and Roghmann, K.J., Parental smoking, presence of older siblings, and family history of asthma increase risk of bronchiolitis, *Am. J. Dis. Child*, 140, 806, 1986.

36. Taylor, B. and Wadsworth, J., Maternal smoking during pregnancy and lower respiratory tract illness in early life, *Arch. Dis. Child.*, 62, 786, 1987.

37. Rylander, E., Pershagen, G., Eriksson, M., and Nordvall, L., Parental smoking and other risk factors for wheezing bronchitis in children, *Eur. J. Epidemiol.*, 9, 517, 1993.

38. Young, S., Le Souëf, P.N., Geelhoed, G.C., Stick, S.M., Turner, K.J., and Landau, L.I., The influence of a family history of asthma and parental smoking on airway responsiveness in early infancy, *N. Engl. J. Med.*, 324, 1168, 1991.

39. Buonocore, G., Zani, S., Tomasini, B., Tripodi, V., Grano, S., and Bracci, R., Serum IgE concentrations in the neonatal period, *Biol. Neonate*, 62, 10, 1992.

40. Chilmonczyk, B.A., Knight, G.J., Palomaki, G.E., Pulkkinen, A.J., Williams, J., and Haddow, J.E., Environmental tobacco smoke exposure during infancy, *Am. J. Public Health*, 80, 1205, 1990.

41. Charlton, A. and Blair, V., Absence from school related to children's and parental smoking habits, *Br. Med. J.*, 298, 90, 1989.

42. Warner, J.A., Little, S.A., Pollock, I., Longbottom, J.L., and Warner, J.O., The influence of exposure to house dust mite, cat, pollen and fungal allergens in the home on primary sensitization in asthma, *Pediatr. Allergy Immunol.*, 1, 79, 1991.

43. Weitzman, M., Gortmaker, S., Walker, D.K., and Sobol, A., Maternal smoking and childhood asthma, *Pediatrics*, 85, 505, 1990.

44. McConnochie, K.M. and Roghmann, K.J., Weezing at 8 and 13 years: changing importance of brochilitis and passive smoking, *Pediatr. Pulmonol.*, 6, 138, 1989.

45. Chen, Y., Li, W., Yu, S., and Qian, W., Chang-Ning epidemiological study of children's health: passive smoking and children's respiratory diseases, *Int. J. Epidemiol.*, 17, 348, 1988.

46. Malo, J.L., Filiatrault, S., and Martin, R.R., Bronchial responsiveness to inhales methacholine in young asymptomatic smokers, *J. Appl. Physiol.*, 52, 1464, 1982.

47. Menon, P., Rando, R.J., Stankus, R.P., Salvaggio, J.E., and Lehrer, S.B., Passive cigarette smoke-challenge studies: increase in bronchial hyperreactivity, *J. Allergy Clin. Immunol.*, 89, 560, 1992.

48. Ekwo, E.E., Weinberger, M.M., Lachenbruch, P.A., and Huntley, W.H., Relationship of parenteral smoking and gas cooking to respiratory disease in children, *Chest*, 84, 662, 1983.

49. From, L.J. et al., The effects of open leaf burning on spirometric measurements in asthma, *Chest*, 101, 1236, 1992.

50. Harries, M.G. et al., Role of bronchial irritant receptors in asthma, *Lancet*, 1, 5, 1981.

51. Musk, A.W., Venables, K.M., Crook, B., et al., Respiratory symptoms, lung function and sensitization to flour in a British bakery, *Br. J. Ind. Med.*, 46, 636, 1989.

52. Burge, P.S., Edge, G., Hawkins, R., White, V., and Newman-Taylor, A.J., Occupational asthma in a factory making cored solder containing colophony, *Thorax*, 36, 828, 1981.

53. Cullinan, P., Lowson, D., Nieuwenhuijsen, M.J., et al., Work-related symptoms, sensitisations and estimated exposure in workers not previously exposed to laboratory rats, *Occup. Environ. Med.*, 57, 589, 1994.

54. Zetterstrom, O., Nordvall, S.L., Bjorksten, B., Ahlstedt, S., and Sterlander, M., Increased IgE antibody responses to rats exposed to tobacco smoke, *J. Allergy Clin. Immunol.*, 75, 594, 1985.

55. Venables, K.M., Toping, M.G., Howe, W., Luczynski, C.M., Hawkins, R., and Newman-Taylor, A.J., Interaction of smoking and atopy in producing specific IgE antibody against a hapten protein conjugate, *Br. Med. J.*, 290, 201, 1985.

56. Zetterstrom, O., Osterman, K., Machado, L., and Johansson, S.G.O., Another smoking hazard revised: serum IgE concentrations and increased risk of occupational allergy, *Br. Med. J.*, 283, 1215, 1981.

57. Venables, K.M., Dally, M.B., Nunn, A., et al., Smoking and occupational allergy in a platinium refinery, *Br. Med. J.*, 299, 939, 1989.

58. Howe, W., Venables, K., Topping, M., et al., Tetrachlorophthalic anhydride asthma: evidence for specific IgE antibody, *J. Allergy Clin. Immunol.*, 71, 5, 1983.

59. Cartier, A., Malo, J., Forest, F., et al., Occupational asthma in snow-crab processing workers, *J. Allergy Clin. Immunol.*, 74, 261, 1984.

60. Calverley, A.E., Rees, D., Dowdeswell, R.J., Linnett, P.J., and Kielkowski, D., Platinium salt sensitivity in refinery workers: incidence and effects of smoking and exposure, *Occup. Environ. Med.*, 52, 661, 1995.

61. Paggiaro, P.L., Loi, A.M., Rossi, O., et al., Follow-up study of patients with respiratory disease due to toluene diisocyanate (TDI), *Clin. Allergy*, 14, 463, 1984.

62. Chang-Yeung, M., Lam, S., and Koerner, S., Clinical features and natural history of occupational asthma due to Western red cedar (Thuja plicata), *Am. J. Med.*, 72, 411, 1982.

63. Carrillo, T., Rodríguez de Castro, F., Cuevas, M., Díaz, F., and Cabrera, P., Effect of cigarette smoking on the humoral immune response in pigeon fanciers, *Allergy*, 46, 241, 1991.

64. Holt, P.G., Immune and inflammatory function in cigarette smokers, *Thorax*, 42, 241, 1987.

65. Burrows, B., Halonen, M., Barbee, R.A., and Lebowitz, M., The relationship of serum immunoglobulin E to cigarette smoking, *Am. Rev. Respir. Dis.*, 124, 523, 1981.

66. Taylor, B., Norman, A.P., Orgel, H.A., Stokes, C.R., Turner, M.W., and Soothill, J.F., Transient IgA deficiency and pathogenesis of infantile atopy, *Lancet*, 1, 111, 1973.

67. Norman, V., An overview of the vapor phase, semivolatile and nonvolatile components of cigarette smoke, in Tobacco Chemists Research Conference, Vol. 30, Greensboro, NC, 1977, 5–7.

68. Lehrer, S.B., McCants, M., and Salvaggio, J.E., Analysis of tobacco leaf allergens by crossed radioimmunoelectrophoresis, *Clin. Allergy*, 56, 616, 1985.

69. Gleich, C.J., Welsh, P.W., Yunginger, J.W., et al. Allergy tobacco: an occupational hazard, *N. Engl. J. Med.*, 13, 617, 1980

70. Ebner, C., Birkner, T., Valenta, R., et al., Common epitopes of birch pollen and apples. Studies by western and northern blot, *J. Allergy Clin. Immunol.*, 88, 588, 1991.

71. Valenta, R., Duchene, M., Ebner, C., et al., Profilin constitute a novel family of functional plant pan-allergens, *J. Exp. Med.*, 175, 377, 1992.

72. Ortega, N., Quiralte, J., Blanco, C., Castillo, R., Alvarez, M.J., and Carrillo, T., Tobacco allergy: demonstration of cross-reactivity with other members of Solanaceae family and mugwort pollen, *Ann. Allergy*, 82, 194, 1999.

73. Hernandez, J., Garcia Selles, F.J., Pagan, J.A., and Negro, J.M., Hipersensibilidad inmediata a frutas, verduras y polinosis, *Allergol. Immunopathol.*, 13, 197, 1985.

6   Environmental Tobacco Smoke and Adult Asthma

Mark D. Eisner and Paul D. Blanc

CONTENTS

6.1   INTRODUCTION

Asthma is a common chronic health condition, affecting 5% of the U.S. adult population.[1] During the past decade, the prevalence of adult asthma has increased more than 50%.[1] The mortality from asthma has approximately doubled. Understanding the factors contributing to asthma morbidity and mortality has important clinical and public health implications.

A complex mixture of over 4000 chemical compounds, environmental tobacco smoke (ETS) contains potent respiratory irritants such as sulfur dioxide, ammonia, formaldehyde, and acrolein.[2] These chemicals may induce asthma through irritant or sensitizing mechanisms.[2] Because persons with established asthma have chronic respiratory disease, they may also be susceptible to adverse health effects from ETS exposure. This chapter evaluates the evidence that ETS exposure (1) is a risk factor for new-onset asthma among adults and (2) exacerbates pre-existing adult asthma, resulting in greater respiratory symptoms and health care utilization.

As shown in Figure 6.1, this chapter focuses on three conceptually related respiratory outcomes. If ETS exposure is a causal factor in adult asthma onset, studies should also demonstrate an elevated risk of respiratory symptoms, which may reflect undiagnosed asthma. Elucidating the relationship between ETS exposure and asthma exacerbation would further support the coherence of the association between ETS and asthma.

0-8493-0311-7/00/$0.00+$.50
© 2001 by CRC Press LLC

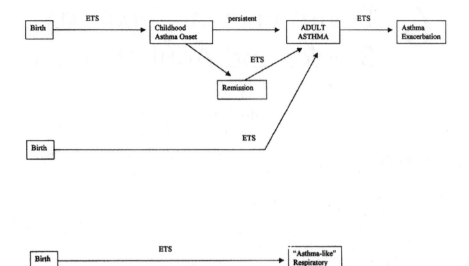

FIGURE 6.1 Conceptual framework of the causal factors of ETS on adult asthma. Several possible pathways are shown for how ETS may affect adult asthma. ETS exposure may cause childhood asthma, which persists into adulthood. In persons whose childhood asthma undergoes remission, ETS may cause recrudescence in adulthood. ETS may induce new-onset asthma in adulthood. In adults with established asthma, ETS exposure may result in exacerbation. ETS exposure may also cause adult respiratory symptoms, perhaps reflecting undiagnosed asthma.

6.2   PREVALENCE OF ETS EXPOSURE

A significant proportion of U.S. adults report exposure to ETS, ranging from 37 to 63%.[2] The third National Health and Nutrition Examination Survey (NHANES), based on a representative sample of the U.S. population, estimated that 37% of non-smoking adults are regularly exposed to ETS.[3] The population-based National Health Interview Survey from 1991 found a similar prevalence of ETS exposure among adults (39%).[4] Using measurements of serum cotinine, the major metabolite of nicotine, the NHANES investigators found that 88% of adults had elevated levels consistent with at least some ETS exposure.

Although they might be expected to avoid ETS, many U.S. adults with asthma experience exposure.[5] Among adult HMO members with asthma, 38% indicated regular exposure.[6] A population-based study from Canada found that 42% of non-smoking children and adults with asthma reported ETS exposure during the previous 24 h, compared with 32% of the general population.[7] In sum, adults with asthma appear to experience significant ETS exposure.

6.3   ETS AND NEW-ONSET ADULT ASTHMA

Extensive data support a causal association between ETS exposure and induction of asthma in children.[2] In a recent meta-analysis of 37 epidemiologic studies, ETS exposure was associated with a greater risk of developing childhood asthma (odds ratio [OR] 1.44; 95% confidence interval [CI] 1.27 to 1.64).[2] The relationship between ETS exposure and adult-onset asthma has received less attention.[8]

The Swiss Study on Air Pollution and Lung Diseases in Adults (SAPALDIA) focused on a random sample of adult never smokers aged 18 to 60 years residing in Switzerland (Table 6.1).[9] In a cross-sectional analysis, investigators observed an association between self-reported ETS exposure during the previous 12 months and a greater risk of self-reported physician diagnosis of asthma (OR 1.39; 95% CI 1.04 to 1.86). Statistically controlling for age, gender, atopy, education, maternal and paternal smoking during childhood, and parental asthma history had no appreciable impact on this relationship. Furthermore, the investigators observed statistically significant exposure-response trends for hours per day of ETS exposure, number of smokers, and years of exposure.

The population-based design strengthens the generalizability of the SAPALDIA results. The investigators controlled for important potential confounding variables, including demographic and socioeconomic factors. However, the cross-sectional design precludes complete elucidation of the causal pathway. Due to retrospective exposure assessment, ETS exposure preceding asthma onset cannot be unambiguously established. If adults with asthma are more likely to remember and report ETS exposure, the association between ETS and asthma would not reflect causal action.

A case-control study from semi-rural Sweden evaluated ETS exposure as a risk factor for asthma.[10] During a 9-month period, cases were identified from all persons filling a prescription for beta-agonist medications in two communities. For each potential case (n = 271), the diagnosis of asthma was confirmed by the study lung specialist (n = 138). An additional 44 subjects were excluded whose asthma began before age 20, leaving 79 cases. From this primary study base, controls were randomly selected from a general population register and matched to cases by age (of asthma diagnosis), gender, and community (n = 304 controls). ETS exposure at both home and work was assessed by a written questionnaire, which was defined as exposure for at least 3 years prior to the age at asthma diagnosis (or comparable age for controls). Using this definition, workplace ETS exposure was associated with an increased risk of asthma (OR 1.5; 95% CI 0.8 to 2.5), but the CI did not exclude no association. Exposure to ETS at home was not associated with the risk of asthma (OR 0.9; 95% CI 0.5 to 1.5).

Using a primary study base, the investigators sampled adults with asthma from the entire population of two communities. Selection of controls from a general population registry likely recruited cases and controls from the same source population. However, the use of beta-agonist dispensing to identify adults with asthma may have missed some persons with mild or intermittent asthma. The study of prevalent asthma cases complicates interpretation of the elevated risk associated with workplace ETS exposure, which could reflect increased risk of asthma or a longer duration of asthma symptoms. Finally, the investigators did not exclude active smokers from study. Because active smoking and ETS exposure are correlated, smoking may also have introduced confounding.

A population-based, case-control study examined the relationship between ETS exposure and obstructive respiratory disease.[11] Using a primary study base consisting

TABLE 6.1
Environmental Tobacco Smoke and New-Onset Adult Asthma

Study	Design	Subjects	N	Participation	ETS exposure measure	Study outcomes	Results OR (95% CI)
Leuenberger, P.[9]	Cross-sxn	Population based 18–60 years Switzerland Never Smokers	4197	59%	Any ETS past 12 months (home, work)	Asthma[a] Wheezing Dyspnea on exertion Chronic bronchitis	1.39 (1.04–1.86) 1.94 (1.39–2.70) 1.45 (1.2–1.79) 1.65 (1.17–2.16)
Flodin, U.[10]	Case control	Population based Sweden Cases: 20-65 years w/asthma Controls: age/gender matched	79 304	N/A	ETS at home, work	Asthma, clinical diagnosis Workplace ETS Home ETS	1.5 (0.8–2.5) 0.9 (0.5–1.5)
Dayal, H.H.[11]	Case control	Population based Cases: self-report obstructive airway disease, never smokers Controls: age/gender/ neighborhood matched	219 657	N/A	Live with smoker	Obstructive airway disease (asthma, chronic bronchitis, or emphysema) Light ETS exposure Heavy ETS exposure	1.16 (0.78–1.7) 1.86 (1.2–2.9)
Hu, F.B.[12]	Cohort	20–22 years Southern California	2041	N/A	Parental smoking	Asthma[a] at 7 years f/u Mother smoking Father smoking	1.8 (1.1–3.0) 1.6 (1.1–2.4)
Greer, J.R.[13] and Robbins, A.S.[15]	Cohort	Adult (≥25 years) non-smoking Seventh-Day Adventists	3917	87%	Duration of workplace ETS exposure (per 10 years of exposure)	Asthma[a] at 10 years f/u Men Women	1.5 (1.12–2.01) 1.5 (1.17–1.92)

TABLE 6.1 (continued)
Environmental Tobacco Smoke and New-Onset Adult Asthma

Study	Design	Subjects	N	Participation	ETS exposure measure	Study outcomes	Results OR (95% CI)
McDonnell, W.F.[14]	Cohort	Adult (≥25 years) non-smoking Seventh-Day Adventists	3091	50%	Duration of workplace ETS exposure (per 10-years of exposure)	Asthma[a], at 15-years f/u	
						Men	N.S. (not reported)
						Women	1.21 (1.04–1.39)

Note: Abbreviations: F/U = f/u; ETS = environmental tobacco smoke; cross-sxn = cross-sectional; N.S. = not significant, and OR = odds ratio.

[a] Self-reported, physician-diagnosed asthma.

of nine neighborhoods in Philadelphia, the investigators randomly selected 4200 households for interview. From this study base, they identified 219 prevalent cases of never smokers with obstructive respiratory disease, defined as self-reported asthma, chronic bronchitis, or emphysema. For each case, three randomly selected controls were matched on age, gender, and neighborhood. In conditional logistic regression analysis, self-reported high-level household ETS exposure (>1 pack per day) was associated with a greater risk of obstructive respiratory disease (OR 1.86; 95% CI 1.2 to 2.86). Lower level household ETS exposure (<1 pack per day) was not significantly associated with the risk of obstructive respiratory disease (OR 1.16; 95% CI 0.78 to 1.72).

Strengths of this study include the population-based sampling methods and the control of confounding by restriction of participants to never smokers and matching by demographic and neighborhood characteristics. The latter also provides statistical control for socioeconomic factors and outdoor air pollution. Unfortunately, the use of a composite case definition limits inferences regarding asthma alone. Because cases were defined as prevalent respiratory disease, the relationship between ETS and disease could be explained by either increased incidence or longer duration of symptoms. Finally, ETS exposure was likely underestimated because work and other sources of exposure were not ascertained.

Hu and colleagues evaluated a cohort of 1469 seventh grade students 7 years after participating in a school-based smoking prevention program in Southern California.[12] At baseline, ETS exposure status was determined by parental reports of personal smoking. Self-reported, physician-diagnosed asthma was ascertained 7 years later, during young adulthood, by a written questionnaire. Exposure to parental ETS at baseline was associated with an increased risk of reporting asthma at the 7-year follow-up. Compared with no maternal smoking or light smoking at baseline (≤ half a pack per day), heavier maternal smoking was associated with an increased risk of reporting asthma after controlling for sex, race, and educational attainment (OR 1.8; 95% CI 1.1 to 3.0). Similarly, heavy paternal smoking was related to a greater risk of asthma (OR 1.6; 95% CI 1.1 to 2.4). The investigators observed an exposure-response relationship between the number of parents smoking at baseline and the risk of asthma 7 years later: OR 1.3 for one smoking parent vs. none (95% CI 0.9 to 2.0) and 2.9 for two smoking parents (95% CI 1.6 to 5.6) (*p* for trend <0.01).

Several limitations complicate interpretation of this study. Because the proportion of subjects completing the 7-year follow-up was not reported, this selection bias cannot be estimated. At baseline, prevalent asthma cases were not excluded, so subjects reporting asthma at follow-up combine incident and prevalent cases. Consequently, the temporal association between ETS exposure and asthma onset is not clear for all subjects. Furthermore, the authors do not evaluate personal smoking. Because adolescents exposed to smoking parents may be more likely to smoke themselves, active personal smoking could confound the relationship between ETS exposure during young adulthood and the subsequent risk of developing asthma.

A prospective cohort study of 3914 adult, non-smoking, Seventh-Day Adventists living in California evaluated the relationship between ETS exposure and the incidence of self-reported, physician-diagnosed asthma during a 15-year period.[13,14] The investigators reported the 10-[13] and 15-year cohort follow-up.[14] The cumulative incidence of asthma at 10 years was 21/1000 for men and 22/1000 for women. At 15 years, the cumulative incidence was 32/1000 and 43/1000, respectively. The duration of working with a smoker was associated with an increased risk of developing asthma (OR 1.5 per 10-year increment; 95% CI 1.2 to 1.8) after controlling for age, education, gender, and ambient ozone concentration. The stratified results by gender were highly similar. At 15-year follow-up, the duration of working with a smoker was associated with an increased risk of incident asthma for women only (OR 1.21; 95% CI 1.04 to 1.39). Although not statistically significant, the results were not reported for men. In both analyses, there was no reported relationship between the duration of residence with a smoker and the risk of asthma, although detailed results were not presented.

In a related publication focusing on the same cohort, the investigators examined long-term exposure to ETS in both workplace and residential settings.[15] Exposure to ETS during adulthood was associated with a greater risk of incident asthma (OR 1.57; 95% CI 0.81 to 2.97), although the CI did not exclude no-relationship. Longer term ETS exposure, including both childhood and adulthood, was associated with an increased risk of developing asthma (OR 1.89; 95% CI 1.13 to 3.15).

The cohort study experienced substantial losses to follow-up, which may have biased the relationship between ETS exposure and risk of asthma. By 15 years, subjects lost to follow-up were more likely to have asthma and higher levels of ETS exposure,[14] potentially biasing the OR toward the null value. The observed reduction in effect size from 10 to 15 year follow-up supports this interpretation. The authors also measured longer term ETS exposure, which ignores the acute effects of ETS exposure on respiratory health.[16–18] Furthermore, the analyses implicitly assume a linear exposure-response relationship between years of ETS exposure and the risk of asthma, which may not be justified.

6.4 ETS AND "ASTHMA-LIKE" RESPIRATORY SYMPTOMS

Currently, there is no widely accepted "gold standard" for defining asthma. Although self-reported, physician-diagnosed asthma is commonly used in survey research, this definition may not detect some individuals with asthma.[19,20] Subject-reported respiratory symptoms, such as wheezing, dyspnea, and cough, may have a greater sensitivity for identifying adults with asthma (albeit a lower specificity).[20]

Several investigators have examined whether ETS exposure is related to developing adult-onset wheezing and other respiratory symptoms that could reflect asthma (Table 6.2). In the SAPALDIA study, ETS exposure during the previous 12 months was cross-sectionally associated with a greater risk of wheezing (OR 1.94; 95% CI 1.39 to 2.70), dyspnea on exertion (OR 1.45; 95% CI 1.2 to 1.79), and chronic bronchitis symptoms of cough or phlegm production (OR 1.65; 95% CI 1.17 to 2.16).[9] As in the analysis focusing on asthma, controlling for demographic and socioeconomic covariates had minimal effect. For these symptom endpoints, investigators observed exposure-response trends for hours per day of exposure, number of smokers, and years of exposure.

In an analysis of 43,732 adults completing the Health Promotion and Disease Prevention supplement of the 1991 National Health Interview Survey, Mannino and

TABLE 6.2
Environmental Tobacco Smoke and "Asthma-Like" Respiratory Symptoms

Study	Design	Subjects	N	Participation	ETS exposure measure	Study outcomes	Results RR (95% CI)[a]
Mannino, D.M.[4]	Cross-sxn	Population based ≥18 years National Health Interview Survey	43,732	N/A	ETS at home, work	Respiratory disease exacerbation	1.44 (1.07–1.95)
Hole, D.J.[21]	Cross-sxn	Population based 45–64 years Scotland	7,997	80%	Live with smoker	Sputum production	1.19 (0.85–1.67)
						Dyspnea	1.09 (0.82–1.45)
Kauffmann, F.[23]	Cross-sxn	Population based Women 24–74 years U.S. and France Never smokers	6,075	N/A	Live with smoker	Chronic cough	
						U.S.	1.14 (0.62–2.09)
						France	1.35 (0.78–2.36)
						Sputum production	
						U.S.	1.65 (0.72–3.78)
						France	0.77 (0.29–2.03)
						Dyspnea	
						U.S.	1.35 (0.68–2.61)
						France	1.17 (0.87–1.57)
						Wheeze	
						U.S.	1.35 (0.97–1.87)
						France	1.03 (0.77–1.38)

TABLE 6.2 (*continued*)
Environmental Tobacco Smoke and "Asthma-Like" Respiratory Symptoms

Study	Design	Subjects	N	Participation	ETS exposure measure	Study outcomes	Results RR (95% CI)[a]
Comstock, G.W.[24]	Cross-sxn	Population based >20 years Maryland	1,802	89–98%	Live with smoker	Chronic cough	
						Men	0.96[b]
						Women	0.17
						Wheeze	
						Men	1.04
						Women	1.45
						Dyspnea	
						Men	1.08
						Women	1.79
Ng, T.P.[25]	Cross-sxn	Population based Women 20–74 years Singapore Never smokers	1,438	79%	Live with heavy smoker	Asthma[c]	1.6 (0.69–3.70)
						Chronic cough	3.01 (1.13–8.03)
						Sputum production	2.29 (0.94–5.59)
						Dyspnea	1.83 (1.30–2.58)
						Wheezing	2.69 (1.23–5.88)
Eisner, M.D.[27]	Case crossover	Bartenders ≥18 years San Francisco Before/after prohibition of workplace smoking	53	81%	Self-reported ETS exposure duration (previous 7 days)	Respiratory symptoms Per 5-h reduction work ETS	0.7 (0.5–0.9)

TABLE 6.2 (continued)
Environmental Tobacco Smoke and "Asthma-Like" Respiratory Symptoms

Study	Design	Subjects	N	Participation	ETS exposure measure	Study outcomes	Results RR (95% CI)[a]
Jaakkola, M.S.[28]	Cohort	15–40 years Never smokers Canada	117	N/A 13% at f/u	Total ETS exposure index (intensity and duration)	Wheezing Dyspnea Cough Sputum production Any symptom	1.15 (0.64–2.06) 2.37 (1.25–4.51) 1.55 (0.61–3.90) 0.69 (0.21–2.26) 1.48 (0.88–2.49)
Schwartz, J.[30]	Cohort	Student nurses	100	N/A	Live with smoker	Cough Sputum production	p = N.S.[b] (data N/A) 1.41 (1.08–1.85)
White, J.R.[31]	Cohort	Adult participants in univ. fitness program Never smokers, no home ETS No repiratory disease	80	80% at f/u	Workplace ETS for >1 year	Cough Sputum production Dyspnea	70% vs. 25%[d] 68% vs. 20% 68% vs. 15%
Withers, N.J.[32]	Cohort	Population based Children (6–8 years) followed into adolescence (14–16 years) U.K.	2,289	71% at f/u	Live with smoker	Current wheeze (at f/u interview) Current cough New-onset wheeze (over f/u) New-onset cough New-onset wheeze — father smoking New-onset wheeze — mother smoking	1.48 (1.17–1.88) 1.47 (1.11–1.95) N.S.[b] N.S. 1.55 (1.03–2.32) N.S.

TABLE 6.2 (continued)
Environmental Tobacco Smoke and "Asthma-Like" Respiratory Symptoms

Study	Design	Subjects	N	Participation	ETS exposure measure	Study outcomes	Results RR (95% CI)[a]
Strachan, D.P.[33]	Cohort	Population based Followed all persons born March 3–9, 1958 into adulthood U.K.	18,559	31% at 33 years f/u	Live with smoker	New-onset wheeze by age 33	0.92 (0.73–1.15)
						Paternal smoking at age 16	1.71 (0.97–3.0)
						Maternal smoking (pregnancy)	1.19 (0.86–1.65)
						Maternal smoking (age 16)	1.40 (1.08–1.82)
						Maternal smoking (pregnancy and age 16)	

Note: Abbreviations: F/U = f/u; ETS = environmental tobacco smoke; RR = relative risk; and cross-sxn = cross-sectional.

[a] RR is the odds ratio, except for White (proportion of subjects with symptoms) and Sippel (RR based on Poisson regression)
[b] 95% CI not available; $p > 0.05$.
[c] Self-reported, physician-diagnosed asthma.
[d] $p < 0.001$, except where otherwise indicated.

colleagues examined the cross-sectional association between self-reported ETS exposure at home or work and the risk of "chronic respiratory disease exacerbation."[4] This study outcome was defined as activity limitation *or* a physician visit due to asthma, chronic bronchitis, emphysema, or chronic sinusitis. Among never smokers, ETS exposure was associated with an increased risk of chronic respiratory disease exacerbation after controlling for age, socioeconomic status (based on income and education), gender, race, and region of the country (OR 1.44; 95% CI 1.07 to 1.95). Although the population-based sampling and careful control of confounding are study strengths, the relationship between ETS exposure and asthma alone cannot be determined.

Four other population-based, cross-sectional studies evaluated the association between living with a smoker and respiratory symptoms in the west of Scotland,[21,22] the U.S.,[23,24] France,[23] and Singapore.[25] In the first two studies,[21–23] residential ETS exposure was associated with an elevated risk of respiratory symptoms such as cough, wheezing, and dyspnea. The CIs, however, span 1.0 (no association) in all cases. Comstock et al. found no association between residential ETS exposure and respiratory symptoms among men, whereas women had an elevated risk of wheeze (OR 1.45) and dyspnea (OR 1.79) that were not statistically significant (p values or CIs were not presented).[24] Among never-smoking women living in Singapore, living with a heavy smoker (more than 20 cigarettes per day) was not statistically related to the risk of self-reported, physician-diagnosed asthma after controlling for age, race, and employment status (OR 1.6; 95% CI 0.69 to 3.70).[25] Residing with a heavy smoker was associated with a greater risk of chronic cough (OR 3.01; 95% CI 1.13 to 8.03), phlegm production (OR 2.29; 95% CI 0.94 to 5.59), dyspnea (OR 1.83; 95% CI 1.30 to 2.58), and wheezing (OR 2.69; 95% CI 1.23 to 5.88). Exposure to light smoking at home was not related to any respiratory symptom endpoint.

In all four studies, ETS exposure outside the home was not assessed, potentially underestimating the relationship between ETS exposure and respiratory symptoms. Furthermore, comparison of several similar respiratory symptom endpoints without statistical adjustment for multiple comparisons may have increased the likelihood of chance findings. Finally, the cross-sectional design may have introduced bias, because symptom status may influence the likelihood of reporting ETS exposure.[2,26]

Using a case-crossover design, we studied the effects of California State Assembly Bill 13, which prohibited tobacco smoking in bars and taverns, on the respiratory health of bartenders.[27] Based on a random sample of all bars and taverns in San Francisco, we interviewed and performed spirometry on 53 bartenders before and after the smoking ban (median length of follow-up was 56 d). Because 45% of participating bartenders reported current personal cigarette smoking, we performed stratified and multivariate analyses to control for direct smoking.

At baseline, all 53 subjects reported ETS exposure while working in bars or taverns during the 7 d prior to interview. After prohibition of smoking, self-reported workplace ETS exposure sharply declined from a median of 28 to 2 h/week. Thirty-nine (74%) of the 53 bartenders reported at least one respiratory symptom at baseline (including cough, dyspnea, and wheezing), while only 17 (32%) were still symptomatic at follow-up. Of the 39 bartenders reporting baseline symptoms, 23 subjects (59%) no longer indicated any respiratory symptoms after prohibition of smoking

(p <0.001). In particular, 70% of the 17 bartenders reporting baseline wheezing noted resolution after workplace smoking prohibition (p = 0.02). In conditional logistic regression analysis, a 5-h reduction of workplace ETS exposure was associated with a lower risk of respiratory symptoms at follow-up (OR 0.7; 95% CI 0.5 to 0.9), after controlling for upper respiratory infections and reduced personal cigarette smoking. In stratified analyses by current smoking status, we observed similar symptom reduction in both current smokers and non-smokers. After prohibition of workplace smoking, we also observed improvement in mean forced vital capacity (FVC) (0.189 L; 95% CI 0.082 to 0.296) and mean forced expiratory volume in 1 second (FEV$_1$) (0.039 L; 95% CI –0.030 to 0.107). Complete cessation of workplace ETS exposure was associated with an even greater pulmonary function improvement.

Although we did not specifically study asthma, the reduction of respiratory symptoms after workplace smoking prohibition suggests that ETS exerts harmful effects on the lower respiratory tract. The reduction in wheezing, in particular, suggests that ETS exposure may cause or exacerbate asthma-like symptoms. Further study limitations include reliance on self-reported ETS exposure and symptoms. The low participation rate by bars (30%) raises the concern of generalizability to all bars and taverns. Although we adhered to a standard spirometry protocol, we cannot exclude the contribution of training to the observed pulmonary function improvement.

In a prospective cohort study, investigators studied the relationship between self-reported ETS exposure and the incidence of respiratory symptoms among 117 young adult never smokers (15 to 40 years).[28] Comparing ETS-exposed subjects with unexposed subjects, the cumulative incidence ratio (CIR) was elevated for wheezing (CIR 1.5), dyspnea (CIR 2.0), and cough (CIR 2.4) (no p values or CIs were reported). In multiple logistic regression controlling for age, gender, and atopy, a total ETS exposure index based on duration and intensity of exposure was associated with a greater risk of developing dyspnea (OR 2.4 per 10-point increment; 95% CI 1.3 to 4.5) or any respiratory symptom (OR 1.5; 95% CI 0.9 to 2.5). There was no statistical association between the ETS exposure index and the risk of wheezing, cough, or phlegm. Examination of ETS exposure categories (none, low exposure, and high exposure) yielded similar results, with high exposure related to an increased risk of dyspnea only (OR 6.8; 95% CI 1.4 to 33.1).

The principal limitation of this study is the small sample size, with limited power to detect clinically significant increases in respiratory symptoms. As a consequence, the investigators were not able to assess the relationship between ETS exposure and incident asthma. Furthermore, less than 15% of subjects originally studied cross-sectionally in 1980 to 1981 were evaluated at an 8-year follow-up,[29] which may have introduced bias. If subjects who developed respiratory symptoms were less likely to complete follow-up, the relationship between ETS exposure and incident symptoms may be biased toward the null value.

A prospective cohort study examined 100 student nurses who completed daily symptom diaries for 3 years.[30] After controlling for personal smoking, living with a smoking roommate was associated with an increased risk of incident phlegm production (relative risk [RR] 1.41; 95% CI 1.08 to 1.85). There was no association between ETS exposure and development of cough. Although the investigators did not explicitly control for demographic covariates, study subjects were similar in gender and education. Study limitations included failure to assess non-residential ETS exposure and other respiratory symptoms. Because the study does not report losses to follow-up, this selection bias cannot be estimated.

In a prospective cohort study, 80 never-smoking adult participants in a university-based physical fitness program were followed for 9 months.[31] Subjects who reported residential ETS exposure or baseline respiratory conditions were excluded. Compared with unexposed subjects, a greater proportion of persons reporting ongoing workplace ETS exposure (for at least 12 months) indicated cough (70 vs. 25%), phlegm production (68 vs. 20%), or dyspnea (68 vs. 15%) ($p \leq 0.0001$ in all cases). Unfortunately, the exclusion of 50% of fitness program participants (for a long list of criteria) may have biased these results in an unpredictable fashion. The investigators also did not control for potential confounding variables such as demographic and socioeconomic indicators.

A population-based prospective cohort study followed U.K. children aged 6 to 8 years into adolescence (age 14 to 16 years) to examine factors associated with respiratory symptoms over time.[32] Subjects who resided with one or more smokers at a follow-up postal questionnaire were classified as ETS exposed. ETS exposure was cross-sectionally associated with current cough (OR 1.47; 95% CI 1.11 to 1.95) and wheeze (OR 1.48; 95% CI 1.17 to 1.88). Among previously asymptomatic persons, paternal smoking was associated with new-onset wheeze during longitudinal follow-up after controlling for demographic and socioeconomic indicators (OR 1.55; 95% CI 1.03 to 2.32). Maternal smoking, however, was not associated with new-onset wheeze. Moreover, there was no apparent relationship between ETS exposure and new-onset cough.

Although this study appears to link childhood ETS exposure to the development of respiratory symptoms, the reliance on parental assessment of subjects' respiratory symptoms may have resulted in misclassification of symptom status. Furthermore, the classification of ETS exposure status based on follow-up interviews assumes no change in exposure during the 8-year follow-up period. Assessment of the association between *baseline* ETS exposure and the subsequent incidence of respiratory symptoms would have enabled stronger etiologic inferences.

Another population-based U.K. cohort study followed 18,559 children born during a single week in March 1958 through age 33, with 31% complete follow-up.[33] The study examined the relationship between household ETS exposure and the subsequent incidence of wheezing after controlling for demographic, socioeconomic, and personal smoking variables. At both age 7 and 33 years, maternal smoking during pregnancy was associated with an increased risk of incident wheezing illness (OR 1.72; 95% CI 1.11 to 2.67 and OR 1.71; 95% CI 0.97 to 3.0, respectively). At age 33, maternal smoking at subject age 16 was associated with an increased incidence of wheezing (OR 1.19; 95% CI 0.86 to 1.65), although the CI overlapped the null value. ETS exposure *both* during pregnancy and at age 16 was related to a greater risk of incident wheezing (OR 1.4; 95% CI 1.08 to 1.82). Unlike the study by Withers et al.,[32] these investigators prospectively assessed ETS exposure and examined the relationship with incident symptoms. The low study participation rate at age 33, however, may have introduced significant bias if ETS exposure or development of wheezing was related to the likelihood of study participation at successive waves.

6.5   ETS EXPOSURE AND EXACERBATION OF PRE-EXISTING ADULT ASTHMA

Among children, ETS exposure has been strongly linked with exacerbation of pre-existing asthma.[2,34] Although adults with asthma commonly report ETS exposure as a trigger for asthma exacerbation, the effect of exposure on adult asthma status has received less research.[35,36] In a cross-sectional study, investigators examined the impact of self-reported ETS exposure on 200 never-smoking adults with asthma attending a · university-based chest clinic in India (Table 6.3).[37] Compared with unexposed patients, adult asthmatics reporting ETS exposure indicated greater reliance on daily bronchodilators (66 vs. 56%, p <0.01) and intermittent corticosteroid use (56 vs. 42%, p <0.01). Although there was no relationship with hospitalization, ETS-exposed subjects had a higher mean number of emergency department visits for asthma during the previous year (0.82 vs. 0.6 visits per person, p <0.01) and more work absence (3.6 vs. 3.0 weeks per person, p <0.01). ETS exposure was also associated with worse pulmonary function, including lower FEV_1 (68.7 vs. 80.8% of predicted), FEV_1/FVC (63.5 vs. 78.4%), and $FEF_{25-75\%}$ (54.3 vs. 75.7%) (p <0.01 in all cases).

ETS exposure and asthma status were assessed cross-sectionally, potentially resulting in misclassification of exposure. If adults with poorly controlled asthma were more likely to report ETS exposure, a bias toward greater association may have resulted. The investigators also made no attempt to control for confounding variables. Moreover, generalizing the results obtained from a university center in India to the general population of asthmatics in the U.S. requires caution.

Blanc and colleagues examined the cross-sectional impact of self-reported regular ETS exposure at work among 2065 adult participants (20 to 44 years) in the Swedish component of the population-based European Community Respiratory Health Survey.[38] In multivariate analysis controlling for age, gender, personal smoking, and work characteristics, regular workplace ETS exposure was associated with a greater risk of respiratory-related work disability (prevalence ratio [RR] 1.8; 95% CI 1.1 to 3.1), defined as self-reported change in job or leaving work due to affected breathing. Moreover, workplace ETS exposure was associated with a greater risk of work-associated symptomatic asthma, defined as self-reported asthma, airway hyperresponsiveness, and work-related chest tightness or wheezing (PR 1.7; 95% CI 0.9 to 3.3). Because this study focused on workplace factors, home and other sources of ETS exposure were not examined.

In a prospective panel study of 164 adult non-smokers with asthma, Ostro and colleagues examined the impact of ETS exposure on asthma status during a 3-month period.[39] Subjects completed daily diaries including ETS exposure (home and work) and respiratory symptoms. During longitudinal follow-up, ETS exposure was associated with subsequent greater risks of cough (OR 1.21; 95% CI 1.01 to 1.46), dyspnea (OR 1.85; 95% CI 1.57 to 2.18), nocturnal asthma symptoms (OR 1.24; 95% CI 1.00 to 1.53), and restricted activity (OR 2.08; 95% CI 1.63 to 2.64). There

TABLE 6.3
Environmental Tobacco Smoke and Exacerbation of Pre-Existing Asthma

Study	Design	Subjects	N	Participation	ETS exposure measure	Study outcomes	Results RR (95% CI)[a]
Jindal, S.K.[37]	Cross-sxn	Adults with asthma 15–50 years Chest clinic, India	200	N/A	≥1 h/day ETS for at least 1 year	Daily bronchodilator use Corticosteroid use ED visits (# per patient/year) Hospitalization (# per patient/year) Work absence (weeks per patient/year)	66% vs. 56%[b] 56% vs. 42% 0.82 vs. 0.6 0.33 vs. 0.34; p = N.S. 3.6 vs. 3.0
Blanc, P.D.[38]	Cross-sxn	Population based 20–44 years Sweden	2065	86%	Regular ETS at work	Respiratory-related work disability Work-associated symptomatic asthma	1.8 (1.1–3.1) 1.7 (0.9–3.3)
Ostro, B.[39]	Cohort	Adults with asthma Denver Non-smokers	164	78%	ETS exposure at home, work	Cough Dyspnea Nocturnal respiratory symptoms Restricted activity	1.21 (1.01–1.46) 1.85 (1.57–2.18) 1.24 (1.00–1.53) 2.08 (1.63–2.64)
Sippel, J.M.[6]	Cohort	15–55 years Kaiser Permanente members with asthma (Northwest U.S.)	619	N/A	Regular ETS exposure, home or work	Hospital-based care (ED, hospital)	2.34 (1.8–3.1)

TABLE 6.3 (*continued*)
Environmental Tobacco Smoke and Exacerbation of Pre-Existing Asthma

Note: Abbreviations: F/U = f/u; ETS = environmental tobacco smoke; RR= relative risk; cross-sxn = cross-sectional; and ED = emergency department.

[a] **RR** is the odds ratio, except where otherwise indicated.

[b] $p < 0.001$, except where otherwise indicated.

[c] Self-reported asthma, airway responsiveness, and work-related chest tightness or wheezing.

was no apparent association between ETS exposure and emergency department visits or hospitalization for asthma (data not reported in publication). Although demographic characteristics were not included in logistic regression models, the use of 1-day lagged ETS exposure makes significant confounding unlikely. In this longitudinal panel study, the close temporal link between ETS exposure and outcome supports a causal relationship between exposure and asthma exacerbation.

A cohort study of 619 adult HMO members with asthma examined the association between ETS exposure and health outcomes.[6] The prevalence of self-reported regular ETS exposure was 38%, and a small proportion of subjects (11%) indicated current personal cigarette smoking. In cross-sectional analysis of baseline data, regular ETS exposure was associated with worse asthma-specific quality of life (QOL) and generic health status (SF-36 physical functioning and general health domains). During longitudinal follow-up, ETS exposure was associated with a greater incidence of hospital-based episodes of asthma care (28 events per 100 person-years vs. 10 events per 100 person-years). After adjusting for age, gender, and asthma severity, ETS exposure remained associated with a greater risk of hospital-based care (RR 2.34; 95% CI 1.8 to 3.1). Excluding current smokers from analysis did not appreciably affect these results.

6.6   ETS EXPOSURE AND PULMONARY FUNCTION

Because asthma is characterized by reversible airway obstruction, the effect of ETS exposure on pulmonary function in the adult general population has relevance for asthma. In children, more than 30 studies have linked domestic ETS exposure with decreased development of lung function.[2] Among adults, most cross-sectional analyses support the association between self-reported ETS exposure and a decrement in pulmonary function.[21,24,25,29,40-43] In two other cross-sectional studies, there was no apparent relationship.[23,44] The fewer prospective investigations have provided mixed results, with some studies demonstrating an association between ETS exposure and decreased pulmonary function over time[45-47] and others finding no association.[48] Finally, a study of 26 Canadian bar workers found significant acute decrements in lung function after a work shift.[49] Although there is significant heterogeneity among study results, the cross-sectional and longitudinal data together support a small deleterious effect of ETS on pulmonary function.

6.7   CONTROLLED HUMAN EXPOSURE STUDIES

Controlled experimental studies support the biologic plausibility of ETS-related asthma exacerbation. In chamber exposure experiments, investigators studied the impact of acute ETS exposure on asthmatic subjects. Dahms and colleagues demonstrated a significant decrement (approximately 20%) in FEV_1 and FVC after ETS exposure for 1 h.[17] Similarly, five of ten subjects with baseline airway hyperresponsiveness experienced a 10% or greater decrement in FEV_1 after exposure.[16] Another study found that one third of asthmatic subjects experienced a substantial decline in FEV_1 after chamber exposure (>20%).[50] The same group demonstrated that pretreatment with bronchodilators prevented the acute decline in FEV_1 in previously reactive subjects.[51] In ten adult subjects with asthma, experimental ETS exposure for 3 h resulted in reduced FEV_1 (5.9%) and FVC (9.1%).[52] Other studies, however, have found no effect of acute chamber ETS exposure on lung function in asthmatic subjects.[53,54] Interpretation of these controlled exposure studies is limited by small sample size, variable subject inclusion criteria, and variation in chamber exposure methodology. Nonetheless, these experimental studies support a modest adverse effect of acute ETS exposure on pulmonary function.

6.8 ETS EXPOSURE AND ADULT ASTHMA: EVIDENCE FOR A CAUSAL ASSOCIATION

In considering whether ETS exposure is causally related to adult asthma induction and exacerbation, confounding factors require consideration. ETS exposure has been associated with younger age, female gender, non-white race, lower education, lower income, blue collar occupation, and personal cigarette smoking.[4,6,21] Many of these factors have also been associated with an increased prevalence of asthma and asthma-related morbidity.[1] As a result, an observed association between ETS exposure and adverse asthma health outcomes could be potentially explained by confounding. In the studies examined, control for confounding by design or statistical strategies ranged from poor[12,31,37] to excellent,[4,10,11,13,14,21,24,25,27,30,32,38,55] with most studies examining at least some potential confounders. Taken together, the studies reviewed suggest that the relationship between ETS exposure and adverse asthma outcomes is not explained by confounding variables.

Ascertainment of ETS exposure by self-report is potentially subject to bias, which limits interpretation of all the studies reviewed. Numerous reports have demonstrated modest correlation between self-reported ETS exposure and biomarkers of tobacco exposure, such as cotinine. In all studies examined, we cannot exclude systematic misclassification of ETS exposure. For example, persons with poor asthma status might be more likely to remember and report ETS exposure, whereas asthmatics with mild disease severity might underreport ETS exposure. This bias would inflate the estimated risk associated with ETS exposure. The prospective data, however, should be less affected by this potential bias.

Examination of the Hill criteria[56] supports a causal association between ETS exposure and adult asthma onset. Several studies demonstrated an exposure-response relationship between ETS exposure and the risk of developing new-onset adult asthma or asthma-like conditions. Investigators observed significant exposure-response relationships for total daily hours of ETS exposure,[9] number of smokers in the environment,[9,12] duration of exposure to smokers,[9] duration of working with a smoker,[13,14] and an ETS exposure index that incorporates both intensity and duration of exposure.[28] Other studies are more consistent with a threshold effect, with heavy ETS exposure associated with a greater risk of asthma or asthma-like conditions.[11,25] Taken together, these studies are consistent with a causal relationship between ETS exposure and adult asthma onset and exacerbation.

Because ETS contains potent respiratory irritants, exposure may adversely affect bronchial smooth muscle tone and airway inflammation.[2] Studies linking ETS exposure with a decrement in pulmonary function support the biologic plausibility of ETS-related asthma onset and exacerbation. Controlled human exposure studies are also consistent with an adverse biologic effect.

The consistency of study findings supports a causal relationship between ETS exposure and asthma morbidity. In samples drawn from different populations, ranging from clinical to population-based samples, investigators have observed an association between ETS exposure and new-onset asthma and asthma exacerbation. Similarly, the relationship between ETS exposure and asthma morbidity has been observed in cross-sectional, case-control, and cohort study designs. Exposure in different environments such as home and work has also been linked with asthma.

The studies reviewed also demonstrate coherence in the association between ETS exposure and asthma morbidity. ETS exposure has been associated with new-onset asthma, whether defined as self-reported, physician-diagnosed asthma or clinical asthma diagnosis. Furthermore, ETS exposure is associated with related health outcomes, including chronic respiratory disease and respiratory symptoms such as dyspnea, cough, and wheezing. In persons with pre-existing asthma, ETS exposure appears to affect a variety of health outcomes, including disease severity, QOL, and health care utilization. The coherence of these findings among diverse respiratory outcomes supports a causal association.

During the past 20 years, the prevalence of cigarette smoking has declined significantly.[57] At the same time, the prevalence of asthma has increased.[1] Although ETS exposure appears causally related to adult asthma induction, other etiologic factors must also be contributing to new asthma onset. These factors may include other environmental exposures such as allergens or ambient pollutants.

6.9   PUBLIC HEALTH IMPLICATIONS AND FUTURE RESEARCH

The long-term health consequences of ETS exposure have been established over the past two decades. Consistent epidemiologic evidence links ETS exposure with lung cancer and atherosclerotic cardiovascular disease.[2,58,59] In this chapter, the evidence suggests a causal relationship between ETS exposure and new-onset asthma and asthma exacerbation among adults. Despite the growing knowledge of ETS-related health effects, smoking is still permitted in many public locations and workplaces.[60,61] Further elucidating the short- and long-term respiratory health consequences of ETS exposure will provide policymakers with the basis for further regulation of public smoking.

To evaluate the impact of ETS exposure on new-onset asthma, future research should focus on incident asthma cases over longitudinal follow-up. Previous studies have examined prevalent asthma cases as the study outcome. As a consequence, these studies cannot separate the effects of ETS exposure on asthma induction from exacerbation of pre-existing disease. Investigating incident asthma cases would enhance etiologic inferences about the role of ETS in asthma initiation.

In future studies, the measurement of ETS should be refined. Because self-reported ETS exposure is subject to bias, assessment of biomarkers should be performed, such as cotinine or nicotine. Questionnaire-based exposure measurement should be validated using biomarker measurement. Moreover, separation of the acute and chronic effects of ETS exposure on adult asthma induction requires further study. Among adults with established asthma, studies should evaluate how short- and long-term ETS exposure affects health outcomes.

Elucidating adult subgroups at high risk for asthma induction or exacerbation would facilitate targeted preventive programs. Future research should explicate the subgroups at high risk for ETS-related asthma morbidity by factors such as atopic status, gender, personal smoking history, socioeconomic status, and occupation. Based on this knowledge, specific educational and other preventive health intervention programs can be designed, implemented, and evaluated.

Given the extensive evidence that ETS has adverse health effects, what are the current barriers to prohibiting smoking in public places? Although they comprise the minority, persons who actively smoke may resist efforts to curtail their behavior. Despite the scientific literature, many members of the lay public also do not believe that ETS affects their health. Among bartenders working in San Francisco, CA, 21% expressed the belief that ETS exposure has no adverse effect on their personal health.[27] Business owners, especially restaurant, bar, and hotel proprietors, are concerned that public smoking prohibition may result in economic losses.[62] Further research into the social and economic barriers to creating smoke-free environments would facilitate efforts to reduce ETS exposure.

6.10 CONCLUSIONS

During the past decade, the U.S. morbidity and mortality from asthma have increased substantially. The evidence indicates that adults who are exposed to ETS have a greater risk of developing asthma. Among adults with pre-existing asthma, ETS appears causally related to adverse health outcomes. Based on these and other health consequences, public policies should prohibit smoking in the workplace and other public locations. Prohibition of public smoking would be expected to have beneficial effects on adult respiratory health.

REFERENCES

1. Mannino, D.M., Homa, D.M., Pertowski, C.A., et al., Surveillance for asthma — United States, 1960–1995. Morb and Mort Wkly Rep. Cdc Surveillance Summaries, 47, 1–27, 1998.
2. California Environmental Protection Agency, Health Effects of Exposure to Environmental Tobacco Smoke, Office of Environmental Health Hazard Assessment, Sacramento, CA, 1997.
3. Pirkle, J.L., Flegal, K.M., Bernert, J.T., Brody, D.J., Etzel, R.A., and Maurer, K.R., Exposure of the U.S. population to environmental tobacco smoke: the Third National Health and Nutrition Examination Survey, 1988 to 1991, *J. Am. Med. Assoc.*, 275, 1233–1240, 1996.

4. Mannino, D.M., Siegel, M., Rose, D., Nkuchia, J., and Etzel, R., Environmental tobacco smoke exposure in the home and worksite and health effects in adults: results from the 1991 National Health Interview Survey, *Tob. Control*, 6, 296–305, 1997.

5. Eisner, M.D., Yelin, E.H., Smith, S., Henke, J., and Blanc, P.D., Impact of environmental tobacco smoke on asthma severity, *J. Invest. Med.*, 46, 97A, 1998.

6. Sippel, J.M., Pedula, K.L., Vollmer, W.M., Buist, A.S., and Osborne, M.L., Associations of smoking with hospital-based care and quality of life in patients with obstructive airway disease, *Chest*, 115, 691–696, 1999.

7. Leech, J.A., Wilby, K., and McMullen, E., Environmental tobacco smoke exposure patterns: a subanalysis of the Canadian Human Time-Activity Pattern Survey, *Can. J. Public Health*, 90, 244–249, 1999.

8. Coultas, D.B., Health effects of passive smoking. 8. Passive smoking and risk of adult asthma and COPD: an update, *Thorax*, 53, 381–387, 1998.

9. Leuenberger, P., Schwartz, J., Ackermann-Liebrich, U., et al., Passive smoking exposure in adults and chronic respiratory symptoms (SAPALDIA Study). Swiss study on air pollution and lung diseases in adults, SAPALDIA team, *Am. J. Respir. Crit. Care Med.*, 150, 1222–1228, 1994.

10. Flodin, U., Jeonsson, P., Ziegler, J., and Axelson, O., An epidemiologic study of bronchial asthma and smoking, *Epidemiology*, 6, 503–505, 1995.

11. Dayal, H.H., Khuder, S., Sharrar, R., and Trieff, N., Passive smoking in obstructive respiratory disease in an industrialized urban population, *Environ. Res.*, 65, 161–171, 1994.

12. Hu, F.B., Persky, V., Flay, B.R., and Richardson, J., An epidemiological study of asthma prevalence and related factors among young adults, *J. Asthma*, 34, 67–76, 1997.

13. Greer, J.R., Abbey, D.E., and Burchette, R.J., Asthma related to occupational and ambient air pollutants in nonsmokers, *J. Occup. Med.*, 35, 909–915, 1993.

14. McDonnell, W.F., Abbey, D.E., Nishino, N., and Lebowitz, M.D., Long-term ambient ozone concentration and the incidence of asthma in nonsmoking adults: the AHSMOG study, *Environ. Res.*, 80, 110–121, 1999.

15. Robbins, A.S., Abbey, D.E., and Lebowitz, M.D., Passive smoking and chronic respiratory disease symptoms in non-smoking adults, *Int. J. Epidemiol.*, 22, 809–817, 1993.

16. Danuser, B., Weber, A., Hartmann, A.L., and Krueger, H., Effects of a bronchoprovocation challenge test with cigarette sidestream smoke on sensitive and healthy adults, *Chest*, 103, 353–358, 1993.

17. Dahms, T.E., Bolin, J.F., and Slavin, R.G., Passive smoking. Effects on bronchial asthma, *Chest*, 80, 530–534, 1981.

18. Magnussen, H., Jeorres, R., and Oldigs, M., Effect of one hour of passive cigarette smoking on lung function and airway responsiveness in adults and children with asthma, *Clin. Invest.*, 70, 368–371, 1992.

19. McWhorter, W.P., Polis, M.A., and Kaslow, R.A., Occurrence, predictors, and consequences of adult asthma in NHANESI and follow-up survey, *Am. Rev. Respir. Dis.*, 139, 721–724, 1989.

20. Toraen, K., Brisman, J., and Jearvholm, B., Asthma and asthma-like symptoms in adults assessed by questionnaires. A literature review, *Chest*, 104, 600–608, 1993.

21. Hole, D.J., Gillis, C.R., Chopra, C., and Hawthorne, V.M., Passive smoking and cardiorespiratory health in a general population in the west of Scotland, *Br. Med. J.*, 299, 423–427, 1989.

22. Gillis, C.R., Hole, D.J., Hawthorne, V.M., and Boyle, P., The effect of environmental tobacco smoke in two urban communities in the west of Scotland, *Eur. J. Respir. Dis. Suppl.*, 133, 121–126, 1984.

23. Kauffmann, F., Dockery, D.W., Speizer, F.E., and Ferris, B.G., Jr., Respiratory symptoms and lung function in relation to passive smoking: a comparative study of American and French women, *Int. J. Epidemiol.*, 18, 334–344, 1989.

24. Comstock, G.W., Meyer, M.B., Helsing, K.J., Tockman, M.S., Respiratory effects on household exposures to tobacco smoke and gas cooking, *Am. Rev. Respir. Dis.*, 124, 143–148, 1981.

25. Ng, T.P., Hui, K.P., and Tan, W.C., Respiratory symptoms and lung function effects of domestic exposure to tobacco smoke and cooking by gas in non-smoking women in Singapore, *J. Epidemiol. Community Health*, 47, 454–458, 1993.

26. Hammond, S.K., Exposure of U.S. workers to environmental tobacco smoke, *Environ. Health Perspect.*, 107(Suppl. 2), 329–340, 1999.

27. Eisner, M.D., Smith, A.K., and Blanc, P.D., Bartenders' respiratory health after establishment of smoke-free bars and taverns, *J. Am. Med. Assoc.*, 280, 1909–1914, 1998.

28. Jaakkola, M.S., Jaakkola, J.J., Becklake, M.R., and Ernst, P., Effect of passive smoking on the development of respiratory symptoms in young adults: an 8-year longitudinal study, *J. Clin. Epidemiol.*, 49, 581–586, 1996.

29. Masi, M.A., Hanley, J.A., Ernst, P., and Becklake, M.R., Environmental exposure to tobacco smoke and lung function in young adults, *Am. Rev. Respir. Dis.*, 138, 296–299, 1988.

30. Schwartz, J. and Zeger, S., Passive smoking, air pollution, and acute respiratory symptoms in a diary study of student nurses, *Am. Rev. Respir. Dis.*, 141, 62–67, 1990.

31. White, J.R., Froeb, H.F., and Kulik, J.A., Respiratory illness in nonsmokers chronically exposed to tobacco smoke in the work place, *Chest*, 100, 39–43, 1991.

32. Withers, N.J., Low, L., Holgate, S.T., and Clough, J.B., The natural history of respiratory symptoms in a cohort of adolescents, *Am. J. Respir. Crit. Care Med.*, 158, 352–357, 1998.

33. Strachan, D.P., Butland, B.K., and Anderson, H.R., Incidence and prognosis of asthma and wheezing illness from early childhood to age 33 in a national British cohort, *Br. Med. J.*, (Clinical Research Ed.) 312, 1195–1199, 1996.

34. Chilmonczyk, B.A., Salmun, L.M., Megathlin, K.N., et al., Association between exposure to environmental tobacco smoke and exacerbations of asthma in children, *N. Engl. J. Med.*, 328, 1665–1669, 1993.

35. Dales, R.E., Kerr, P.E., Schweitzer, I., Reesor, K., Gougeon, L., and Dickinson, G., Asthma management preceding an emergency department visit, *Arch. Int. Med.*, 152, 2041–2044, 1992.

36. Abramson, M.J., Kutin, J.J., Rosier, M.J., and Bowes, G., Morbidity, medication and trigger factors in a community sample of adults with asthma, *Med. J. Aust.*, 162, 78–81, 1995.

37. Jindal, S.K., Gupta, D., and Singh, A., Indices of morbidity and control of asthma in adult patients exposed to environmental tobacco smoke, *Chest*, 106, 746–749, 1994.

38. Blanc, P.D., Ellbjar, S., Janson, C., et al., Asthma-related work disability in Sweden, *Am. J. Respir. Crit. Care Med.*, 160, 2028–2033, 1999.

39. Ostro, B.D., Lipsett, M.J., Mann, J.K., Wiener, M.B., and Selner, J., Indoor air pollution and asthma. Results from a panel study, *Am. J. Respir. Crit. Care Med.*, 149, 1400–1406, 1994.

40. O'Connor, G.T., Weiss, S.T., Tager, I.B., and Speizer, F.E., The effect of passive smoking on pulmonary function and nonspecific bronchial responsiveness in a population-based sample of children and young adults, *Am. Rev. Respir. Dis.*, 135, 800–804, 1987.

41. Svendsen, K.H., Kuller, L.H., Martin, M.J., and Ockene, J.K., Effects of passive smoking in the Multiple Risk Factor Intervention Trial, *Am. J. Epidemiol.*, 126, 783–795, 1987.

42. White, J.R. and Froeb, H.F., Small-airways dysfunction in nonsmokers chronically exposed to tobacco smoke, *N. Engl. J. Med.*, 302, 720–723, 1980.

43. Xu, X. and Li, B., Exposure-response relationship between passive smoking and adult pulmonary function, *Am. J. Respir. Crit. Care Med.*, 151, 41–46, 1995.

44. Schilling, R.S., Letai, A.D., Hui, S.L., Beck, G.J., Schoenberg, J.B., and Bouhuys, A., Lung function, respiratory disease, and smoking in families, *Am. J. Epidemiol.*, 106, 274–283, 1977.

45. Brunekreef, B., Fischer, P., Remijn, B., van der Lende, R., Schouten, J., and Quanjer, P., Indoor air pollution and its effect on pulmonary function of adult non-smoking women: III. Passive smoking and pulmonary function, *Int. J. Epidemiol.*, 14, 227–230, 1985.

46. Abbey, D.E., Burchette, R.J., Knutsen, S.F., McDonnell, W.F., Lebowitz, M.D., and Enright, P.L., Long-term particulate and other air pollutants and lung function in nonsmokers, *Am. J. Respir. Crit. Care Med.*, 158, 289–298, 1998.

47. Carey, I.M., Cook, D.G., and Strachan, D.P., The effects of environmental tobacco smoke exposure on lung function in a longitudinal study of British adults, *Epidemiology*, 10, 319–326, 1999.

48. Jaakkola, M.S., Jaakkola, J.J., Becklake, M.R., and Ernst, P., Passive smoking and evolution of lung function in young adults. An 8-year longitudinal study, *J. Clin. Epidemiol.*, 48, 317–327, 1995.

49. Dimich-Ward, H., Lawason, J., and Chan-Yeung, M., Work shift changes in lung function in bar workers exposed to environmental tobacco smoke, *Am. J. Respir. Crit. Care Med.*, 157, A505, 1998.

50. Stankus, R.P., Menon, P.K., Rando, R.J., Glindmeyer, H., Salvaggio, J.E., and Lehrer, S.B., Cigarette smoke-sensitive asthma: challenge studies, *J. Allergy Clin. Immunol.*, 82, 331–338, 1988.

51. Menon, P.K., Stankus, R.P., Rando, R.J., Salvaggio, J.E., and Lehrer, S.B., Asthmatic responses to passive cigarette smoke: persistence of reactivity and effect of medication, *J. Allergy Clin. Immunol.*, 88, 861–869, 1991.

52. Nowak, D., Jeorres, R., Schmidt, A., and Magnussen, H., Effect of 3 hours' passive smoke exposure in the evening on airway tone and responsiveness until next morning, *Int. Arch. Occup. Environ. Health*, 69, 125–133, 1997.

53. Wiedemann, H.P., Mahler, D.A., Loke, J., Virgulto, J.A., Snyder, P., and Matthay, R.A., Acute effects of passive smoking on lung function and airway reactivity in asthmatic subjects, *Chest*, 89, 180–185, 1986.

54. Shephard, R.J., Collins, R., and Silverman, F., "Passive" exposure of asthmatic subjects to cigarette smoke, *Environ. Res.*, 20, 392–402, 1979.

55. Strachan, D.P. and Cook, D.G., Health effects of passive smoking. 6. Parental smoking and childhood asthma: longitudinal and case-control studies, *Thorax*, 53, 204–212, 1998.

56. Hill, A.B., *Principles of Medical Statistics*, Oxford University Press, New York, 1971.

57. Anon., Achievements in public health, 1900–1999: tobacco use, United States, 1900–1999. Mmwr., *Morbidity Mortality Weekly Rep.*, 48, 986–993, 1999.

58. Hackshaw, A.K., Law, M.R., and Wald, N.J., The accumulated evidence on lung cancer and environmental tobacco smoke [see comments], *Br. Med. J.* (Clinical Research Ed.), 315, 980–988, 1997.

59. Kawachi, I., Colditz, G.A., Speizer, F.E., et al., A prospective study of passive smoking and coronary heart disease, *Circulation*, 95, 2374–2379, 1997.
60. Emmons, K.M., Marcus, B.H., Abrams, D.B., et al., Use of a 24-hour recall diary to assess exposure to environmental tobacco smoke, *Arch. Environ. Health*, 51, 146–149, 1996.
61. Gerlach, K.K., Shopland, D.R., Hartman, A.M., Gibson, J.T., and Pechacek, T.F., Workplace smoking policies in the United States: results from a national survey of more than 100,000 workers, *Tob. Control*, 6, 199–206, 1997.
62. Macdonald, H.R. and Glantz, S.A., Political realities of statewide smoking legislation: the passage of California's Assembly Bill 13, *Tob. Control*, 6, 41–54, 1997.

7 Policies to Reduce Involuntary Exposure to Environmental Tobacco Smoke: Prevalence, Nature, Effects, and Future Directions

Elizabeth A. Klonoff and Hope Landrine

CONTENTS

0-8493-0311-7/00/$0.00+$.50
© 2001 by CRC Press LLC

ABSTRACT

Many countries have implemented some form of tobacco-control policy. These typically require health warnings on packages of tobacco, increase tobacco surtaxes, and prohibit tobacco sales to youth, but do little to control involuntary exposure to environmental tobacco smoke (ETS). Only 12 countries in the world have nationwide ETS policies (i.e., with no regional exceptions), and these tend to restrict smoking to a confined area (rather ban it) within enclosed spaces. Even in the U.S. where tobacco-control policies are somewhat tougher than in much of the world, only 5 states have statewide ETS policies, and only 259 (of thousands of) cities have citywide policies. Furthermore, U.S. policies tend to protect Whites, adults, and white-collar workers, leaving their Black, Latino, young, and blue-collar cohorts still exposed to high levels of ETS. Children ($\leq$5 years of age) are particularly vulnerable to the deleterious effects of ETS, and their exposure is widespread and involuntary. Half of all children $\leq$5 years of age in the U.S. and in the world — 700 million children — live with at least one adult smoker and are exposed to ETS each year. Hence, The World Health Organization (WHO) recently suggested adding warnings to tobacco packs regarding the dangers of exposing others (children in particular) to ETS. Given the paucity of ETS policies in the world, the inability to legislate people's behavior in their homes, and the widespread acceptance of health warnings on packs of tobacco, we reiterate WHO's suggestion. Adding warnings about the dangers of ETS to existing warnings on packages of tobacco is an inexpensive strategy for reducing adult and child ETS exposure in developed and developing nations alike, insofar as this strategy may circumvent the tobacco industry's efforts to block implementation of ETS policies that curtail smoking in public places. We add to WHO's suggestion that such future, potential ETS warning labels should be strong and rotated (like current health warnings on cigarette packs in Canada and Thailand) and should occupy at least 20% of the package (like current health warnings on cigarette packs in Poland and South Africa). Finally, efforts must be made in the U.S. to overcome ethnic, racial, and social class inequality in ETS exposure.

7.1 INTRODUCTION

ETS (secondhand smoke) is a Group A carcinogen, a substance proven to be a cause of cancer in humans.[1] In recent years, the health hazards of exposure to ETS have been increasingly recognized.[1-2] Hence, public policies to eliminate or reduce ETS exposure similarly have increased. In this chapter, we briefly review the evidence on the health consequences of ETS exposure and then examine gender, social class, age, ethnicity and culture, and international differences in level and frequency of exposure to ETS. We follow this with a detailed examination of public policies to reduce/eliminate ETS exposure in the U.S. as well as in other nations. This examination highlights the frequency and scope of such policies and the extent to which they eliminate/reduce the ETS exposure of those with the highest levels of exposure. Finally, we highlight some future directions for policies designed to reduce/eliminate exposure to ETS. Throughout this chapter, our focus is on involuntary exposure to ETS — i.e., on the exposure of non-smoking adults and children to the ETS of smokers ("passive smoking" or "involuntary smoking").

7.2 HEALTH CONSEQUENCES OF ETS EXPOSURE

ETS consists of approximately 5000 chemicals, 200 of which are poisons and 43 of which are known carcinogens.[3-4] Numerous studies have firmly established the health hazards of exposure to ETS. These include a comprehensive review of the evidence published by the U.S. Environmental Protection Agency (EPA) in 1992,[1] a comprehensive review by the California EPA in 1997,[5] reports from the U.S. Surgeon General[6] and from the National Academy of Sciences,[7] and several other studies.[8-16] These reports and studies indicate that exposure to ETS plays a significant, causal role in lung cancer and heart disease among non-smoking adults and in Sudden Infant Death Syndrome, pneumonia, bronchitis, asthma, learning disabilities, behavioral problems, low birthweight, chronic respiratory symptoms, and middle-ear disease in non-smoking children. Specifically, where U.S. adults are concerned, these studies indicate that the risk of lung cancer is 30% higher for the non-smoking spouses of smokers compared to the non-smoking spouses of non-smokers, and that the risk of heart disease in non-smokers exposed to ETS is 20% higher than that of those not exposed. For non-smoking adults in the U.S., exposure to ETS accounts for 3000 lung cancer deaths and 62,000 deaths from ischemic heart disease each year, making involuntary exposure to ETS the third leading, preventable cause of death among adults in the U.S. Where children in the U.S. are concerned, these studies indicate that ETS exposure accounts for 150,000 to 300,000 cases of respiratory disease each year; for 8000 to 26,000 new asthma cases each year; and for 400,000 to 1 million cases annually of respiratory disease, asthma, and asthma exacerbation. The WHO[16] estimated that worldwide, involuntary exposure to ETS is associated with increased morbidity and mortality for youth and adults at an annual health and economic cost of $1000 million. Other estimates (see later) are higher.

7.2.1   U.S. Status Differences in Levels of ETS Exposure

Numerous studies indicate that the majority of U.S. non-smokers' exposure to ETS is in the workplace and at home.[17] Data from the 1988 National Health Interview Survey indicated that 28.9 million U.S. non-smokers worked in environments that permitted smoking in some area of the workplace.[18] While this exposure is decreasing in the U.S. in response to various ETS-control policies, social class and other status differences in levels of ETS exposure in the U.S. persist. **Race/Ethnicity** — Blacks and Hispanics are significantly less likely than Whites to work in smoke-free work-places and, hence, have higher levels of exposure to ETS than their White working counterparts.[17,19–21] Similar ethnic differences have been found in New Zealand.[22] **Education and Income** — Similarly, U.S. workers with less than a high-school education are less likely to work in smoke-free workplaces than their educated counterparts and have higher levels of ETS exposure.[20] This is also the case for people of low socioeconomic status in the U.S.[17] and elsewhere in the world.[22] Likewise, blue-collar (e.g., factory) workers are significantly less likely than their white-collar (e.g., professional, managerial, health) cohorts to be employed in smoke-free work-places.[17,21] **Age** — Young children (birth to 5 years) are particularly vulnerable to the effects of ETS exposure,[23] and their exposure is both widespread and involuntary; the vast majority of children's exposure to ETS is the result of smoking among adults and occurs at home and in automobiles.[2,24] Children's ETS exposure is also higher than that of adults. Compared with adults, children have higher relative ventilation rates leading to higher internal exposure to ETS despite the same level of external exposure as adults; this finding holds using urinary[25] or serum[26,27] cotinine. Children in single-parent families have higher exposure than those in two-parent families.[26–28] Likewise, children in low-income families have extremely high levels of exposure to ETS. Low-income parents are more likely to smoke and to have friends who smoke; are more likely to live in apartments with limited windows, limited outdoor access, and shared ventilation systems; and typically have no balcony or garage to use for smoking outdoors.[26,29] In the U.S., 49% of all children ages 5 years and younger are exposed to ETS at home due to the smoking of parents.[2,24] Hence, parental and other adult smoking is a major, preventable cause of morbidity and mortality among U.S. children and results in annual direct medical expenditures of $4.6 billion and loss of life costs of $8.2 billion.[30] Similarly, internationally, 50% of all young children live with at least one adult smoker, and hence, 700 million children are exposed to ETS each year.[16] Teenage workers (ages 15 to 19 years) in the U.S. also are particularly vulnerable to ETS exposure insofar as their exposure in the workplace significantly exceeds that of other workers; this is because teenage workers are concentrated in food-service jobs where smoke-free workplace policies generally are absent.[21]

Thus, ethnic minorities, the poor, and young children in the U.S. have the highest levels of exposure to ETS. The question is, do ETS policies protect these groups?

7.2.2   International Differences in Levels of ETS Exposure

Studies of ETS exposure in other nations are somewhat rare; it can be assumed, however, that exposure levels are high in the many nations that lack any form of ETS policy at any (nationwide, regional, city) level (see later). Most studies of ETS exposure have been conducted in northern developed (industrialized) countries such as the U.S., Canada, and the Scandinavian/Nordic nations; none have been conducted in tropical environments until recently.[31] The sole study of ETS exposure in a tropical environment was conducted in Puerto Rico; results revealed that 86% of Puerto Rican children (compared to 49% of U.S. children) are exposed to ETS. Exposure levels as high as those found in Puerto Rico may hold for other tropical countries that similarly lack ETS policies.[31] There appear to be few studies of status differences in levels of ETS exposure in countries other than the U.S. and Canada. One study, however, was conducted in New Zealand and examined ETS exposure levels by social class and ethnicity among a population of 7725 New Zealanders.[22] Results revealed that lower and working-class people had higher levels of ETS exposure than their middle-class cohorts and that Maori (ethnic minority) people had significantly higher levels of exposure than Whites (Europeans) in the workplace and at home.[22]

7.3   ETS-CONTROL POLICIES IN THE U.S.

U.S. policies to reduce ETS exposure by and large began in 1986 when ETS was demonstrated to be a cause of lung cancer.

7.3.1   FEDERAL POLICIES

The only federal laws to reduce ETS exposure (nationwide) are (1) bans on smoking on domestic airline flights, enacted in 1988 and 1989; (2) bans on smoking in federal office buildings, implemented in August 1997; and (3) bans on smoking in federally funded facilities for children (including health care, day care, and public schools), implemented in December 1994.[17] There are no federal policies to reduce or ban smoking nationwide in public or private sector workplaces;[17] the U.S. Occupational Safety and Health Administration has one that remains under consideration.[21]

7.3.2   STATE POLICIES

Over the past decade, states have moved to the forefront of tobacco control. Starting with California in 1988, and followed by Massachusetts, Arizona, Oregon, and others, states have increased tobacco taxes and dedicated a fraction of the revenues to reducing tobacco use.[32] Legislatures in Alaska, Hawaii, Maryland, Michigan, New Jersey, New York, and Washington likewise have increased tobacco taxes and created programs to reduce youth and adult smoking.[32] Similarly, policies to restrict youth access to tobacco are in effect in the entire U.S.[33,34] The best evidence for the effectiveness of these state tobacco-control efforts comes from comparing states with different intensities of tobacco control, as measured by funding levels and aggressiveness.[32] For example, California, Massachusetts, and Oregon implemented programs that were more intense than those of other states, and they showed greater decreases in tobacco use than other states.[32–34]

Hence, the majority (n = 47) of U.S. states have some sort of policy to reduce ETS, but these vary widely in their degree of restrictiveness; most ban smoking only in hospitals, child care facilities, and on public transportation.[17–18,35] While these are important steps, nonetheless, most adult exposure to ETS occurs in the workplace, and hence, workplace policies are our focus. As of March 1999, only three of 50 states (6%) — California, Maryland, and Washington State — had enacted statewide policies that ban smoking entirely in public and private sector workplaces statewide.[8]

7.3.3 LOCAL AND WORKPLACE POLICIES

Numerous cities have passed policies to entirely ban smoking in workplaces and/or restaurants, but the number of these is small relative to the number of cities in the U.S. As shown in Table 7.1, as of March 1999, only 148 (of tens of thousands of) U.S. cities and counties had banned smoking in workplaces (public and private sector) and restaurants; 115 of these cities/counties (78%) are in California, which banned smoking in all public buildings (excluding bars) statewide in 1994 and then banned smoking in all bars statewide in 1998. Likewise, 33 additional cities/counties and two states (Maryland and Washington State) have banned smoking entirely in workplaces statewide, but restaurants are excluded from that ban (see Table 7.2). An additional 74 cities/counties and two states (Utah and Vermont) have banned smoking entirely in restaurants, but not in other workplaces (Table 7.3). Thus, totaling these reveals that only 259 (of tens of thousands of) U.S. cities/counties have 100% smoke-free workplaces (private and public sector, including or excluding restaurants).

Although only 259 cities/counties have entirely banned smoking in some or all workplaces, many additional cities have passed policies to restrict smoking in workplaces to specific, designated areas. Brownson et al.[17] reported that there are approximately 700 such local ordinances entailing thousands of workplaces. Likewise, Gerlach et al.[21] analyzed the National Cancer Institute's data from a 1992–1993 cross-sectional survey of 100,561 employed, indoor workers (ages 15 and older) regarding the presence and nature of ETS policies in their workplaces. Results revealed that 81.6% of workers reported that their workplace had an official ETS policy that restricted smoking in some manner.[21] This finding at first suggests that the U.S. has met the U.S. Public Health Service goal of banning or severely restricting smoking in 75% of all workplaces nationwide by the year 2000.[37] However, only 46% of workers reported that the policy in question banned smoking in both the working and the common (restrooms and cafeteria) areas; most policies only banned smoking in the former and, hence, were not restrictive[21] in the manner desired by the U.S. Public Health Service.

In addition, this national survey found that Black workers (men in particular) were less likely than Whites and Asians to be employed in workplaces with an ETS policy.[21] This was also the case for young (ages 15 to 19 years) workers — who were the least likely of all workers to be employed in a workplace with an ETS policy.[21] Similarly, white-collar (i.e., professional, managerial) workers were 1 1/2 times more likely than their blue-collar counterparts (i.e., skilled and unskilled laborers) to be employed in workplaces with an ETS policy. For example, only 22% of factory and food-service workers were employed in a smoke-free/smoking-restricted workplace, compared to 50% of sales/finance workers, 60% of college professors, 70% of public school teachers, and 80% of professionals in health-related

TABLE 7.1
U.S. Cities/Counties with 100% Smoke-Free Workplace Ordinances (Including Restaurants) 1999: N = 148

Bethel, AK	Larkspur, CA	San Rafael, CA	Amherst, MA
	Lathrop, CA	San Ramon, CA	Arlington, MA
Mesa, AZ	Lemon Grove, CA	Santa Barbara Co., CA	Brookline, MA
	Livermore, CA	Santa Clara, CA	Easthampton, MA
Albany, CA	Long Beach, CA	Santa Clara Co., CA	Falmouth, MA
Anderson, CA	Los Gatos, CA	Santa Cruz, CA	Foxborough, MA
Antioch, CA	Marin Co., CA	Santa Cruz Co., CA	Greenfield, MA
Arcata, CA	Martinez, CA	Santa Rosa, CA	Haverhill, MA
Auburn, CA	Mendocino Co., CA	Saratoga, CA	Holyoke, MA
Belmont, CA	Mill Valley, CA	Sausalito, CA	Lexington, MA
Belvedere, CA	Millbrae, CA	Scotts Valley, CA	Medfield, MA
Berkeley, CA	Modesto, CA	Sebastopol, CA	Montague, MA
Burlingame, CA	Monterey Co., CA	Shafter, CA	Northampton, MA
Butte Co., CA	Moorpark, CA	Shasta Co., CA	Norwell, MA
Calistoga, CA	Mountain View, CA	Shasta Lake, CA	Orange, MA
Camarillo, CA	Napa, CA	Solana Beach, CA	Wakefield, MA
Capitola, CA	Napa Co., CA	Solano Co., CA	Westborough, MA
Ceres, CA	Novato, CA	Sonoma Co., CA	
Chico, CA	Oakland, CA	Stanislaus Co., CA	Austin, TX
Chino Hills, CA	Ojai, CA	Tiburon, CA	Fort Worth, TX
Clayton, CA	Palo Alto, CA	Tracy, CA	Wichita Falls, TX
Colfax, CA	Paradise, CA	Tuolumne Co., CA	
Concord, CA	Pasadena, CA	Ukiah, CA	Benton Co., OR
Contra Costa Co., CA	Patterson, CA	Union City, CA	Corvallis, OR
Cotati, CA	Petaluma, CA	Vallejo, CA	
Cupertino, CA	Pittsburg, CA	Ventura, CA	Marquette, MI
Danville, CA	Placer Co., CA	Ventura Co., CA	
Davis, CA	Placerville, CA	Visalia, CA	Erie Co., NY
Del Mar, CA	Pleasanton, CA	Walnut Creek, CA	Monroe Co., NY
Duarte, CA	Redding, CA	Watsonville, CA	Niagara Co., NY
Dublin, CA	Richmond, CA	Westlake Village, CA	
El Cerrito, CA	Roseville, CA	Whittier, CA	Boulder, CO
El Dorado Co., CA	Sacramento, CA	Yountville, CA	
Fairfax, CA	Sacramento Co., CA	California[a]	Albany, GA
Folsom, CA	Salinas, CA		
Fort Bragg, CA	San Anselmo, CA		Talbot Co., MD
Foster City, CA	San Bernardino, CA	New Hanover, NC	
Fremont, CA	San Carlos, CA	New Hanover Co., NC	
Gilroy, CA	San Francisco, CA		
Healdsburg, CA	San Jose, CA		
Hercules, CA	San Juan Bautista, CA		
Hughson, CA	San Luis Obispo, CA		
Laguna Hills, CA	San Mateo, CA		

[a] Although local ordinances in California are listed, it is important to note that all cities and counties in California have smoke-free ordinances under state law as of 1994.

Source: Table freely reprinted from Americans for Non-Smokers' Rights, American Nonsmokers' Rights Foundation, 1999.

TABLE 7.2
U.S. Cities/Counties with 100% Smoke-Free Workplace Ordinances (Excluding Restaurants) 1999: N = 35[b]

Tempe, AZ	Honolulu, HI	Highland Park, NJ
Alpine Co., CA	Attleboro, MA	Zanesville-Muskingum, OH
Lafayette, CA	Saugus, MA	
Lindsay, CA	Williamstown, MA	Arlington, TX
Merced, CA		
Orinda, CA	Buncombe Co., NC	Brooke Co., WV
Pinole, CA	Burke Co., NC	Lincoln Co., WV
Pleasant Hill, CA	Catawba Co., NC	Wyoming, Co., WV
San Diego, CA	Chatham Co., NC	
San Mateo Co., CA	Craven Co., NC	Maryland[a]
Santa Ana, CA	Henderson Co., NC	
Santa Clarita, CA	Montreat, NC	Washington State[b]
Sonoma, CA	Orange Co., NC	
South Lake Tahoe, CA	Wake Co., NC	
Turlock, CA	Wilkes Co., NC	

Note: Some of the above ordinances allow smoking in enclosed, separately ventilated rooms.

[a] Maryland's regulations prohibiting smoking in enclosed workplaces can be found in the Code of MD Regs §9.12.23 (1994).

[b] Washington's state law only applies to office workplaces, WAC 296-62-12000 et seq. (1994).

Source: Table freely reprinted from Americans for Non-Smokers' Rights, American Non-smokers' Rights Foundation, 1999. (www.no-smoke.org)

jobs.[21] Hence, higher levels of ETS exposure persist among minorities, the poor, the uneducated, and the young despite the flurry of new ETS workplace policies.

7.3.4 Two Difficulties in Enacting Local ETS Policies

As cities and counties increase their efforts to enact local ETS policies, so too has the tobacco industry increased its efforts to block these. In at least 17 states thus far, the tobacco industry has successfully countered such local efforts by supporting passage of "preemptive" state laws that prohibit local jurisdictions from implementing ETS policies that are more restrictive than state law.[35] Hence, many cities and counties that may desire to pass total bans on smoking in workplaces and restaurants may find it difficult to do so. Thus, for example, in 1989, the city of Greensboro, NC passed the first ETS policy in the state.[38] Several communities in North Carolina took steps to do the same, which led the tobacco industry to sponsor HB 957, a law that preempts local jurisdictions from enacting ETS policies more stringent than North Carolina state law. HB 957 was passed in July 1993 and contained no measures whatsoever to protect the public from ETS exposure, but it did allow

TABLE 7.3
U.S. Cities/Counties with 100% Smoke-Free Restaurant (Only) Ordinances 1999: N = 76

Homewood, AL	Acushnet, MA	Howard Co., MD
	Andover, MA	
Flagstaff, AZ	Bedford, MA	Portland, ME
	Bellingham, MA	
Agoura Hills, CA	Belmont, MA	Guilford Co., NC
Buellton, CA	Boston, MA	
Calabasas, CA	Boxborough, MA	Glassboro, NJ
Carpinteria, CA	Chicopee, MA	
Dana Point, CA	East Longmeadow, MA	Las Cruces, NM
El Cajon, CA	Gardner, MA	
Fresno, CA	Holden, MA	Livingston, Co., NY
Grass Valley, CA	Lee, MA	Nassau Co., NY
Hollister, CA	Lenox, MA	New York City, NY
Huntington Beach, CA	Leominster, MA	Suffolk Co., NY
Imperial Co., CA	Middleton, MA	Westchester Co., NY
Laguna Beach, CA	Needham, MA	
Lodi, CA	Newton, MA	West Lake Hills, TX
Loma Linda, CA	Oak Bluffs, MA	
Los Angeles, CA	Pittsfield, MA	Madison, WI
Menlo Park, CA	Plainville, MA	Middleton, WI
Orange Co., CA	Sharon, MA	Shorewood Hills, WI
Rancho Palos Verdes, CA	Somerville, MA	
Rohnert Park, CA	South Hadley, MA	Gilmer, WV
Ross, CA	Stockbridge, MA	Mercer Co., WV
San Jose, CA	Tewksbury, MA	Morgan Co., WV
San Juan Capistrano, CA	West Springfield, MA	Upshur Co., WV
Thousand Oaks, CA	Westfield, MA	
West Hollywood, CA	Westwood, MA	Utah[a]
Willows, CA	Wilbraham, MA	
		Vermont[b]
Aspen, CO		
Pitkin Co., CO		

Note: Some of the above ordinances allow smoking in enclosed, separately ventilated rooms.

[a] Utah's regulations prohibiting smoking in restaurants can be found in UT Code Ann. §26-28-1 (1994)

[b] Vermont's regulations prohibiting smoking in restaurants can be found in VT Stat. Ann. Title 18, §1744 (1993).

Source: Table freely reprinted from Americans for Non-Smokers' Rights, American Nonsmokers' Rights Foundation, 1999. (www.no-smoke.org)

North Carolina communities 3 months to adopt their own policies prior to its implementation; 89 ETS regulations were passed in that 3 months,[38] suggesting that communities can sometimes circumvent preemptive state laws. A second difficulty in enacting local ETS ordinances in workplaces and restaurants is the tobacco industry's campaign to convince local governments that such ordinances will harm business and tourism.[39] Evidence indicates that this is not the case: one study found growth in hotel revenues and tourists,[39] and another[40] found no difference in these despite the presence of smoke-free restaurant and/or bar ordinances.[39] A study in North Carolina demonstrated that the aforementioned 1993 ETS policies did not decrease local revenues.[38]

7.3.5 ENFORCEMENT OF U.S. LOCAL POLICIES

The agency responsible for enforcing local policies varies from city to city. In 50% it is the health department/board of health, but in others it is city managers (29%), police/sheriffs (5%), or any assortment of agencies such as the fire department.[17] The extent to which local ETS policies are enforced remains unknown, but it is clear that few public resources are devoted to that purpose.[17]

7.3.6 U.S. AIRPORT POLICIES

Although federal law bans smoking on domestic airline flights, there is no federal policy to ban smoking in airports. Banning smoking in airports continues to be a local (state, county, or city) issue. Hence, of the ten busiest airports in the U.S. (Chicago-O'Hare, Atlanta International, Los Angeles International, Dallas/Fort Worth International, San Francisco International, Miami International, Denver International, JFK International, Detroit Metropolitan, and Phoenix Sky Harbor), only four are 100% smoke-free indoors.[36] Those four are Chicago-O'Hare, Los Angeles International, Dallas/Fort Worth, and JFK International. In the other six busiest airports, smoking is permitted in designated lounges that may or may not be separately ventilated and may or may not be enclosed.[36]

7.3.7 U.S. HOSPITAL AND PRIVATE SECTOR POLICIES

Hospitals are the only business in the U.S. to voluntarily implement a nationwide ban on smoking.[41] This was announced in 1991, with full nationwide implementation required in all public and private hospitals by 1993.[41] As of 1994, 96% of hospitals were in compliance with the policy.[41]

Both the U.S. Bureau of National Affairs[42] and the U.S. Department of Health and Human Services National Survey of Worksite Health Promotion Activities[43] reported that approximately 85% of U.S. private sector businesses/workplaces had some sort of ETS-control policy in place by 1991 to 1992. As of 1992, however, 66% of these policies restricted smoking to designated areas that may or may not be separately ventilated or even enclosed, and only 34% of them banned it entirely in the workplace.[42,43] Given data on the health hazards of ETS published post-1992, many more policies banning smoking in private sector workplaces may have been introduced.

7.4 EFFECTS OF U.S. ETS POLICIES

7.4.1 EFFECTS ON EXPOSURE TO ETS

Smoking bans and restrictions reduce ETS exposure significantly, irrespective of whether they are workplace policies or local ordinances.[17,55,56] Non-smokers in workplaces that lack workplace or local policies (that cover workplaces) are eight times more likely to be exposed to ETS than their cohorts in policy-covered workplaces;[19] direct measures of nicotine vapor concentrations in the workplace have found significantly lower levels in workplaces that ban or restrict smoking.[45,57] In the cases of restricting smoking, such policies reduce ETS exposure only if the designated smoking area has separate ventilation.[58] For example, Hedge et al.[59] examined the effects of five different ETS policies on indoor air quality in 27 air-conditioned buildings. Although the type of smoking policy had no effect on carbon monoxide, carbon dioxide, relative humidity, formaldehyde, air temperature, or illumination for open office areas, differences based on restrictiveness of the policy were found for levels of respirable suspended particulates, ultraviolet particulate matter, and nicotine. These investigators concluded that, for ETS-associated pollutants, ambient indoor air quality in non-smoking areas of air-conditioned office buildings which spatially restrict smoking is comparable to that in buildings which prohibit smoking.[59]

However, real-time concentrations of nicotine vary greatly over hours or days. For example, Muramatsu et al.[60] measured real-time concentrations of particle and 15-min average nicotine concentrations in an office throughout the workday and found particle concentrations ranging from 20 to 200 $\mu g/m^3$ and 30-min average concentrations ranging from 2 to 26 $\mu g/m^3$. Because of this variability, Hammond[61] recently reviewed only those studies in which the airborne concentration of nicotine in the workplace was sampled for 1 h or more. Nonetheless, Hammond found that workplace smoking policies make a substantial difference in the concentration of ETS; nicotine concentrations were much higher in both office and non-office areas of work sites where smoking was allowed than in those workplaces that either restricted smoking to a relatively few areas or banned it completely.[61] While policies that restricted smoking to certain locations had uneven efficacy in limiting ETS exposure (probably due to differing levels of enforcing the restriction), complete smoking bans were effective, with nicotine concentrations in those settings generally lower than 1 $\mu g/m^3$. By contrast, Hammond[61] reported that where smoking was not banned, the mean nicotine concentrations in offices was generally between 2 and 6 $\mu g/m^3$; in restaurants it was between 3 and 8 $\mu g/m^3$; in bars it was between 10 and 40 $\mu g/m^3$, and in other diverse blue-collar occupations it was between 1 and 6 $\mu g/m^3$. Hammond[61] contrasted these values with nicotine concentrations in the homes of smokers, which generally average between 1 and 3 $\mu g/m^3$, and concluded that the average workplace concentrations of nicotine in areas where smoking is allowed are generally greater than concentrations in the homes of smokers. In addition, however, status differences in ETS exposure persist despite these policies, with Blacks, Latinos, young people, and those of low socioeconomic status still more likely to work in places without ETS policies.[17,19] Hence, to date, ETS policies do not adequately cover those with the highest levels of exposure.

7.4.2   Effects on Smoking

Numerous studies have found that making workplaces smoke free increases rates of smoking cessation (quit rates) and reduces the number of cigarettes that smokers consume.[44-52] On average, smoke-free workplaces reduce the number of smokers by 5% (i.e., as smokers quit) and reduce the number of cigarettes smoked (by those who continue to smoke) by 10%.[53] Hence, some have attributed the large tobacco consumption drop in Australia (1988 to 1995) and in the U.S. (1988 to 1994) to smoke-free workplace policies.[54] For example, Browson et al.[17] reviewed 22 such studies that were published between 1983 and 1996. Sample sizes ranged from 349 to 10,579 employees, and settings ranged from hospitals (most frequent setting) to small businesses and large government agencies. These 22 studies found a decrease in the number of cigarettes smoked per day by those employed in workplaces with ETS policies and, in some cases, found decreases in the overall prevalence of smoking among employees.[17] A few of these studies also found long-term (at 6- to 20-month follow-ups) decreases in the number of cigarettes smoked per day by employees.[17] Population-based studies similarly suggest that ETS policies decrease smoking and increase quitting.[17] Likewise, at least one study indicates that policies that restrict smoking in public places decrease smoking among youth more than among adults; this may be because youth are somewhat more sensitive than adults to how they are viewed by others — to peer censure.[62]

7.5   Tobacco-Control Policies around the Globe

The World Health Assembly is the governing body of the WHO and is made up of representatives of all 191 WHO member states. Since 1970, the World Health Assembly has adopted numerous resolutions calling for the global implementation of comprehensive tobacco-control policies and for international collaboration in achieving global tobacco control.[63] These resolutions were passed unanimously (without dissent) and, hence, indicate international consensus among the WHO nations on the need for tobacco control. Specifically, the World Health Assembly outlined ten steps needed to achieve international tobacco control.[63] These included increasing taxes on cigarettes; using tobacco taxes to finance tobacco control; eliminating tobacco advertising, promotion, and sponsorship; strong warnings on cigarette packs; reducing youth access to tobacco and smoking initiation; and reducing or eliminating involuntary exposure to ETS.

Hence, in the 1980s and early 1990s, several countries implemented policies that require health warnings on packs of cigarettes; increased tobacco surtaxes; and restricted, curtailed, or banned tobacco advertising. These include Brazil, Cyprus, Nepal, China, Costa Rica, Cuba, India, Mongolia, Poland, Turkey, Slovenia, Lithuania, the U.S., South Africa, and the 15 nations of the European Union.[63] That such policies are effective is most clear in the case of Kuwait. In the period of reconstruction following its invasion and subsequent liberation, Kuwait temporarily suspended health warnings and import duties on packs of cigarettes. Consequently, for a short period in the early 1990s, packs of cigarettes bore no health warnings and were incredibly cheap. Smoking prevalence among men increased from 34% in 1989 to 52% in 1992, while among women it doubled — from 6 to 12% in the same period.[63]

7.6 INTERNATIONAL ETS-CONTROL POLICIES

7.6.1 COUNTRIES WITH NATIONWIDE ETS POLICIES

Although several countries have restricted tobacco advertising, increased tobacco surtaxes, and required health warnings on packs of cigarettes, only a few (i.e., 12) countries in the world have implemented nationwide (i.e., with no regional or state exceptions) policies to restrict or ban smoking indoors. These are Nepal, Thailand, France, New Zealand, Sweden, Finland, Iceland, Norway, Portugal, Canada, Singapore, and (in 1995) Kuwait.[63] Of these, Canada provides an example of a model comprehensive tobacco-control policy that includes reduction in exposure to ETS.

7.6.2 THE CANADIAN EXAMPLE

In 1989, Canada adopted the Non-Smokers' Health Act, banning smoking in buses, airplanes, all federal government offices, other forms of public transportation, other public places, and private sector workplaces that come under federal jurisdiction for health and safety matters. Canada also passed The Tobacco Products Control Act in 1989, banning most forms of tobacco advertising and promotion and requiring strong health warnings on cigarette packs and the mandatory labeling of the tar, nicotine, and carbon monoxide in tobacco products.[63] In addition, Canada implemented a system of payments to tobacco farmers who agreed to permanently leave tobacco growing and find an alternative business.[63] Other aspects of Canada's tobacco-control policy include financial support for smoking cessation, education and prevention programs, and sharp increases in tobacco taxes.[63] Devoting public funds to smoking cessation, decreasing tobacco production, and simultaneously raising the price of all tobacco products may be reasons that Canada's ETS reduction policy has been uniquely successful.

The success of Canada's ETS policies is clear: consumption of tobacco in Canada decreased immediately in response to the 1989 policies, with a 6% decrease in 1989 and another 6% decrease in 1990.[63] Similar declines were found in France (which banned smoking in workplaces and public places in 1991 and 1993). Comparable declines in smoking were found in nations (i.e., Australia, Norway, Finland, Canada, and New Zealand) that have banned or curtailed smoking indoors.[63]

7.6.3 COUNTRIES WITH POLICIES THAT HAVE REGIONAL/STATE EXCEPTIONS

Although only a handful of countries have nationwide policies to reduce exposure to ETS, many countries nonetheless have less global ETS policies. Such policies ban smoking entirely in workplaces and public buildings, but the policies have been implemented only in certain environments (e.g., hospitals, schools, and day care centers) or implemented only in certain states or regions.[63] The ETS-control policies of the U.S. and Australia are examples of such strong, but context-specific and

TABLE 7.4
Flights to and from the U.S. by Carriers of Indicated Region

Region	Total annual flights	Percent non-smoking
Asia	26,000	93
Europe	47,000	91
Middle East	2,500	60
South America	8,500	53

Source: Table freely reprinted from Americans for Non-Smokers' Rights, American Nonsmokers' Rights Foundation, 1999. (www.no-smoke.org)

limited, policies. Other countries that fall into this category are those that have not banned, but have restricted smoking to areas in workplaces, public places, and public transportation. The ETS control policies of New Zealand, Nepal, Thailand, Sweden, and Kuwait are examples of such policies.[63]

7.6.4 COUNTRIES CONSIDERING NATIONWIDE ETS POLICIES

Although only a few countries have nationwide policies to reduce exposure to ETS, several have such policies currently under consideration; these include Poland, Lithuania, Slovakia, and Slovenia.[63] These policies would ban or restrict smoking in public buildings, workplaces, and on all forms of public transportation throughout the country, with no regional or state exceptions.[63]

7.6.5 INTERNATIONAL AIRLINE ETS POLICIES

International flights constitute 51% of all flights to and from the U.S.[36] Smoking is banned or restricted on many foreign carriers. Specifically, approximately 50 foreign carriers have banned or nearly completed banned smoking; this includes the flights of African, Caribbean, North American, and Oceanian carriers serving the U.S. The percentage of smoke-free flights of foreign carriers serving the U.S. from the remaining regions of the globe (as of 1999) is shown in Table 7.4.[36] As shown, most flights from Europe and Asia are smoke free, whereas only some flights from the Middle East and Latin America are smoke free.

7.7 FUTURE DIRECTIONS FOR ETS-CONTROL POLICIES

7.7.1 IMPROVING AND ENFORCING POLICIES

As noted here, only 34% of U.S. ETS workplace policies ban smoking in the workplace; the majority (66%) of policies restrict smoking to a designated area that may not be separately ventilated or even enclosed, and most policies do not restrict smoking in the common areas of the workplace. The restrictiveness (i.e., quality) of ETS policies must be assessed using our instrument[72] or some other, and then the quality of the policies must be improved. Furthermore, the extent to which local ETS policies are enforced remains unknown. Clearly, ETS workplace policies must be improved, and public funds must be devoted to the enforcement of local, state, workplace, and other ETS policies. Without enforcement, such policies serve no purpose.

7.7.2 Achieving Equality in Policies

As noted in this chapter, there are few statewide ETS policies in the U.S. (n = 5), few citywide policies in the U.S. (n = 259), few federal policies in the U.S., and few policies around the globe. Such policies, where they exist, are unequally distributed by ethnicity, race, and social class, such that ethnic minorities and the poor continue to have higher levels of exposure to ETS than their whiter and more privileged counterparts in the U.S. and abroad. Hence, studies are needed to assess the extent to which the higher ETS exposure of disadvantaged groups is the result of the stratified distribution of ETS policies. Likewise, studies are needed to assess the percentage of excess morbidity and mortality among the disadvantaged that results from inequities in ETS exposure; the possibility that ETS policies not only reflect social status, but also may perpetuate status differences in health must be examined. Similarly, it is clear that efforts must be made by federal, state, and local governments (as well as by health departments and agencies) to assure equality in coverage by ETS policies. To date, groups with the highest ETS exposure levels are the least affected by ETS policies.

7.7.3 Protecting Children from ETS

As noted in this chapter, young children's exposure to ETS exceeds that of adults in the U.S. and abroad, is widespread and involuntary, and poses serious threats to their health. Children's exposure occurs almost exclusively at home and in automobiles, is the result of smoking by adults, and (in the U.S.) results in annual direct medical expenditures of $4.6 billion and loss of life costs of $8.2 billion. Clearly, specific efforts must be made to protect this large, but powerless (disadvantaged) population from ETS.[26] This means that efforts to reduce ETS in homes and automobiles must be made. Some policy efforts to legislate smoking in homes and cars are currently underway. For example, after learning that smoking in a car is 23 times more toxic than in a house and 8.5 times more toxic than in an aircraft because of the smaller enclosed space, Colorado State Senator Dorothy Rupert filed a bill to impose a $56 fine on adults caught smoking in a car where a child under age 16 is present.[26] Whether this will pass or not remains to be seen. Likewise, given concern about exposing non-smoking apartment tenants to ETS via the shared ventilation system, the housing authority of Fort Pierce, FL enacted a policy requiring all new tenants to sign an agreement not to smoke in apartments under its jurisdiction. This is the first time that people have been banned from smoking at home by a public body.[26] We believe that Senator Rupert and the city of Fort Pierce, FL can serve as models for household and automobile ETS policies.

In general, however, it is extremely difficult to enact policies to reduce ETS in homes and automobiles because both are considered off limits to governmental and other outside intervention,[64] with the home in particular viewed as one's "castle."

Consequently, efforts may need to encourage the public to voluntarily reduce household and automobile ETS.[26,64]

7.7.3.1 Public Education

Hence, one future direction for reducing ETS in homes and automobiles is public education about the dangers of exposing others (children in particular) to ETS and encouraging smokers to smoke outdoors.[64] Such efforts have begun in some countries (e.g., Canada, Scandinavia/Nordic countries[65]), have been effective, and so must be expanded.[64] We believe that Canada and Iceland are among the best models for such efforts.

7.7.3.2 Encouraging Adult Smokers to Quit

A second direction is greater funding for smoking cessation programs for adults that are designed specifically to reduce household ETS. Several such individual/family[66,67] and community[68,69] interventions have been tried, and these have had variable success. For example, the intervention of Emmons et al.[66] was successful in reducing ETS in homes by 60%, but Greenberg et al.[68] and Chilmonczyk et al.[70] obtained no significant differences in measured ETS postintervention. Clearly, new interventions, conducted in homes with families, are needed. We believe that the work of Emmons et al.[66] can serve as a model for such efforts.

7.7.3.3 Litigation

Clearly, however, educating people and encouraging them to smoke outdoors may not be sufficient, and legal remedies eventually may need to be sought.[64] This indeed seems to already be occurring: in the U.S., courts have handed down several decisions to protect children from ETS using the legal umbrella/standard of the "best interest of the child."[64] For example, in a recent case in Tennessee, a couple's divorce decree included the rule that neither parent would expose their children to tobacco smoke in enclosed places or allow others to expose them to it. A few days after the divorce, the mother filed a motion for contempt against the father in which she alleged that he smoked around the child and exposed her to ETS during his visitation. The court found the father guilty of criminal contempt for exposing his child to cigarette smoke, sentenced him to 2 days in jail, and suspended his visitation rights.[64] Thus, ETS is now increasingly used in divorce and child custody cases involving children who have asthma.[26,64] Parents who refuse to provide smoke-free environments for asthmatic children have been denied access or custody.[26] In some cases, custody was granted to the non-smoking parent on ETS grounds alone; in other cases, both parents were forbidden by the court to smoke in the child's presence; in other cases, the court banned smoking in the home and in automobiles.[26] Litigation is thus an alternative to policy and may lead to policy.[26]

Such litigation has not been limited to children, but, instead, is increasingly being used by individuals to protect themselves from ETS exposure in households, automobiles, and public places.

7.7.4   ETS LITIGATION AS AN ALTERNATIVE TO POLICY

A recent case in Ohio revealed how litigation can be used to reduce household ETS.[64] A non-smoking tenant acquired a lease to reside on the top floor of a two-family dwelling. During the lease, the landlord, a smoker, moved into the first-floor unit, sending smoke into the tenant's apartment through the heating/cooling systems. The tenant sent letters to the landlord complaining about this, but the landlord ignored these. The tenant moved and filed a successful suit against the landlord on the grounds that the landlord had breached their rental contract insofar as it stated that the landlord would take all reasonable steps to keep the premises safe and habitable.[64]

Successful ETS suits also have been filed against McDonald's and Burger King using the Americans With Disabilities Act. Under U.S. federal law, establishments open to the public cannot have policies that discriminate against people with disabilities.[64] These lawsuits claimed that by permitting smoking in their restaurants, McDonald's and Burger King were discriminating against people with chronic respiratory illnesses (children with asthma included) and denying them equal access to the fast-food chain.[64] Likewise, a class action suit was filed by flight attendants exposed to ETS and was settled in October 1997 for $300 million. Such suits suggest that ETS home and automobile legislation may eventually be necessary not only to reduce involuntary ETS exposure, but also to reduce such lawsuits.[26,64]

7.7.5   WHO's RECOMMENDATIONS FOR FUTURE ETS POLICIES

Given the paucity of countries that have implemented policies to reduce involuntary exposure to ETS, and given the danger that ETS exposure represents for young children in particular, in May 1996, the World Health Assembly of WHO planned an international forum for discussing and recommending global policies to reduce children's exposure to ETS. Hence, WHO convened the "International Consultation on Environmental Tobacco Smoke (ETS) and Child Health" in Geneva, Switzerland from January 11 to 14, 1999. The consultation brought together experts from developed and developing countries to examine the effects of ETS on child health and to recommend interventions (to be implemented internationally) to reduce these harmful effects and eliminate children's exposure. The consultation reviewed numerous empirical studies conducted by the environmental protection and public health agencies of several nations to reach its unanimous conclusions. These were that young children's involuntary exposure to ETS is causally associated with respiratory illness, asthma, middle-ear disease, Sudden Infant Death Syndrome (SIDS), learning difficulties, behavioral problems, language impairment, and physiological changes in children that may increase the risk of cardiovascular disease in adulthood.[16] Hence, the consultation declared that children have the right to grow up in an environment free from tobacco smoke under Article 19 of the WHO Constitution (protecting children's health) and declared that this right must be safeguarded by national and local governments, voluntary bodies, community leaders, health workers, educators, and parents.[16] The consultation drew the following conclusions:

1. Governments must pass laws to control ETS in public spaces. Foremost among such laws must be those that ban smoking in places frequented by children (schools, child and health care facilities, sports clubs, restaurants, shopping centers, and public transportation).
2. Such laws, however, cannot reduce children's exposure to ETS at home, and this is the source of the majority of children's exposure. Therefore, governments must use public education strategies that emphasize the risks to children that ETS represents, and these efforts must target the household decisionmaker in the culture/nation in question.
3. Similar educational strategies are needed to protect children from ETS exposure in automobiles. One educational strategy (for influencing adults to cease exposing children to ETS in homes and automobiles) is to add health warnings to cigarette packages advising smokers that their tobacco smoke is injurious to their children and to others in their presence.

The numerous countries that participated in the consultation agreed to these suggestions. Hence, the governments and health agencies of the participating nations are expected to take WHO's suggested steps in the future; this may mean that new and more comprehensive ETS policies will be enacted in the U.S., Canada, England, and elsewhere in the next 2 years.

7.7.6 ONE POLICY STRATEGY: ETS WARNINGS ON TOBACCO PACKAGES

WHO's[16] suggestion that future warnings on packages of tobacco include the dangers of ETS is a sound one for several reasons. First, adding warnings about the dangers of ETS to existing warnings on packages of tobacco is an easy, inexpensive, public education strategy that can be adopted in developing and developed nations alike to raise consciousness about ETS. Thus far, at least 40 countries have laws requiring cigarette warning labels, and numerous others have voluntary labeling agreements with the tobacco industry.[71] Although such warnings are (with a few exceptions) more common in developed than in developing countries, the warnings *are* present in many countries and (importantly) in many nations that lack ETS policies. Second, studies indicate that cigarette warning labels can and do influence smokers' behavior in numerous countries,[71] and hence, ETS warnings may not only provide information, but may alter smokers' behavior — particularly if they emphasize the dangers to children that ETS represents. Finally, using warnings on packages of tobacco to encourage the public to voluntarily reduce ETS may be a strategy that circumvents the tobacco industry's efforts to block implementation of ETS policies that curtail smoking in public places.

Thus, in addition to the future directions for ETS control outlined here as well as elsewhere,[26] we suggest, in line with WHO,[16] that ETS warning labels be added to packages of tobacco. In addition, we add to WHO's[16] suggestion that such future, potential ETS warning labels should be strong and rotated (like current health warnings on cigarette packs in Canada and Thailand) and should occupy at least 20% of the package (like current health warnings on cigarette packs in Poland and South Africa).

ACKNOWLEDGMENTS

This chapter was supported by funds provided by the Tobacco-Related Disease Research Program Grant Nos. 6RT-0081 and 8RT-0013 and by the California Department of Health Services Tobacco Control Section Grant 94-20962.

REFERENCES

1. U.S. Environmental Protection Agency, Respiratory Health Effects of Passive Smoking: Lung Cancer and Other Disorders, EPA 600/6-90/006F, U.S. Environmental Protection Agency, Office of Research and Development, Washington, D.C., December 1992.
2. Samet, J.M., Lewit, E.M., and Warner, K.E., Involuntary smoking and children's health, *Crit. Issues Child. Youth*, 4, 94–114, 1994.
3. Repace, J.L., Tobacco smoke pollution, in *Nicotine Addiction*, Orleans, C.T. and Slade, J., Eds., Oxford University Press, New York, 1993, 129–142.
4. International Agency for Research on Cancer, *Environmental Carcinogens: Methods of Analysis and Exposure Measurement, Vol. 9: Passive Smoking*, International Agency for Research on Cancer, Lyon, France, 1987.
5. California Environmental Protection Agency, Health Effects of Exposure to Environmental Tobacco Smoke: Final Report, California Environmental Protection Agency, Sacramento, CA, 1997.
6. U.S. Department of Health and Human Services, *The Health Consequences of Involuntary Smoking: A Report of the Surgeon General*, U.S. Government Printing Office, Washington, D.C., 1986.
7. National Research Council, *Committee on Passive Smoking: Environmental Tobacco Smoke. Measuring Exposures and Assessing Health Effects*, National Research Council, Board on Environmental Studies and Toxicology, National Academy Press, Washington, D.C., 1986. (www.no-smoke.org)
8. Americans for Nonsmokers' Rights, Protecting Nonsmokers from Secondhand Smoke, Americans Nonsmokers' Rights Foundation, October 15, 1998.
9. Glantz, S.A. and Parmley, W., Passive smoking and heart disease: epidemiology, physiology, and biochemistry, *Circulation*, 83, 1–12, 1991.
10. Taylor, A., Johnson, D., and Kazemi, H., Environmental tobacco smoke and cardiovascular disease, *Circulation*, 86, 699–702, 1992.
11. Wells, A.J., Passive smoking as a cause of heart disease, *J. Am. Coll. Cardiol.*, 24, 546–554, 1994.
12. Brownson, R.C., Alavanja, M.C.R., Hock, E.T., and Loy, T.S., Passive smoking and lung cancer in non-smoking women, *Am. J. Public Health*, 82, 1525–1530, 1992.
13. Kabat, G.C., Stellman, S.D., and Wynder, E.L., Relation between exposure to environmental tobacco smoke and lung cancer in lifetime nonsmokers, *Am. J. Epidemiol.*, 142, 141–148, 1995.
14. Jinot, J. and Bayard, S., Respiratory health effects of passive smoking — EPA's weight-of-evidence analysis, *J. Clin. Epidemiol.*, 47, 339–349, 1994.
15. Scoendorf, K.C. and Kiely, J.L., Relationship of Sudden Infant Death Syndrome to maternal smoking during and after pregnancy, *Pediatrics*, 90, 905–908, 1992.
16. WHO, *WHO's International Consultation on Environmental Tobacco Smoke (ETS) and Child Health*, World Health Organization, Geneva, 1999.

17. Brownson, R.C., Eriksen, M.P., Davis, R.M., and Warner, K.E., Environmental tobacco smoke: health effects and policies to reduce exposure, *Annu. Rev. Public Health*, 18, 163–185, 1997.

18. Centers for Disease Control, Discomfort from environmental tobacco smoke among employees at worksites with minimal smoking restrictions, United States, 1988, *MMWR*, 41, 351–354, 1992.

19. Borland, R., Pierce, J.P., Burns, D., Gilpin, E., Johnson, M., and Bal, D., Protection from environmental tobacco smoke in California: the case for a smoke-free workplace, *J. Am. Med. Assoc.*, 268, 749–752, 1992.

20. Patten, C.A., Pierce, J.P., Cavin, S.W., Berry, C., and Kaplan, R.M., Progress in protecting nonsmokers from environmental tobacco smoke in California workplaces, *Tob Control*, 4, 139–144, 1995.

21. Gerlach, K.K., Shopland, D.R., Harman, A.M., Gibson, J.T., and Pechacek, T.F., Workplace smoking policies in the U.S. results from a national survey of over 100,000 workers, *Tob. Control*, 6, 199–206, 1997.

22. Whitlock, G., MacMahon, S., Vander Hoorn, S., Davis, P., Jackson, R., and Norton, R., Association of environmental tobacco smoke exposure with socioeconomic status in a population of 7725 New Zealanders, *Tob. Control*, 7, 276–280, 1998.

23. Koop, C.E. and Kessler, D.A., Final Report of the Advisory Committee on Tobacco Policy and Public Health, *Tob. Control*, 6, 254–261, 1997.

24. Overpeck, M.D. and Moss, A.J., Children's exposure to environmental cigarette smoke before and after birth. Advance data from Vital and Health Statistics of the National Center for Health Statistics, #202, U.S. Department of Health and Human Services, DHHS Publication No. 91-1250, Hyattsville, MD, 1991.

25. Willers, S., Skarping, G., Dalene, M., et al., Urinary cotinine in children and adults during and after semi-experimental exposure to environmental tobacco smoke, *Arch. Environ. Health*, 50, 130–138, 1995.

26. Ashley, M.J. and Ferrence, R., Reducing children's exposure to tobacco smoke in homes: issues and strategies, *Tob. Control*, 7, 61–65, 1998.

27. Pirkle, J.L., Flegal, K.M., Bemert, J.T., Brody, D.J., Etzel, R.A., and Maurer, K.R., Exposure of the U.S. population to environmental tobacco smoke, *J. Am. Med. Assoc.*, 275, 1233–1240, 1996.

28. Greenberg, R.A., Bauman, K.E., Glover, L.H., et al., Ecology of passive smoking by young infants, *J. Pediatr.*, 114, 774–780, 1989.

29. Kleinschmidt, I., Hills, M., and Elliott, P., Smoking behaviour can be predicted by neighbourhood deprivation measures, *J. Epidemiol. Community Health*, 49(Suppl. 2), S72–S77, 1995.

30. Aligne, C.A. and Stoddard, J., Tobacco and children: an economic evaluation of the medical effects of parental smoking, *Arch. Pediatr. Adolesc. Med.*, 151, 648–653, 1997.

31. Preston, A.M., Ramos, L.J., Calderon, M.D., and Sahal, H., Exposure of Puerto Rican children to environmental tobacco smoke, *Preventive Med.*, 26, 1–7, 1997.

32. Institute of Medicine, State Programs Can Reduce Tobacco Use, National Cancer Policy Board, Institute of Medicine, Washington, D.C., 2000.

33. Klonoff, E.A., Landrine, H., and Alcaraz, R., An experimental analysis of sociocultural variables in sales of cigarettes to minors, *Am. J. Public Health*, May, 87(5), 823–826, 1997.

34. Centers for Disease Control, Decline in cigarette consumption following implementation of a comprehensive tobacco prevention and education program-Oregon, 1996-1998, *MMWR*, 48(7), 140–143, 1999.

35. Centers for Disease Control, CDC Surveillance summaries: state laws on tobacco control — United States, 1995, *MMWR*, 44(SS-6), 1995.
36. Americans for Nonsmokers' Rights, Smoking Policies in the Ten Busiest U.S. Airports, Americans Nonsmokers' Rights Foundation, January 1999. (www.no-smoke.org)
37. U.S. Department of Health and Human Services, *Healthy People, 2000*, U.S. DHHS Publication No.(PHS) 91-50212, U.S. Government Printing Office, Washington, D.C., 1990.
38. Goldstein, A.O. and Sobel, R.A., Environmental tobacco smoke regulations have not hurt restaurant sales in North Carolina, *N.C. Med. J.*, 59(5), 284–287, 1998.
39. Glantz, S.A. and Charlesworth, A., Tourism and hotel revenues before and after passage of smoke-free restaurant ordinances, *J. Am. Med. Assoc.*, 281, 1911–1918, 1999.
40. Glantz, S.A. and Smith, L.R.A., The effect of ordinances requiring smoke-free restaurants and bars on revenues: a follow-up, *Am. J. Public Health*, 87(10), 1687–1693, 1997.
41. Longoa, D.R., Brownson, R.C., and Kruse, R.L., Smoking bans in U.S. hospitals: results of a national survey, *J. Am. Med. Assoc.*, 274, 488–491, 1995.
42. Bureau of National Affairs, Smoking in the Workplace, 1991, SHRM-BNA Survey No. 55, Bureau of National Affairs, Washington, D.C., 1991.
43. U.S. Department of Health and Human Services, 1992 Survey of Worksite Health Promotion Activities, U.S. Department of Health and Human Services, Office of Health Promotion, Washington, D.C., 1993.
44. Glantz, S.A., Back to basics: getting smoke-free workplaces back on track, *Tob. Control*, 6, 164–166, 1997.
45. Stillman, F.A., Becker, D.M., Swank, R.T., Hantula, D., Moses, H., et al., Ending smoking at the Johns Hopkins Medical Institutions, *J. Am. Med. Assoc.*, 264, 1565–1569, 1990.
46. Grigham, J., Gross, J., Stitzer, M.L., and Felch, L.J., Effects of a restricted work-site smoking policy on employees who smoke, *Am. J. Public Health*, 84, 773–778, 1994.
47. Kinne, S., Kristal, A.R., White, E., and Hunt, J., Work-site smoking policies: their population impact in Washington State, *Am. J. Public Health*, 93, 1031–1033, 1993.
48. Brenner, H. and Fleischle, B., Smoking regulations at the workplace and smoking behavior: a study from southern Germany, *Prev. Med.*, 23, 230–234, 1994.
49. Woodruff, T.J., Rosbrook, B., Pierce, J., and Glantz, S.A., Lower levels of cigarette consumption found in smoke-free workplaces in California, *Arch. Int. Med.*, 153(12), 1485–1493, 1993.
50. Jeffery, R.W., Kelder, S.H., Forster, J.L., French, S., et al., Restrictive smoking policies in the workplace: effects on smoking prevalence and cigarette consumption, *Prev. Med.*, 23, 78–82, 1994.
51. Farrelly, M.C., Evans, W.N., and Sfekas, E.S., The impact of workplace smoking bans: Results from a national survey, *Tob. Control*, 8, 272–277, 1999.
52. Farkas, A.J., Gilpin, E.A., Distefan, J.M., and Pierce, J.P., The effects of household and workplace smoking restrictions on quitting behaviors, *Tob. Control*, 8, 261–265, 1999.
53. Evans W.N., Farrelly, M.C., and Montgomery, E., Do Workplace Bans Reduce Smoking?, NBER Working Paper No. W5567, National Bureau of Economic Research, Cambridge, MA, 1996.
54. Chapman, S., Borland, R., Scollo, M., Brownson, R.C., Dominello, A., and Woodward, S., The impact of smoke-free workplaces in declining cigarette consumption in Australia and the United States, *Am. J. Public Health*, 89, 1018–1023, 1999.

55. Brownson, R.C., Davis, J.R., Jackson-Thompson, J., and Wilkerson, J.C., Environmental tobacco smoke awareness and exposure, *Tob. Control*, 4, 132–138, 1995.

56. Pierce, J.P., Shanks, T.G., Pertschuk, M., Gilipin, E., et al., Do smoking ordinances protect non-smokers from environmental tobacco smoke?, *Tob. Control*, 3, 15–20, 1994.

57. Hammond, S.K., Sorenson, G., Youngstrom, R., and Ockene, J.K., Occupational exposure to environmental tobacco smoke, *J. Am. Med. Assoc.*, 274, 956–960, 1995.

58. Repace, J.L., Risk management of passive smoking at work and at home, *St. Louis Univ. Public Law Rev.*, 14, 763–785, 1994.

59. Hedge, A., Erickson, W.A., and Rubin, G., The effects of alternative smoking policies on indoor air quality in 27 office buildings, *Ann. Occup. Hyg.*, 38, 265–278, 1994.

60. Muramatsu, M., Umemura, S., Okada, T., and Tomita, H., Estimation of personal exposure to tobacco smoke with a newly developed nicotine personal monitor, *Environ. Res.*, 35, 218–227, 1984.

61. Hammond, S.K., Exposure of U.S. workers to environmental tobacco smoke, *Environ. Health Perspect.*, 107, 329–339, 1999.

62. Wasserman, J., Manning, W.G., and Newhouse, J.P., The effects of excise taxes and regulations on cigarette smoking, *J. Health Econ.*, 10, 43–64, 1991.

63. Collishaw, N.E., *An International Framework Convention of Tobacco Control*, World Health Organization, Geneva, 1997.

64. Sweda, E.L., Gottlieb, M.A., and Porfiri, R.C., Protecting children from exposure to environmental tobacco smoke (Editorial), *Tob. Control*, 7, 1–2, 1998.

65. Biener, L., Cullen, D., Di, Z., and Hammond, S., Household smoking restrictions and adolescents' exposure to environmental tobacco smoke, *Prev. Med.*, 26, 358–363, 1997.

66. Emmons, K.M., Hammond, S.K., and Abrams, D.B., Smoking at home: the impact of smoking cessation on nonsmokers' exposure to environmental tobacco smoke, *Health Psych.*, 13, 516–520, 1994.

67. Severson, H.H., Zoref, L., and Lichenstein, E., Reducing environmental tobacco smoke (ETS) exposure for infants: a cessation intervention for mothers of newborns. *Am. J. Health Promotion*, 8, 252, 1994.

68. Greenberg, R.A., Strecher, V.J., Bauman, K.E., Boat, B.B., Fowler, M.G., Keyes, L.L., Denny, F.W., Chapman, R.S., Stedman, H.S., LaVange, L.M., Glover, L.H., Haley, N.J., and Loda, F.A., Evaluation of a home-based intervention program to reduce infant passive smoking and lower respiratory illness, *J. Beh. Med.*, 17, 273–290, 1994.

69. Fisher, E.B., Auslander, W.F., Munro, J.F., Arfken, C.L., Brownson, R.C., and Owens, N.W., Neighbors for a smokefree north side, *Am. J. Public Health*, 88, 1658–1663, 1998.

70. Chilmonczyk, B.A., Palmaki, G.E., Knight, G.J., et al., An unsuccessful cotinine-assisted intervention strategy to reduce ETS exposure during infancy, *Am. J. Dis. Child.*, 146, 357–360, 1992.

71. Aftab, M., Kolben, D., and Lurie, P., International cigarette labeling practices, *Tob. Control*, 8, 368–372, 1999.

72. Klonoff, E.A., Landrine, H., Alcaraz, R., Campbell, R.R., Lang, D., McSwan, K., and Parekh, B., An instrument for assessing the quality of tobacco-control policies, *Prev. Med.*, 27(6), 808–814, 1998.

8 Animal Models of Environmental Tobacco Smoke-Induced Lung Cancer

Hanspeter Witschi

CONTENTS

8.1 THE TOXICOLOGY OF ENVIRONMENTAL TOBACCO SMOKE

In 1992, the U.S. Environmental Protection Agency (EPA) acknowledged environmental tobacco smoke as a proven human carcinogen.[1] A few years later, the state of California declared environmental tobacco smoke (ETS) to be a toxic air contaminant. This decision was based on a thorough analysis done by California's Environmental Protection Agency.[2] The agency not only confirmed the conclusions that had been drawn by the U.S. EPA, but reinforced them by taking into account the results of the newest available epidemiological data. Exposure to ETS can result in a lifetime risk of lung cancer death in non-smokers of about 0.7% (relative risk 1.2). Although such an increase in risk is comparatively small, the large number of people exposed worldwide makes ETS an important public health problem. While in the U.S. during the last decades smoking has generally decreased and lung cancer rates have seemed to flatten out,[3] smoking remains one of the major health concerns worldwide, particularly in the large populations of Asia.[4–6]

However, cancer was not the only important adverse health effect caused by ETS. Compromised lung function in children and an increase in morbidity and

0-8493-0311-7/00/$0.00+$.50
© 2001 by CRC Press LLC

129

mortality from cardiovascular disease were recognized to be two other major public health concerns of ETS.[2] These possible implications of widespread exposure to ETS for public health spawned multiple new research interests and initiatives in the early 1990s. Within a comparatively short time, a substantial amount of information on the toxicology of ETS became available.[7] In several experimental studies, it could be demonstrated that exposure to ETS at comparatively low concentrations (from 0.5 to 5 mg of total suspended particulate [TSP] matter per cubic meter [m^3]) adversely affected lung development in newborn rats and guinea pigs.[8-11] ETS was also found to have cardiovascular toxicity. In cockerels exposed for 16 weeks, ETS at concentrations around 2 to 3 mg TSP/m^3 enhanced the development of arteriosclerotic plaques in arteries.[12,13] In rabbits fed a cholesterol-rich diet and exposed to even higher concentrations of ETS, similar observations were made.[14] In a coronary artery ligation-reperfusion model, ETS increased myocardial infarct size.[15] On the other hand, several studies, conducted at comparatively low concentrations of ETS (up to 4 mg TSP/m^3), showed little effect on the tissues of the respiratory tract of rats or hamsters.[16-18]

8.2   EXPERIMENTAL TOBACCO SMOKE CARCINOGENESIS

Developing an animal model of ETS-induced pulmonary carcinogenesis proved to be a challenge. Between 1950 and 1980, numerous attempts had been made to duplicate in experimental animals the well-known carcinogenic properties of tobacco smoke. An analysis of the data available until 1986 revealed, however, that animal experimentation had been mostly unsuccessful in producing tobacco smoke-induced lung cancer in experimental animals.[19] The International Agency for Research on Cancer (IARC) documents review altogether five mouse studies, four of which were negative. Out of four rat studies, two were negative, one was questionable, and only one was positive. In hamsters, laryngeal tumors were produced by tobacco smoke, but no tumors in the studies of deep lung of dog and rabbit were questionable or negative. A more recent review on animal carcinogenesis studies with tobacco mainstream smoke concluded that animal experiments had essentially failed to duplicate what was known from human epidemiological studies. Tobacco smoke would not produce lung tumors in experimental animals.[20]

Although these observations were not very encouraging, nevertheless there were some compelling reasons to reinvestigate the carcinogenic potential of ETS. One reason relates to dosimetry. Until quite recently, most animal studies had not been done under well-characterized exposure conditions with analytical data on constituents of the chamber atmospheres. In practically all studies dealing with tobacco smoke exposure, information was usually limited to a description of how many puffs of how many cigarettes the animals had been exposed. Practically no data on the concentrations of tobacco smoke constituents within the inhalation chambers are available in the older literature. This makes it difficult, if not impossible, to relate any outcome of a carcinogenesis study with actual exposure data, information that is needed in any quantitative risk assessment for a given hazard. With the renewed interest in the toxicity and possibly carcinogenicity of ETS, a new generation of well-designed and well-executed inhalation studies became available in which animals were exposed to thoroughly characterized atmospheres of ETS.[21–23]

A second reason to examine possible ETS carcinogenesis was the suspicion that ETS would be a more potent carcinogen than full tobacco smoke. Differences in the composition between mainstream smoke, the smoke inhaled by active smokers, and ETS, a mixture of exhaled mainstream smoke and of sidestream smoke, have been described in detail.[24] Cigarette sidestream smoke, generated at lower burning temperatures than mainstream smoke, has higher concentrations of certain carcinogenic components of cigarette smoke and could thus be a more potent carcinogen. Filters do not remove toxic and carcinogenic chemicals from sidestream smoke.[25] In a skin painting study, it was found that sidestream smoke condensate had considerably more carcinogenic potential than mainstream smoke condensate.[26] On the other hand, such differences become irrelevant, considering the fact that exposure to ETS occurs at concentrations that are orders of magnitude lower than tobacco smoke concentrations inhaled by active smokers.[27]

Not unexpectedly, this difference in exposure has been the subject of considerable controversy and debate. Human health risk assessment of ETS, based on epidemiological studies, suggests an increased relative risk of about 1.2. Translated into annual increase in tobacco smoke-related lung cancer deaths per year, such a risk factor might account for an additional 3000 lung cancer deaths in the U.S.[2] Since quite a few epidemiological studies failed to indicate a significantly increased risk, the validity and particularly the significance of the number have been vigorously challenged.[28] If the risk is calculated based on dosimetric considerations, i.e., taking into account how much tobacco smoke an active smoker might inhale and comparing this with what then could be calculated for involuntary smoking, exposure to ETS might account at most for 1 death per 4.7 million never smokers or about 11 deaths in the 1992 population of 52 million U.S. never smokers.[29] Well-controlled animal studies with ETS could help estimate ETS potency, compare it to the potency of full tobacco smoke, and incorporate the potency factor into risk estimates.

8.3   TOBACCO SMOKE AND LUNG TUMORS IN MICE

From a practical consideration, laboratory rodents (mice, rats, and hamsters) are the species of choice for a study of the carcinogenic potential of ETS. As mentioned earlier, previous assays conducted with full tobacco smoke had produced only a comparatively weak carcinogenic response. The 1986 IARC monograph on tobacco smoking analyzes six chronic exposure studies.[19] If the data from the various studies are combined, while disregarding possible differences in exposure or strain sensitivities, it can be seen that out of an overall total of 1703 mice exposed to tobacco smoke, only 108 animals (6.3%) developed lung tumors. In the corresponding control groups, tumors were found in 39 (3.9%) out of 998 animals. While this is a small difference, the large number of animals makes the difference significant, showing that it is possible to induce lung tumors in mice with tobacco smoke. In another large study involving some 1600 mice, lung tumor incidence in the exposed animals was again comparatively low, 5% and in controls 4.3%. While tumor incidence was not significantly higher in the exposed animals, smoke shortened the time to tumor development.[30,31] Nevertheless, in most mouse strains tobacco smoke seems to have only weak carcinogenic activity.

Therefore, it was of interest to examine whether better results could be obtained with a mouse strain that was particularly prone to developing lung tumors; A/J mice are such a sensitive strain. When exposed to a chemical carcinogen by inhalation or systemic administration, these mice will develop multiple lung tumors. This fact that has been well known for several decades (see Reference 32 for a review of the early literature). Thus, strain A mice can be used to rapidly screen for carcinogens.[32,33] The assay has two major advantages. First, a response is obtained within 6 to 10 months rather than, as is the case with conventional rodent assays, within 2 years. Second, the number of lung tumors that develop following exposure to a carcinogen is strictly dose dependent. It is for this reason that lung tumor multiplicity, rather than lung tumor incidence, is used to gauge the carcinogenic effect.

The criteria for a positive lung tumor assay have been rigorously defined.[33] A positive carcinogenic response in the strain A/J mouse lung tumor assay is obtained when lung tumor multiplicity in the treated animals is significantly increased, preferably to one or more lung tumors per mouse. The mean number of lung tumors in the appropriate controls should approximate the anticipated number for untreated mice of the same age. Additionally, for marginally positive compounds, positive results in an initial test must be repeatable in a second test.

Mouse lung tumors offer an additional advantage for the study of human lung carcinogenesis in an experimental animal model. During the last two decades, there has been a noticeable shift in the types of human lung cancers. Squamous cell cancers have decreased in numbers, whereas adenocarcinomas, often of peripheral origin, are definitely on the increase,[34] an observation made first in 1961.[35] The phenomenon is usually explained by the "changing" cigarette.[36,37] The design of effective filters removes much of the tar from inhaled tobacco smoke. The tar fraction contains most of the polycyclic aromatic hydrocarbons, carcinogens known to produce squamous cell lung cancer in animals. Filters also retain some nicotine. In order to satisfy a nicotine craving, smokers inhale more deeply and retain the smoke, rich in tobacco smoke-specific nitrosamines, longer in the deep lung.[38] Nitrosamines produce adenocarcinomas in experimental animals.

This shift in tumor types makes mouse lung tumors highly relevant for the study of human adenocarcinomas. The lung tumors observed in A/J mice are in many respects similar to human adenocarcinoma. They originate from the same cells, the type II alveolar epithelial cells and Clara cells. They share many of the molecular changes found in cell cycle and cell–cell communicating genes. Striking similarities exist in many biochemical processes, such as the activities of certain enzymes or signaling pathways. Thus, mouse lung tumors and human adenocarcinoma claim many similarities in genetic and epigenetic traits. The relevant literature that analyzes and documents these similarities has been extensively reviewed.[39–41] Thus, according to the current state of knowledge, it can be ascertained that mouse lung tumors are a highly relevant model for human pulmonary adenocarcinoma.

Finally, it must be mentioned that the strain A mouse lung tumor model has become the animal model of choice for an evaluation of agents that might prevent tobacco smoke-induced lung cancer. Such studies are actively pursued in many laboratories.[42–44] Some of the agents that were found to be effective in the mouse lung have been proposed for or are already being evaluated for chemopreventive action in humans. This is further evidence that mouse lung tumors are considered to be an excellent animal model to study human lung adenocarcinoma.

8.4 ETS AND LUNG TUMORS IN STRAIN A/J MICE

During the last few years, we have conducted several experiments on the carcino-genicity of ETS in strain A/J mice. The results clearly show that under appropriate experimental conditions mice exposed to ETS develop substantially more lung tumors than mice kept in filtered air. It can be concluded that ETS is a pulmonary carcinogen in the mouse and thus, by inference, may cause lung cancer in humans.

We have used strain A/J mice in all experiments, purchased from Jackson Laboratories, Bar Harbor, ME. After 2 weeks of acclimatization, the now about 10-week-old animals are exposed to ETS in a stainless steel inhalation chamber (volume, 0.44 m^3). The control animals are placed into a chamber of identical dimensions that is ventilated with filtered air. The ETS atmosphere consists of a mixture of 89% cigarette sidestream smoke (SS) and 11% mainstream smoke (MS). SS is generated by burning standard research Kentucky 1R4F cigarettes in a smoking machine with standard puffs of 35 ml volume of 2 s duration, once every minute, for a total of 8 puffs per cigarette. The smoke curling off the tip of the cigarette between puffs, the so called SS, is drawn into a larger chamber for aging. Once every 58, s the SS is reinforced with a single puff of MS. The final mixture is then drawn into the inhalation chambers, and this represents ETS. Within the inhalation chamber, three constituents of ETS are routinely monitored. Concentrations of carbon monoxide are measured every 30 min. Total suspended particulates (TSP) are collected on filters and weighed after 1, 3, and 5 h of exposure. The daily measurements are averaged for each exposure day. Nicotine was initially measured every day. It then became apparent that the system is stable and nicotine measurements are now done once a week. Within the chambers there is an average temperature of 22 to 25°C and an average relative humidity of 33 to 43%. The chambers are kept on a 12-h light–dark cycle. The smoke generating system, the engineering specifications of the exposure chamber, a detailed characterization of the chamber atmosphere, and the analytical methods we used have been described in detail.[22]

The animals are usually exposed to smoke 6 h/day, 5 days/week. They are observed daily and weighed weekly. At the end of the experiment, the animals are killed, and the lungs are fixed in Tellyesniczky's fluid, a mixture of ethanol, formalin, phosphate buffer, and glacial acetic acid.[45] After a few days in fixative, the lungs are inspected under a magnifying glass, and the tumor nodules visible on the lung surface are counted. Counting surface tumors in Tellyesniczky's fluid-fixed lung accurately reflects tumor count in the entire lung.[45] Paraffin sections are prepared from selected tumors and examined under a light microscope. The tumor counts are presented in two ways: as lung tumor multiplicity and as lung tumor incidence. Tumor multiplicity is defined as the average number of lung tumors per animal, and this includes non-tumor bearing animals. The inclusion of non-tumor bearing animals explains why in the control groups, where many animals do not have a lung tumor, the average number of tumors is less than one. Lung tumor incidence is defined as the percentage of tumor bearing animals in a given experimental group. Statistical tests applied are *t* test or Analysis of Variance (ANOVA) followed by an appropriate post hoc test for lung tumor multiplicity and Fisher's exact test for lung tumor incidence. A *p* value of 0.05 or less is considered significant.

In the first experiment, mice were exposed for a total of 6 months to an ETS concentration with the chambers of 4 mg/m³.[46] Tumors were counted after 6 months. No increase in lung tumor multiplicity or incidence was found. However, there was one interesting observation: in lung tumors harvested from tobacco smoke-exposed animals, there was a shift in K-ras mutations from codon 12/13 to codon 61. This was inasmuch unusual as it has been found that in spontaneously developing lung adenomas, K-ras mutations occur with approximately equal frequency in codon 12 and codon 61.[47] Exposure of animals to methylating or alkylating agents, including the potent lung carcinogen 4-(methylnitrosamino)-1-(3-pyridyl)-1-butanone (NNK), or to polycyclic aromatic hydrocarbons leads to mostly codon 12 mutations, a pattern consistent with the known adduct-forming properties of the carcinogens.[48,49] On the other hand, treatment of A/J mice with urethane results mostly in codon 61 mutations.[48] Therefore, it was surprising to find that ETS, rich in NNK, benzo(a)pyrene, and other polycyclic aromatic hydrocarbons, shifted the mutation spectrum from codon 12 toward codon 61. It was also noteworthy that in SS-exposed animals, the prominent K-ras mutation was the Arg substitution, a feature that has been considered to be a marker for malignancy in mouse lung tumors.[50] The observations on K-ras mutations associated with exposure to ETS need to be expanded in future studies.

The negative data on lung tumor multiplicity called for a modification of the exposure protocol. The concentration of ETS within the chamber was increased to 55 to 90 mg/m³ of TSP and exposure to ETS was maintained for 5 months.[51] After 5 months, the animals were removed from the ETS, and lung tumors were counted in half of the animals. No difference was found between animals exposed to ETS and animals kept in filtered air. The remaining animals were transferred to a conventional animal facility with a temperature of 21 to 22°C and 40 to 70% humidity where they were allowed to recover for another 4 months in filtered air. When they were killed, it was found that lung tumor multiplicities and incidences were significantly higher in the ETS-exposed animals. Two separate experiments were conducted under these exposure conditions, and lung tumor multiplicities were around 1.3.[51,52] It was then decided to increase the concentration of tobacco smoke within the chamber to approximately 130 to 140 mg/m³ of TSP. This substantially increased the tumor yield, as was found in subsequent experiments.[53]

Table 8.1 provides an overview of six independent experiments conducted during the last 5 years. The aggregate data show that in each single experiment the criteria for a positive lung tumor assay were met; lung tumor multiplicities were invariably higher than 1.0, whereas controls were in the range that could be expected for A/J mice approximately 1 year old, where an average of 0.5 to 1.0 spontaneous lung tumor can usually be expected.[32,33,54] In animals exposed to a concentration of 80 to 90 mg/m³ of TSP, the average number of tumors per lung was approximately half the number counted in animals exposed to the higher concentration of 130 to 140 mg/m³ of TSP. This indicates a dose response. In addition, the data also show

TABLE 8.1
Lung Tumor Multiplicity

TSP (mg/m³)	No. of tumors per lung[a]		Ref.
	Filtered air	ETS	
87	0.5 ± 0.2 (24)	1.4 ± 0.2 (24) *	Witschi et al. 1997b[51]
79	0.5 ± 0.1 (24)	1.3 ± 0.3 (26) *	Witschi et al. 1997a[52]
83	0.9 ± 0.2 (29)	1.3 ± 0.2 (33) *	Witschi et al. 1998[62]
132	0.6 ± 0.1 (30)	2.1 ± 0.3 (35) *	Witschi et al. 1999[53]
137	0.9 ± 0.2 (30)	2.8 ± 0.2 (38) *	Witschi et al. 2000[77]
137	1.0 ± 0.1 (54)	2.4 ± 0.3 (28) *	Witschi et al. 2000[77]

[a] All values are given as mean ± SEM with the number of animals in parenthesis. Values significantly higher than for animals kept in filtered air are labeled with an asterisk (*).

TABLE 8.2
Lung Tumor Incidence

TSP (mg/m³)	Percentage of animals with tumors[a]		Ref.
	Filtered air	ETS	
87	38%	83% *	Witschi et al. 1997b[51]
79	42%	58%	Witschi et al. 1997a[52]
83	69%	73%	Witschi et al. 1998[62]
132	50%	86% *	Witschi et al. 1999[53]
137	60%	100% *	Witschi et al. 2000[77]
137	65%	89% *	Witschi et al. 2000[77]

[a] All values are expressed as the percentage of tumor bearing animals. The total number of animals that were at risk are listed in Table 8.1. Values significantly higher than for animals kept in filtered air are labeled with an asterisk (*).

that in 4 of the 6 experiments, the lung tumor incidence (percentage of tumor bearing animals) was significantly higher in the ETS-exposed animals when compared to filtered air controls (Table 8.2).

In all experiments, selected tumors were examined under the light microscope to confirm the diagnosis. Proliferative pulmonary lesions were morphologically similar between control mice and those exposed to ETS and were typical of pulmonary lesions commonly observed in strain A mice. Lesions were categorized as focal alveolar epithelial hyperplasia, alveolar/bronchiolar adenomas, and alveolar/bronchiolar adeno-carcinomas. The majority of adenomas had a solid growth pattern with contiguous neoplastic cells filling alveolar spaces. There was no difference in the relative distribution of adenocarcinomas and adenomas between the two groups. In the ETS-exposed animals, 17% of all tumors had a pattern consistent with the diagnosis of adenocarcinoma, and in the control group 20% did so. Of the four carcinomas observed, two had a solid growth pattern, one had a mixed solid and papillary growth pattern, and one was a papillary neoplasm arising within a solid adenoma. Diagnosis of malignancy was based primarily on the presence of nuclear atypia (hyperchromatic and pleomorphic nuclei) and cellular crowding. Tumor progression appeared to occur along a continuum from hyperplasia to adenoma to carcinoma.

The importance of the recovery period was documented in an additional experiment as follows. Half of the ETS-exposed mice were removed from smoke after 5 months and allowed to recover for the usual 4 months in filtered air. The other half were kept for the full 9 months in smoke. When tumors were counted, it was found that the animals allowed a recovery period had an average of 2.4 lung tumors, significantly higher than animals who had been kept in ETS for 9 months (average lung tumor multiplicity of 1.5 tumors per lung, not significantly higher than lung tumor multiplicity of 1.0 found in the filtered air-exposed animals [Witschi, unpublished data]). There are several explanations for this observation. The most likely one appears to be that ETS is not only carcinogenic, but also has substantial cytotoxic activity. In previous experiments, it was found that ETS significantly inhibited the development of urethane- or methylcholanthrene-induced lung tumors in the lungs of A/J mice.[51] The cytotoxic effects brought about by some as yet unidentified constituents of ETS might be partially offset by tumor growth. Likely cytotoxic agents in tobacco smoke might be oxides of nitrogen, which were found in other experiments to interfere with the growth of chemically induced lung tumors,[55] or other known cytotoxic agents such as acrolein or formaldehyde. Whatever, it seems that the recovery period is instrumental in allowing the full growth of tobacco smoke-induced tumors in the lung. The observation also might explain why in a recent study with A/J mice, where there was no prolonged recovery period, no increase in tumor incidence or multiplicity was produced by quite high concentrations of tobacco smoke.[56] In practically all older studies on tobacco smoke carcinogenesis, the animals were not allowed to recover in filtered air, a factor that might account for the mostly negative data.

The mouse lung tumor model allowed researchers to reexamine another question important to tobacco carcinogenesis: to what extent has the gas phase carcinogenic potential, i.e., components of cigarettes smoke not associated with tar. Some 30 years ago, considerable efforts were spent to develop a "safe" cigarette or at least a cigarette that would represent a smaller risk than the then high tar-high nicotine cigarettes.[57] Removal of the tar from the inhaled smoke by passing it through a filter was a key strategic approach toward this goal. The rationale behind it was that cigarette smoke tar, painted on the back of mice, produced multiple tumors.[58] We examined the contribution of the gas phase in our experimental system. Mice were exposed as described previously, except that the ETS, prior to entering the inhalation chamber, was passed though an absolute high efficiency particulate air (HEPA) filter. Instead of reaching 80 to 90 mg/m^3, the TSP concentration in the chamber was only about 0.1 mg/m^3. The chamber atmosphere was analyzed with gas chromatography/mass spectrometry for several known carcinogens.[52] It was found that the concentrations of several carcinogens, particularly of benzo(a)pyrene and the tobacco smoke-specific nitrosamine NNK (4-(methylnitrosamino)-1-(3-pyridyl)-1-butanone), were greatly reduced. Despite the reduction of "tar" in the chamber atmosphere, the animals developed the same number of lung tumors as did animals kept in the complete smoke atmosphere. Dosimetrric calculations showed that the animals inhaled 3 to 4 orders of magnitude less of benzo(a)pyrene or NNK than would have been necessary to produce one tumor per lung in strain A/J mice. In other words, removal of more than 95% of these two agents from cigarette smoke had no major impact on lung tumor development. It was concluded that the gas phase might contain some, as of yet unidentified, highly potent carcinogens.

This observation is not new; it already had been reported in 1972 that the gas phase of tobacco smoke alone is sufficient to cause lung tumors in mice.[59-61] The observation seems to have been overlooked later on. It has several implications. It is possible that the role of polycyclic aromatic hydrocarbons as causative agents for lung cancer may have been overemphasized. This could explain why the introduction of filter cigarettes had only a minimal effect on the occurrence of lung cancer.[63,64] As far as ETS is concerned, filtering the air in smoke-contaminated rooms might be without any beneficial effect in reducing lung cancer risk.

8.5 CHEMOPREVENTION

Efficient chemoprevention might help to reduce the cancer risk.[65,66] As far as chemoprevention of tobacco smoke-induced lung cancer is concerned, it is of course obvious that complete cessation of smoking would be the most effective way to prevent this burden on the public health. However, since it is recognized that a substantial number of smokers are unable to quit, chemoprevention offers another way to mitigate the impact of smoking.[42,43,67-69] Chemopreventive agents, such as drugs or naturally occurring constituents of the diet, could help protect, to some extent, the active smokers. Perhaps more importantly, they might help to reduce further the risk of developing lung cancer in individuals who have quit smoking or in non-smokers exposed to tobacco smoke in their environment.

The successful production of lung tumors by ETS in the strain A mouse allowed researchers to examine the effectiveness of chemopreventive agents. The search for such agents is currently an active research area, and several animal models have been or continue to be evaluated.[44] We have used several chemopreventive agents to examine their efficacy against tobacco smoke. In a series of experiments, strain A/J mice were given the following agents while being exposed to tobacco smoke and also during the recovery phase: green tea extract in the drinking water and phenethyl isothiocyanate (PEITC), a mixture of phenethyl and benzyl isothiocyanate, N-acetylcysteine, aspirin, D-limonene, and p-XSC (1,4-phenylene-*bis*(methylene)selenoisocyanate) in the diet. All these agents have been found to effectively prevent lung tumor development in strain A/J mice produced by carcinogens found to occur in tobacco smoke, most often NNK.[70-74] However, none of these agents was effective against tobacco smoke. The failure of agents which are demonstrably highly effective against NNK carcinogenesis, such as PEITC, p-XSC, or D-limonene, lends further support to the hypothesis, formulated in our studies with filtered smoke, that NNK is not necessarily the most important or most determining carcinogen in tobacco smoke.

One agent, however, was highly effective. Wattenberg had shown previously that a mixture of myoinositol and dexamethasone protected strain A mice against NNK or benzo(a)pyrene.[75,76] The important observation was that a combination of the two agents not only provided chemoprotection when fed to the animals during carcinogen exposure,[53] but also when fed in the postinitiation phase, e.g., after exposure had ceased. In our first experiment, we tried a feeding mixture of myoinositol-dexamethasone during the entire 9 months. Complete protection against the carcinogenic activity of smoke was seen.[53] In a second experiment, the animals were placed onto the myoinositol-dexamethasone diet only when they were removed from the smoke. Again, it was found that they had complete protection against the carcinogenic effects of tobacco smoke.[77]

The observation that myoinositol-dexamethasone is effective in "quitters" might have some implications for human tobacco carcinogenesis. In several epidemiological studies, it has been found that, in the first few years after quitting, ex-smokers are actually at a higher risk to develop lung cancer than current smokers.[78,79] This has been explained by the fact that, in all likelihood, smokers quit because they are already ill from lung cancer or are plagued by chronic cough and respiratory illnesses. However, not all quitters have, at this stage, obvious lung cancer that calls for aggressive treatment. Chemoprevention initiated after quitting might, in this case, be a beneficial intervention. The success of smoking cessation programs might be enhanced if chemoprevention could be added as an incentive.

8.6 INCIDENTAL PULMONARY CHANGES

Aside from the tumors, the lungs of mice exposed for several months to ETS showed surprisingly few changes. Particularly striking was the complete absence of macrophages carrying ETS particles; the macrophage volume in the lung of ETS-exposed animals did not increase, and the ones that were found had normal morphological features and no inclusion bodies.[51,80] For the time being, it is not known why strain A/J mice are apparently quite resistant to the inflammatory effects of ETS exposure; in rats and hamsters exposed to considerably lower concentrations of ETS than used in our experiments, some mild histopathological changes were nevertheless readily observed.[16,18] However, it is noteworthy that a 9-month exposure of strain A/J mice to 1 ppm of ozone also produced only minimal alterations in the lung and that lower doses had no effect whatsoever.[81] Analysis of the lungs of mice exposed to ETS for the presence of cytochrome P450 enzymes showed an increased staining for CYP 1A1, even at concentrations below those necessary for tumor induction. Upon removal from the smoke, staining disappeared rabidly.[51,80] Exposure to filtered ETS did not increase staining for CYP 1A1.[52] Finally, it was found in labeling studies with bromodeoxyuridine that inhalation of ETS increased cell turnover in the lung parenchyma, the airways, and the nasal passages. Cell turnover was markedly increased in all these regions while the animals were exposed to smoke, particularly during the first 6 weeks of smoke exposure. Upon returning the animals to filtered air, cell proliferation rates fell to normal levels in both the lung and the nasal passages.[51,82] Similar observations were made in the lungs of hamsters exposed to ETS.[83]

In conclusion, animal studies have given biological plausibility to the assertion that ETS causes lung cancer in humans. Hopefully, in the future, the studies with animals exposed to ETS will advance our understanding about underlying mechanisms and help us find and develop effective chemopreventive agents.

ACKNOWLEDGMENTS

This publication was made possible by grants ES07908 and ES07499 from the National Institute of Environmental Health Sciences (NIEHS). Its contents are solely the responsibility of the authors and do not necessarily represent the official views of the NIEHS or the NIH.

REFERENCES

1. U.S. Environmental Protection Agency, Respiratory Health Effects of Passive Smoking: Lung Cancer and Other Disorders, (EPA/600/6-90/006F), U.S. EPA, Office of Health and Environmental Assessment, Washington, D.C., 1992.
2. National Cancer Institute, Health Effects of Exposure to Environmental Tobacco Smoke: The Report of the California Environmental Protection Agency, Smoking and Tobacco Control Monograph No. 10, NIH Pub. No. 99-4645, U.S. Department of Health and Human Services, National Institutes of Health, National Cancer Institute, Bethesda, MD, 1999.
3. Wingo, P.A., Ries, L.A., Giovino, G.A., Miller, D.S., Rosenberg, H.M., Shopland, D.R., Thun, M.J., and Edwards, B.K., Annual report to the nation on the status of cancer, 1973-1996, with a special section on lung cancer and tobacco smoking, *J. Natl. Cancer Inst.*, 91, 675, 1999.
4. Pisani, P., Parkin, D.M., Bray, F., and Ferlay, J., Estimates of the worldwide mortality from 25 cancers in 1990, *Int. J. Cancer*, 83, 18, 1999.
5. Liu, B.Q., Peto, R., Chen, Z.M., Boreham, J., Wu, Y.P., Li, J.Y., Campbell, T.C., and Chen, J.S., Emerging tobacco hazards in China: 1. Retrospective proportional mortality study of one million deaths, *Br. Med. J.*, 317, 1411, 1998.
6. Yang, G., Fan, L., Tan, J., Qi, G., Zhang, Y., Samet, J.M., Taylor, C.E., Becker, K., and Xu, J., Smoking in China: findings of the 1996 National Prevalence Survey, *J. Am. Med. Assoc.*, 282, 1247, 1999.
7. Witschi, H.P., Joad, J.P., and Pinkerton, K.E., The toxicology of environmental tobacco smoke, *Annu. Rev. Pharmacol. Toxicol.*, 37, 29, 1997a.
8. Joad, J.P., Pinkerton, K.E., and Bric, J.M., Effects of sidestream smoke exposure and age on pulmonary function and airway reactivity in developing rats, *Pediatr. Pulmonol.*, 16, 281, 1993.
9. Joad, J.P., Ji, C., Kott, K.S., Bric, J.M., and Pinkerton, K.E., In utero and postnatal effects of sidestream cigarette smoke exposure on lung function, hyperresponsiveness, and neuroendocrine cells in rats, *Toxicol. Appl. Pharmacol.*, 132, 63, 1995.
10. Joad, J.P., Bric, J.M., and Pinkerton, K.E., Sidestream smoke effects on lung morphology and C-fibers in young guinea pigs, *Toxicol. Appl. Pharmacol.*, 131, 289, 1995.
11. Bonham, A.C., Kappagoda, C.T., Kott, K.S., and Joad, J.P., Exposing young guinea pigs to sidestream tobacco smoke decreases rapidly adapting receptor responsiveness, *J. Appl. Physiol.*, 78, 1412, 1995.
12. Penn, A., Chen, L.C., and Snyder, C.A., Inhalation of steady-state sidestream smoke from one cigarette promotes arteriosclerotic plaque development, *Circulation*, 90, 1363, 1994.
13. Penn, A. and Snyder, C.A., Inhalation of sidestream cigarette smoke accelerates development of arteriosclerotic plaques, *Circulation*, 88, 1820, 1993.
14. Zhu, B.Q., Sun, Y.P., Sievers, R.E., Isenberg, W.M., Glantz, S.A., and Parmley, W.W., Passive smoking increases experimental atherosclerosis in cholesterol-fed rabbits, *J. Am. Coll. Cardiol.*, 21, 225, 1993.

15. Zhu, B.Q., Sun, Y.P., Sievers, R.E., Glantz, S.A., Parmley, W.W., and Wolfe, C.L., Exposure to environmental tobacco smoke increases myocardial infarct size in rats, *Circulation*, 89, 1282, 1994.

16. Coggins, C.R., Ayres, P.H., Mosberg, A.T., Sagartz, J.W., and Hayes, A.W., Subchronic inhalation study in rats using aged and diluted sidestream smoke from a reference cigarette, *Inhal. Toxicol.*, 5, 77, 1993.

17. Coggins, C.R., Ayres, P.H., Mosberg, A.T., Ogden, M.W., Sagartz, J.W., and Hayes, A.W., Fourteen-day inhalation study in rats, using aged and diluted sidestream smoke from a reference cigarette. I. Inhalation toxicology and histopathology, *Fundam. Appl. Toxicol.*, 19, 133, 1992.

18. von Meyerinck, L., Scherer, G., Adlkofer, F., Wenzel-Hartung, R., Brune, H., and Thomas, C., Exposure of rats and hamsters to sidestream smoke from cigarettes in a subchronic inhalation study, *Exp. Pathol.*, 37, 186, 1989.

19. IARC (International Agency for Research on Cancer), Biological data relevant to the evaluation of carcinogenic risk to humans. 1. Carcinogenicity studies in animals, in *IARC Monographs on the Evaluation of the Carcinogenic Risk of Chemicals to Humans. Tobacco Smoking. Volume 38*, WHO IARC, Ed., IARC, Lyon, 1986, 127.

20. Coggins, C.R.E., A review of the chronic inhalation studies with mainstream cigarette smoke in rats and mice, *Toxicol. Pathol.*, 26, 307, 1998.

21. Ayres, P.H., Mosberg, A.T., and Coggins, C.R., Design, construction, and evaluation of an inhalation system for exposing experimental animals to environmental tobacco smoke, *Am. Ind. Hyg. Assoc. J.*, 55, 806, 1994.

22. Teague, S.V., Pinkerton, K.E., Goldsmith, M., Gebremichael, A., Chang, S., Jenkins, R.A., and Moneyhun, J.H., Sidestream cigarette smoke generation and exposure system for environmental tobacco smoke studies, *Inhal. Toxicol.*, 6, 79, 1994.

23. Haussmann, H.J., Anskeit, E., Becker, D., Kuhl, P., Stinn, W., Teredesai, A., Voncken, P., and Walk, R.A., Comparison of fresh and room-aged cigarette sidestream smoke in a subchronic inhalation study on rats, *Toxicol. Sci.*, 41, 100, 1998.

24. Guerin, M.R., Jenkins, R.A., and Tomkins, B.A., *The Chemistry of Environmental Tobacco Smoke. Indoor Air Research Series*, Eisenberg, M., Ed., Lewis Publishers, Boca Raton, FL, 1992.

25. Adams, J.D., O'Mara-Adams, K.J., and Hoffmann, D., Toxic and carcinogenic agents in undiluted mainstream smoke and sidestream smoke of different types of cigarettes, *Carcinogenesis*, 8, 729, 1987.

26. Mohtashamipur, E., Mohtashamipur, A., Germann, P.G., Ernst, H., Norpoth, K., and Mohr, U., Comparative carcinogenicity of cigarette mainstream and sidestream smoke condensates on the mouse skin, *J. Cancer Res. Clin. Oncol.*, 116, 604, 1990.

27. Smith, C.J., Sears, S.B., Walker, J.C., and DeLuca, P.O., Environmental tobacco smoke: current assessment and future directions, *Toxicol. Pathol.*, 20, 289, 1992.

28. Gori, G.B. and Luik, J.C., Passive Smoke: The EPA's Betrayal of Science and Policy, The Fraser Institute, Vancouver, BC, 1999.

29. Rosenbaum, W.L., Sterling, T.D., and Weinkam, J.J., Linear extrapolation models of lung cancer risk associated with exposure to environmental tobacco smoke, *Regul. Toxicol. Pharmacol.*, 28, 106, 1998.

30. Henry, C.J. and Kouri, R.E., Chronic Exposure of Mice to Cigarette Smoke. Final Report, The Council for Tobacco Research — USA, Inc. Contract CTR-030, "Smoke Inhalation in Mice," Field, Rich & Associates, New York, 1984.

31. Henry, C.J. and Kouri, R.E., Chronic inhalation studies in mice. II. Effects of long-term exposure to 2R1 cigarette smoke on (C57BL/Cum × C3H/AnfCum)F1 mice, *J. Natl. Cancer Inst.*, 77, 203, 1986.

32. Shimkin, M.B. and Stoner, G.D., Lung tumors in mice: application to carcinogenesis bioassay, *Adv. Cancer Res.*, 21, 1, 1975.
33. Stoner, G.D. and Shimkin, M.B., Lung tumors in strain A mice as a bioassay for carcinogenicity, in *Handbook of Carcinogen Testing*, Milman, H. and Weisburger, E.K., Eds., Noyes Publications, Park Ridge, NJ, 1985, 179.
34. Thun, M.J., Lally, C.A., Flannery, J.T., Calle, E.E., Flanders, W.D., and Heath, C.W., Jr., Cigarette smoking and changes in the histopathology of lung cancer, *J. Natl. Cancer Inst.*, 89, 1580, 1997.
35. Herman, D.L.C.M., Distribution of primary lung carcinomas in relation to time as determined by histochemical techniques, *J. Natl. Cancer Inst.*, 27, 1227, 1961.
36. Hoffmann, D., Hoffmann, I., and Wynder, E.L., Lung cancer and the changing cigarette, *IARC Sci. Publ.*, 105, 449, 1991.
37. Wynder, E.L. and Hoffmann, D., Re: Cigarette smoking and the histopathology of lung cancer, *J. Natl. Cancer Inst.*, 90, 1486, 1998.
38. Djordjevic, M.V., Hoffmann, D., and Hoffmann, I., Nicotine regulates smoking pattern, *Prev. Med.*, 26, 435, 1997.
39. Malkinson, A.M., Primary lung tumors in mice: an experimentally manipulable model of human adenocarcinoma, *Cancer Res.*, 52, 2670s, 1992.
40. Malkinson, A.M., Molecular comparison of human and mouse pulmonary adeno-carcinomas, *Exp. Lung Res.*, 24, 541, 1998.
41. Herzog, C.R., Lubet, R.A., and You, M., Genetic alterations in mouse lung tumors: implications for cancer chemoprevention, *J. Cell. Biochem. Suppl.*, 28–29, 49, 1997.
42. Hecht, S.S., Approaches to chemoprevention of lung cancer based on carcinogens in tobacco smoke, *Environ. Health Perspect.*, 105(Suppl. 4), 955, 1997.
43. Stoner, G.D., Morse, M.A., and Kelloff, G.J., Perspectives in cancer chemoprevention, *Environ. Health Perspect.*, 105(Suppl. 4), 945, 1997.
44. You, M. and Bergman, G., Preclinical and clinical models of lung cancer chemo-prevention, *Hematol. Oncol. Clin. North Am.*, 12, 1037, 1998.
45. Witschi, H.P., Enhancement of tumor formation in mouse lung by dietary butylated hydroxytoluene, *Toxicology*, 21, 95, 1981.
46. Witschi, H.P., Oreffo, V.I.C., and Pinkerton, K.E., Six month exposure of strain A/J mice to cigarette sidestream smoke: cell kinetics and lung tumor data, *Fundam. Appl. Toxicol.*, 26, 32, 1995.
47. You, M., Candrian, U., Maronpot, R.R., Stoner, G.D., and Anderson, M.W., Activation of the Ki-ras protooncogene in spontaneously occurring and chemically induced lung tumors of the strain A mouse, *Proc. Natl. Acad. Sci. U.S.A.*, 86, 3070, 1989.
48. Belinsky, S.A., Devereux, T.R., Maronpot, R.R., Stoner, G.D., and Anderson, M.W., Relationship between the formation of promutagenic adducts and the activation of the K-ras protooncogene in lung tumors from A/J mice treated with nitrosamines, *Cancer Res.*, 49, 5305, 1989.
49. Ronai, Z.A., Gradia, S., Peterson, L.A., and Hecht, S.S., G to A transitions and G to T transversions in codon 12 of the Ki-ras oncogene isolated from mouse lung tumors induced by 4-(methylnitrosamino)-1-(3-pyridyl)-1-butanone (NNK) and related DNA methylating and pyridyloxobutylating agents, *Carcinogenesis*, 14, 2419, 1993.
50. Nuzum, E.O., Malkinson, A.M., and Beer, D.G., Specific Ki-ras codon 61 mutations may determine the development of urethan-induced mouse lung adenomas or adeno-carcinomas, *Mol. Carcinog.*, 3, 287, 1990.
51. Witschi, H.P., Espiritu, I., Peake, J.L., Wu, K., Maronpot, R.R., and Pinkerton, K.E., The carcinogenicity of environmental tobacco smoke, *Carcinogenesis*, 18, 575, 1997b.

52. Witschi, H.P., Espiritu, I., Maronpot, R.R., Pinkerton, K.E., and Jones, A.D., The carcinogenic potential of the gas phase of environmental tobacco smoke, *Carcinogenesis*, 18, 2035, 1997c.

53. Witschi, H., Espiritu, I., and Uyeminami, D., Chemoprevention of tobacco smoke-induced lung tumors in A/J strain mice with dietary myo-inositol and dexamethasone, *Carcinogenesis*, 20, 1375, 1999.

54. Lindenschmidt, R.C., Tryka, A.F., and Witschi, H.P., Inhibition of mouse lung tumor development by hyperoxia, *Cancer Res.*, 46, 1994, 1986.

55. Witschi, H., Breider, M.A., and Schuller, H.M., Modulation of N-nitrosodimethylamine induced hamster lung tumors by nitrogen dioxide, *Inhal. Toxicol.*, 4, 33, 1992.

56. Finch, G.L., Nikula, K.J., Belinsky, S.A., Barr, E.B., Stoner, G.D., and Lechner, J.F., Failure of cigarette smoke to induce or promote lung cancer in the A/J mouse, *Cancer Lett.*, 99, 161, 1996.

57. Gori, G.B., Low-risk cigarettes: a prescription, *Science*, 194, 1243, 1976.

58. Wynder, E.L. and Hoffmann, D., *Tobacco and Tobacco Smoke*, Academic Press, New York, 1967, 623.

59. Leuchtenberger, C. and Leuchtenberger, R., Differential response of Snell's and C57 black mice to chronic inhalation of cigarette smoke. Pulmonary carcinogenesis and vascular alterations in lung and heart, *Oncology*, 29, 122, 1974.

60. Leuchtenberger, C. and Leuchtenberger, R., Effects of chronic inhalation of whole fresh cigarette smoke and of its gas phase on pulmonary tumorigenesis in Snell's mice. in *Morphology of Experimental Respiratory Carcinogenesis. Proceedings of a Biology Division, Oak Ridge National Laboratory, conference held in Gatlinburg, Tennessee, May 13–16, 1970*, Nettesheim, P., Hanna, M.G., Jr., and Deatherage, J.W., Jr., Eds., U.S. Atomic Energy Commission, Division of Technical Information, Washington D.C., 1970, 329.

61. Leuchtenberger, C. and Leuchtenberger, R., Einfluss von frischem Zigarettenrauch auf die Entwicklung von Lungentumoren und auf Lungenkulturen bei der Snell-Maus, *Schweiz. Med. Wochenschr.*, 101, 1374, 1971.

62. Witschi, H., Espiritu, I., Yu, M., and Willits, N.H., The effects of phenethyl isothiocyanate, N-acetylcysteine and green tea on tobacco smoke-induced lung tumors in strain A/J mice, *Carcinogenesis*, 19, 1789, 1998.

63. American Thoracic Society, Cigarette smoking and health, *Am. J. Respir. Crit. Care Med.*, 153, 861, 1996.

64. Tang, J.L., Morris, J.K., Wald, N.J., Hole, D., Shipley, M., and Tunstall-Pedoe, H., Mortality in relation to tar yield of cigarettes: a prospective study of four cohorts, *Br. Med. J.*, 311, 1530, 1995.

65. Wattenberg, L., Chemoprevention of cancer, *Cancer Res.*, 45, 1, 1985.

66. Wattenberg, L.W., Prevention — therapy — basic science and the resolution of the cancer problem: presidential address, *Cancer Res.*, 53, 5890, 1993.

67. Benner, S.E., Lippman, S.M., and Hong, W.K., Chemoprevention of lung cancer, *Chest*, 107, 316S, 1995.

68. Karp, D.D., Lung cancer chemoprevention and management of carcinoma in situ, *Semin. Oncol.*, 24, 402, 1997.

69. van Zandwijk, N., N-acetylcysteine for lung cancer prevention, *Chest*, 107, 1437, 1995.

70. Hecht, S.S., Chemoprevention by isothiocyanates, *J. Cell. Biochem. Suppl.*, 22, 195, 1995.

71. Duperron, C. and Castonguay, A., Chemopreventive efficacies of aspirin and sulindac against lung tumorigenesis in A/J mice, *Carcinogenesis*, 18, 1001, 1997.

72. De Flora, S., Cesarone, C.F., Balansky, R.M., Albini, A., D'Agostini, F., Bennicelli, C., Bagnasco, M., Camoirano, A., Scatolini, L., and Rovida, A., Chemopreventive properties and mechanisms of N-acetylcysteine. The experimental background, *J. Cell. Biochem. Suppl.*, 22, 33, 1995.

73. el-Bayoumy, K., Upadhyaya, P., Desai, D.H., Amin, S., Hoffmann, D., and Wynder, E.L., Effects of 1,4-phenylenebis(methylene)selenocyanate, phenethyl isothiocyanate, indole-3-carbinol, and d-limonene individually and in combination on the tumorigenicity of the tobacco-specific nitrosamine 4-(methylnitrosamino)-1-(3-pyridyl)-1-butanone in A/J mouse lung, *Anticancer Res.*, 16, 2709, 1996.

74. Wattenberg, L.W., Inhibition of carcinogenic effects of polycyclic hydrocarbons by benzyl isothiocyanate and related compounds, *J. Natl. Cancer Inst.*, 58, 395, 1977.

75. Estensen, R.D. and Wattenberg, L.W., Studies of chemopreventive effects of myo-inositol on benzo[a]pyrene-induced neoplasia of the lung and forestomach of female A/J mice, *Carcinogenesis*, 14, 1975, 1993.

76. Wattenberg, L.W. and Estensen, R.D., Chemopreventive effects of myo-inositol and dexamethasone on benzo[a]pyrene and 4-(methylnitrosoamino)-1-(3-pyridyl)-1-butanone-induced pulmonary carcinogenesis in female A/J mice, *Cancer Res.*, 56, 5132, 1996.

77. Witschi, H., Uyeminami, D., Moran, S., and Espiritu, I., Chemoprevention of tobacco smoke lung carcinogenesis in mice after cessation of smoke exposure, *Carcinogenesis*, 21, 977, 2000.

78. Wynder, E.L. and Stellman, S.D. Comparative epidemiology of tobacco-related cancers, *Cancer Res.*, 37, 4608, 1977.

79. Postmus, P.E., Epidemiology of Lung Cancer, in *Fishman's Pulmonary Diseases and Disorders*, Fishman, A.P., Elias, J.A., Fishman, J.A., Grippi, M.A., Kaiser, L.R., and Senior, R.M., Eds., McGraw-Hill, New York, 1998, 1707.

80. Pinkerton, K.E., Peake, J.L., Espiritu, I., Goldsmith, M., and Witschi, H.P., Quantitative histology and cytochrome P450 immunocytochemistry of the lung parenchyma following six months exposure of strain A/J mice to cigarette sidestream smoke, *Inhal. Toxicol.*, 8, 927, 1996.

81. Witschi, H., Espiritu, I., Pinkerton, K.E., Murphy, K., and Maronpot, R., Ozone carcinogenesis revisited, *Toxicol. Sci.*, 52, 162, 1999.

82. Witschi, H.P., Tobacco smoke as a mouse lung carcinogen, *Exp. Lung Res.*, 24, 385, 1998.

83. Witschi, H.P. and Rajini, P., Cell kinetics in the respiratory tract of hamsters exposed to cigarette sidestream smoke, *Inhal. Toxicol.*, 6, 321, 1994.

9 Reducing Children's Exposure to Environmental Tobacco Smoke: A Review and Recommendations

Melbourne F. Hovell, Dennis R. Wahlgren, Joy M. Zakarian, and Georg E. Matt

CONTENTS

0-8493-0311-7/00/$0.00+$.50
© 2001 by CRC Press LLC

ABSTRACT

This chapter summarizes issues and empirical evidence for the reduction of children's residential environmental tobacco smoke (ETS) exposure. Interventions designed for residential/child ETS exposure control have included policy, minimal clinical advice, and counseling services. Divorce court and adoption services have limited custody to protect children from ETS exposure. Controlled trials of clinicians' one-time counseling services have generally shown null results. One-time minimal interventions appear ineffective, but large-scale studies may be warranted to determine possible minimal effects from which substantive public health benefits might be derived. Four trials found that repeated counseling/shaping procedures based on Social Learning Theory reduced children's ETS exposure. Insufficient numbers of controlled studies of repeated session counseling procedures have been completed to determine efficacy for ETS exposure reduction, but evidence is promising and ongoing trials will add to the empirical evidence in the next few years. Measurement issues and their use for both assessment and intervention purposes are discussed. Suggestions for future research concerning ETS exposure and means of reducing exposure are provided.

9.1   INTRODUCTION

Cigarette smoking is a leading cause of morbidity/mortality,[1] and environmental tobacco smoke (ETS) exposure is associated with lung cancer.[2] ETS has been classified as a carcinogen[3] and is linked to pulmonary and ear diseases and sudden death in infants.[3–9] ETS exposure has been estimated as the third leading preventable cause of death.[10]

The public health consequences of ETS exposure may be enormous. The Centers for Disease Control (CDC) reported the prevalence of children's ETS exposure in the home ranging from 11.7 to 34.2% of children across states, based upon numbers of homes with an adult smoker where smoking was allowed in some or all areas.[11] Most smokers are of child-rearing age, leading to as many as 50% of children exposed in their homes.[8,12–17] National Health and Nutrition Examination Survey (NHANES) data indicated that 43% of U.S. children lived in a home with at least one smoker.[15] Huss and associates[18] found that 56% of families with an asthmatic child included a smoker. Prevalence estimates vary, yet suggest residential exposure to a substantial number of children.

These rates may be underestimates due to measurement error.[19] For instance, Repace[20] and Löfroth[21] have suggested that true protection for ETS exposure is not likely unless smoking takes place exclusively outside the residence or unless the smokers in the home quit. Thus, children living with a smoker who smokes in the home may insure exposure, even if the child is not present while smoking takes place. The true prevalence rate of ETS exposure in residences may be closer to 50%, given these qualifications. The true morbidity/mortality burden attributable to ETS exposure may be greater than currently recognized, as well. This is especially likely if growing up exposed to ETS increases a child's risk of becoming a smoker as an adolescent. Finally, to our knowledge, there is no benefit to the child from ETS exposure. Thus, the cost-to-benefit ratio likely for ETS exposure consequences is weighted strongly toward the cost side. While additional research is needed to illuminate the exact risks, evidence to date supports policies and interventions for control of ETS exposure, including control within private residences. This chapter summarizes interventions to reduce children's residential ETS exposure and provides recommendations for future research.

9.2   REDUCING CHILDREN'S ETS EXPOSURE

The World Health Organization (WHO) recommended legislation and education to protect children from ETS exposure,[8] but few studies have demonstrated efficacious interventions. Obviously, getting smokers to quit smoking should protect children from ETS exposure.[22] However, Wahlgren et al. found that 67% of parents were unable to quit or reduce their asthmatic child's ETS exposure after physician advice.[23] Other studies have shown limited cessation rates, and most smokers who quit restart smoking.[24–26] Thus, more effective technology for effecting smoking cessation and/or interventions aimed specifically at reducing ETS exposure independent of smoking cessation are needed to achieve adequate child protection. Indeed, development of new technology directed to ETS exposure control may contribute to cessation interventions.

9.2.1   THEORY

Borland reviewed theoretical models underpinning health behavior interventions, including those for reducing ETS exposure.[22] He concluded that Social Learning Theory (SLT) provides the best model for understanding ETS exposure practices and for informing means of changing them. This model emphasizes reinforcement and modeling, where behavior is promoted by seeing models and sustained by reinforcing consequences. We have extended this model to a Behavioral Ecological Model that emphasizes cultural contingencies of reinforcement.[27,28]

The Behavioral Ecological Model is based on SLT and cultural anthropology by Glenn and others.[29] The model presumes that tobacco smoking is reinforced by almost immediate relief and pleasure that is biologically mediated by the addictive properties of nicotine. To these consequences can be added complex social consequences. Often, these are reinforcing. This can be seen when a stranger asks for a light and initiates a new conversation, possibly making a new friend. However, the social consequences are no longer exclusively reinforcing. Smokers are shunned by non-smokers and may be formally criticized or asked to leave or put out a cigarette. These aversive consequences tend to be less reliable than the many social reinforcers, mostly because the smoker seeks reinforcing social networks and the larger community has not developed a sufficiently well-distributed cultural practice of shunning cigarette smoking to reliably criticize smoking in all instances. Still, the growing counter-control culture is making a difference. With states and communities passing regulations for taxing cigarette sales and with new funds going to anti-tobacco media and research programs,[30] the power of the counter-control contingencies is increasing. The most recent settlement reached by the attorneys general is expected to yield billions of dollars, some of which will be directed toward tobacco control. Collectively, these efforts can be viewed as cultural contingencies aimed at reducing cigarette smoking, and early evidence suggests efficacy for reducing the prevalence of adult smoking rates.[31]

These efforts also have emphasized that youth not start smoking, often with messages of disease or social unpopularity.[32] The prevention effort remains to be demonstrated as efficacious, with some programs showing little effect and others showing promising effects (for examples of the latter, details and results from Florida's,[33] California's,[34] and Massachusetts'[35] programs are currently available online). The effort to get smokers to quit has emphasized aversive control, with social penalties, tax-based financial costs, and restrictions on where smoking can take place in public. The emphasis on aversive control tends to suppress behavior as long as the aversive contingencies are in place. However, aversive control does not work for conditions not under aversive contingencies, such as one's home, or when the aversive contingencies have been removed, such as visiting a country with more liberal smoking policies. In short, the anti-tobacco program has not provided reinforcing consequences for alternatives to smoking.

Tobacco control programs have focused on ETS exposure only as a minor part of the overall epidemic. ETS has been addressed as an occupational risk and has resulted in restrictions in public buildings, airplanes, and other public transportation vehicles.[36] Again, this has taken an aversive control approach, with penalties, real or implied, for smoking in restricted areas. Much less emphasis has been directed to protecting children or adults living in private homes with a smoker. The potential for doing so with much less emphasis on aversive control is quite high, but not consistent with cultural norms. Early policy-based interventions tend to emphasize aversive control, while early clinical interventions have been more positive.

Use of SLT and our Behavioral Ecological Model can be directed to ETS exposure control in residences with considerable confidence for success. Smoking around children can be broken down into component behaviors that can be targeted for change, where new "protective" practices can be reinforced instead of older smoking practices punished. These approaches can include smoking outside, smoking only when the child is not present, smoking fewer cigarettes, chewing gum instead of smoking, or chewing nicotine gum or using the patches designed as part of nicotine replacement therapy to reduce dependence on cigarettes as a source of nicotine. The model asserts levels of environmental control. At the individual level, biological feedback systems (i.e., withdrawal symptoms) prompt smoking and doing so provides relief and pleasure. These reinforcing consequences sustain smoking, including smoking around children. However, as family, friends, or others criticize smoking and ETS exposure or encourage alternative behavior (e.g., smoking outdoors), the rate, timing, and context of smoking can change. Social contingencies can compete with biological contingencies.

Influence can come from social agencies more distant than family and friends. These include clinicians, the broader society, and the media. These sources of influence range from minimal to intensive in nature. Physician advice might be considered minimal, and ongoing counseling might be considered more intensive intervention. Repeated counseling services offer the opportunity to employ SLT-based procedures such as individualized shaping techniques. These involve identifying the unique behavior patterns to be changed and unique targets to be acquired for a given individual (e.g., smoking parent) and the use of individualized prompting and reinforcing consequences to gradually teach new behaviors and extinguish others. Smoking only in one or two rooms of the house may be reinforced, then in only one room, and then only outside. Over time, the complexity of the final behavior to be established is increased. Likewise, the child may be trained to avoid the parent while she/he is smoking or to provide prompts to the parent to "recall their agreement" to smoke elsewhere, in an increasing number of contexts. In this process, it is critical for reinforcing consequences or feedback to be delivered as immediately after the behavior as possible. It is equally important that consequences that are truly reinforcing be employed. Visiting with family members might be reinforcing for some people and relatively aversive for others. Thus, the target behavior, reinforcing consequences, and sequence of successive changes must be selected as unique to the individual/family to maximize shaping effects. Use of these procedures should increase the power and thereby the efficiency of clinical interventions directed to ETS exposure control.

Borland[22] has noted that clinical interventions are unlikely to be cost-effective for tobacco control. However, this judgment may be premature and may miss the larger principle. Tobacco control may require many available interventions simultaneously in order to reduce both smoking and ETS exposure to near zero levels, especially in light of the addictive properties and ongoing tobacco industry's effort to promote smoking. Cost effectiveness concerns should be postponed until efficacious interventions, including shaping procedures and multi-level community-wide interventions, have successfully offset the effects of the tobacco industry.

Interventions employed for ETS exposure control have been policy-based restrictions in public buildings, legal sanctions for residential/child exposure, media/educational influences, and clinical services. Almost all of these interventions have evolved based on limited use of learning theories. None has yet been perfected or applied in an extensive manner throughout a community. Studies of these interventions have been few and usually limited to surveys. Future intervention development and research should be based on learning theory, including integration of cultural contingencies and combined interventions.

9.2.2 POLICY AND REGULATIONS OF ETS IN PUBLIC BUILDINGS

Ordinances that ban smoking in public buildings are effective in reducing ETS in controlled environments, may influence ETS exposure in residences, and may serve as models for residential policies. Thompson and colleagues[37] surveyed over 20,000 adults and found that ETS exposure was related to the degree of regulation at work. Moskowitz and co-workers found that more restrictive community ordinances were associated with less ETS exposure.[38] Eisner and colleagues studied bartenders before and after smoking prohibition and found ETS exposure was reduced from 28 to 2 h/week.[39] Respiratory symptoms decreased and pulmonary function increased after the policy was in effect, suggesting improved health for bartenders. Based on particle size and density, Repace[20] has computed estimates of ETS exposure under varying degrees of ventilation and room size. He found that smoke particles were distributed throughout a building and concluded that the most reliable means of reducing ETS exposure is to ban smoking indoors.[20] Recently, Löfroth[21] demonstrated that the dispersion of ETS through a home is likely to expose residents, including children, even if they are relatively distant from the smoker at the time of smoking. Their conclusions support those from Repace: to control ETS exposure, smoking should be restricted to outdoors. Winkelstein and co-authors[40] found that children's urine cotinine (a major metabolite of nicotine) was negatively associated only with smoking outside. This suggests that a ban (or other means of decreasing ETS) on all in-residence smoking may be required to decrease children's ETS exposure.

9.2.3 AIRFLOW, ETS MARKERS, AND EXPOSURE COMPLEXITIES

However, these outcomes have been based on airflow and particle dosimeters. Other research suggests that nicotine specifically does not disperse to the degree that small smoke particles might. Both chamber studies[41–43] and indoor air measurements[21] suggest that nicotine disappears from air faster than other ETS constituents, and hence, its use as a marker may underestimate the relative concentrations of other constituents.

These studies also suggest that nicotine can be deposited on surfaces and may be "off-gassed" even when no smoking is taking place. This means that nicotine exposure is possible from physical contact with contaminated surfaces or by breathing air from a previously contaminated room. Recent work by Apte et al.[44,45] has shown that the indoor behavior of nicotine is so complex that it may not be a good representation of ETS particles. For instance, nicotine is much less mobile than other ETS compounds and is not distributed as much or as quickly into adjacent rooms. Moreover, walls and surfaces soak up (i.e., sorb) nicotine, but not particles, until the surface is saturated. When the nicotine air concentration is low, the nicotine absorbed in the wall is slowly released into the air. This means nicotine can be released into the air from smoking that took place much earlier. Thus, researchers may detect nicotine/cotinine in the air, blood, or urine even though no indoor smoking has taken place recently. Similarly, researchers may not detect nicotine/cotinine in the air, blood, or urine even though a person was exposed to other ETS compounds. This is due to nicotine's tendency to stay within the room in which the smoking took place, whereas other ETS compounds travel to adjacent rooms more freely.

These studies illustrate the complexities involved in both measuring and intervening to control ETS exposure. These studies lead to questions about the degree to which distance from a smoker can or cannot protect a non-smoker from various components of ETS. Early studies suggest that distance from a smoker may protect a child from nicotine, but less so from small particle and other toxic agents. Early studies also suggest that nicotine exposure may take place simply by spending time in or touching contaminated surfaces of a room where smoking has taken place. These complexities imply different types of risks for infants and children potentially exposed in their home. Infants, for example, may be more likely to contact surfaces such as floors and furniture and infants may be more likely to ingest nicotine from surfaces, given their hand to mouth practices. Older children may be less at risk for some exposures, such as that from hand to mouth, but they are likely to be exposed to dispersed particles (and related toxins) even if they are always at a distance in their own home from the place of smoking.

Additional research is needed to illuminate these issues more fully. In the meantime, to avoid ETS exposure the smoker must quit or smoke exclusively out-doors and children or other special populations might be advised to frequent as little as possible settings where smoking has taken place. To our knowledge, no studies have formally set smoking outdoors as the ultimate standard to be achieved in order to protect children from residential ETS exposure.

9.2.4 BANS IN PUBLIC BUILDINGS MIGHT GENERALIZE TO RESIDENCES

Abernathy and associates used air dosimeters in public places for one week prior to and following a ban on smoking.[46] A 67% decrease in nicotine levels was observed. They suggested that bans for public buildings might decrease ETS exposure in homes (i.e., generalization). This possibility is supported by data showing that bans in pubic settings are associated with fewer adults remaining smokers and lower levels of smoking.[47] This should reduce ETS exposure in residences. A recent survey of over 48,000 adults suggested that smokers who lived or worked in a home or worksite with a ban on smoking were more likely to quit and were more likely to be light smokers, respectively.[48] Odds ratios for quitting were larger for those with a ban on smoking in their residences compared to those with a workplace ban. This was significant for families with only one smoker, but not so for families with two or more smokers. Again, Winkelstein and co-authors[40] found that children's urine cotinine was negatively associated only with smoking outside. Borland and colleagues assessed the effects of ordinances and educational/media campaigns on smoking practices of six different samples of 2500 adults from 1989 through 1997.[49] Results showed an increase in the proportion (14 to 33%) of respondents who did not smoke near children and an increase in the proportion (20 to 28%) who smoked outdoors. Adults who lived with non-smokers, with children in the house, and whose worksite banned cigarette smoking were most likely to restrict indoor smoking. This study implies that public building ordinances and media may influence residential smoking prac-tices. However, these results also suggest that the majority of smokers did not reduce children's ETS exposure as a result of these interventions.

While this study demonstrates that public ordinances are unlikely to reduce ETS exposure in a majority of children, it provides a very important model. The fact that results were attenuated is not surprising. The nature of the interventions is weak and diluted across a society. For this evaluation to have found any rela-tionships is quite remarkable. Until the level of media and public bans for smoking have become so prevalent as to be virtually ubiquitous, the true cumulative effects of these programs may not be visible. They also may be important backgrounds to clinical interventions, with the possibility that community-wide bans and media could synergistically increase the effects of clinical programs, and vice versa. Borland and colleagues[49] provide the evidence needed to justify far more aggressive bans for public settings and media campaigns, continuously and community-wide. Such expansion of community-wide controls is consistent with our Behavioral Ecological Model and may provide the general background needed to support tobacco control in general and ETS exposure control in particular.

The surveys that have focused on bans in public buildings and considered possible effects on residential exposure suggest that public bans may generalize to reduced smoking in the home or to cessation, thereby reducing ETS exposure in homes. However, this effect has not been demonstrated definitively, and it appears limited to homes that include only one smoker.

9.2.5   POLICING RESIDENTIAL BANS

These surveys also suggest that a ban on all in-residence smoking may be required to decrease children's ETS exposure. However, banning home exposure is not the same as banning smoking in public buildings. The increasing reference in the literature to residential bans or policies against smoking in the home is a direct analogy to public policies. However, this analogy is misleading. Residential "policies" are formal statements, signs, and social instructions from a parent/adult to others not to smoke or to stop when in the home. These actions are not different than any other efforts an individual might take to influence others' smoking in the home. They are not equivalent to community laws and community ordinances that obligate police and employers to take (usually aversive control) actions to enforce public policies. Since the use of public policing actions is supported by whole systems of legislation, court actions (e.g., fines on businesses), and police, the analogy is all the more concerning. If community ordinances are created to ban smoking in homes (or around children), how will the community enforce such a ban? How will parents enforce residential policies that may apply to family members? What types of consequences will be forthcoming for compliance or non-compliance?

Community ordinances banning smoking in homes are summarized by Ezra.[50] He points out that competition exists between individuals' rights to privacy and the community's obligation to protect children's health. Although no laws have been enacted to restrict children's ETS exposure in homes, legal sanctions have been employed in special cases. Ezra[50] and Ashley and Ferrence[51] point out that ETS exposure has limited child custody/visitation privileges in divorce cases and disqualified adoption applicants. These cases set precedence and may be forerunners of community ordinances to protect children in all homes.

Policies restricting visitation rights if parents expose their children to ETS have face validity. However, no research has verified that restrictions actually result in lower ETS exposure. Parents restricted from smoking around their children as a condition of custody might still smoke when their child is present, especially when in private. Enforcing court orders is difficult, and exposure will depend on degree of enforcement. Effective enforcement may require development of real-time objective measures as well as ethical considerations about penalties. This is especially important since the former spouse may falsely accuse the smoker of exposing the child to ETS for personal reasons. Even true exposure that results in loss of visitation rights may result in more social damage than health damage likely from ETS exposure. If ETS exposure bans are extended to private residences, these trade-offs need to be considered; empirical evidence of efficacy and cost/benefits are needed.

Sanctions to control ETS exposure in residential settings could have unintended consequences, including counteraggression from smokers. Ezra cites a case in which a smoker was asked to not smoke in a restaurant and returned and killed the non-smoker with a shotgun.[50] Research is needed to determine the social factors that cause a parent to set anti-smoking policies and that determine his/her enforcement practices and the reinforcing consequences that sustain them. It is equally important to study the possible reactions of family and friends who are the focus of possibly aversive means of enforcing a residential ban on smoking in order to determine the social costs (if any), such as family strife. Finally, it cannot be assumed that no-smoking policies in the home are enforced successfully, and future research needs to determine the degree to which such actions by parents result in lower ETS exposure.

Future research also should target positive means of reducing exposure, especially since smokers themselves are victims of the tobacco industry. Sweda[52] suggests that policy restrictions in residences will be necessary. If so, research is needed to define efficacious and acceptable policies and means by which communities can enforce residential policies that do not infringe on traditional constitutional rights, that involve reinforcement for alternative practices instead of exclusively aversive control of exposure practices, and that provide adequate clinical services to smokers to assist them in quitting.

9.2.6 CLINICAL SERVICES FOR ETS EXPOSURE REDUCTION

A number of studies have tested clinical/educational interventions for ETS exposure reduction among children. These have ranged from brief interventions to repeated counseling over months. Table 9.1 shows a listing of recent clinical interventions designed to reduce ETS exposure to children in private residences. These have been reviewed with regard to methodological rigor and the degree to which their interventions have been consistent with the SLT/Behavioral Ecological Model.

Woodward and associates provided women with a pamphlet that instructed them to protect their infants from smoke exposure and/or quit smoking.[53] The pamphlets were supplemented with a physician follow-up telephone call 1 month later. Only four women quit smoking, and there were no differences among cohorts/groups. About 50% of the control and experimental groups reported not smoking in the same room with the child. Again, differences between groups were not significant. No differences were seen in child cotinine assays, either. Chilmonezyk and colleagues assigned mother/child to groups at random.[54] Physicians phoned mothers to report their baby's cotinine level and advised them to avoid exposing their baby to tobacco smoke. No significant differences were obtained in cotinine measures at post-test. Vineis and co-investigators studied infants receiving vaccinations by nurses who counseled parents for 15 min and provided booklets concerning ETS exposure and accident prevention.[55] Again, this study reported null outcomes.

Murray and Morrison[56] conducted surveys of two cohorts of parents. Compared to the first cohort, they found a significantly lower rate of smoking in the room with an asthmatic child and fewer symptoms from the asthmatic child in a second cohort. They suggested this effect was due to physician advice, but numerous confounding variables could explain the observed association, including secular changes in tobacco acceptability in the community. Most importantly, the study was not designed as an experiment, and the presumed advice from physicians was not

TABLE 9.1
Clinical Interventions for ETS Control in Residences

Author and date	Sample	Design	Intervention	Primary measures	Results
Woodward et al. 1987[53]	184 women who smoked during pregnancy	Non-random 3 cohorts (N = 61–62) 1 received tobacco counseling; others served as controls 3 month follow-up	Pamphlet on tobacco control and advice to not smoke near newborn infants; telephone follow-up at 1 month	Smoking rates, smoking in room with infant. Infant's cotinine	"Low" exposure Tx = 61% Ctrl 1 = 50% Ctrl 2 = 51% (not signif.) Cotinine Tx: Mdn = 10 µg/l Ctrl: Mdn = 11 µg/l (not signif.)
Chilmonczyk et al. 1992[54]	103 smoking moms attending well-baby pediatric visits	Randomized Counseling (N = 52) Control (N = 51) 2-month follow-up	M.D. telephoned to report child's urine cotinine and sent a letter with recommendations for ETS exposure control; did not stress cessation	Infant's cotinine, reported as a ratios of post-test to pre-test values × 100	Mean pre-post cotinine ratio: Tx = 2.05 Ctrl = 2.17 (not signif.)
Vineis et al. 1993[55]	Families attending 3-month well-baby vaccination visits	Non-random Tx = Cohort 1 (N = 402) Ctrl = Cohort 2 (N = 613) 2-year follow-up	15 min counseling by nurse plus booklets on accident prevention and health effects of smoking and ETS	Mailed questionnaires emphasizing cessation	No signif. differences in cessation
Murray and Morrison 1993[56]	Families with an asthmatic child and a smoking parent	Non-random, sequential cohorts (prior to and after implementation of Tx) Ctrl = Cohort 1 (N = 415) Tx = Cohort 2 (N = 392)	M.D. advice to parents to not expose child to ETS Implemented for all parents of asthmatic children patients since July 1986	Parents' smoking rate; # cigs smoked near child	No effect on smoking rate Significant reduction in number smoked near child (mean cigs/day): Mothers: Ctrl = 7 Tx = 3 (p = .005) Fathers: Ctrl = 5 Tx = 2 (p = .001)

TABLE 9.1 (*continued*)
Clinical Interventions for ETS Control in Residences

Author and date	Sample	Design	Intervention	Primary measures	Results
Meltzer et al. 1993[67]	5 families with an asthmatic child obtaining specialty medical care	Multiple baseline quasi-experimental design	5 counseling sessions over 4 weeks; ≈30 min each Behavior modification procedures to shape ETS exposure reduction	Proportion of each cig smoked in child's presence rated (0 = absent, 4 = whole cigs) and summed across cigs for daily score	Each child's ETS exposure decreased by 40–80% 4/5 maintained by 1-month follow-up
McIntosh et al. 1994[58]	92 families with an asthmatic child	Random assignment "Usual care" control: N = 44 Tx: N = 48	Both groups received M.D. information on effects of ETS, advice to quit or smoke outside. Tx added cotinine feedback letter from M.D. 1 month later and self-help manual	Cessation of smoking inside the home Child's cotinine Post measures 4–6 months after baseline questionnaire	Tx (35%) vs. Ctrl (17%) smoked outside ($p = .10$) Cotinine levels not reported by group but reported as non-significant
Hovell et al. 1994[69]	91 families with an asthmatic child exposed to at least 1 cig/day by a parent	Randomly assigned No Tx control Monitoring control Counseling Interviews at baseline, 1 and 2 months.; follow-up at 6, 9, and 12 months	6 counseling sessions (≈30 min each, using behavior modification shaping procedures) 5 sessions over 2 months, final session at 6 months	ETS exposure recalled for previous week (a cotinine-validated measure)	Decrease in reported ETS exposure ($p < .001$) Counseling: 79% Monitoring: 34% Control: 42%

TABLE 9.1 (*continued*)
Clinical Interventions for ETS Control in Residences

Author and date	Sample	Design	Intervention	Primary measures	Results
Greenberg et al. 1994[68]	Mothers and newborn infants from 933 families (583 had complete data at follow-up)	Solomon 4-group. However, analyses emphasized only Ctrl vs. Tx, stratified by mother's smoking status. BL at 18 days of child age; follow-up at 7 and 12 months.	4 home counseling sessions (45 min each) by RN over about 6 months; shaped ETS exposure changes	1 week recall of ETS (cigs/day) measured as child in the same room with smoker; urine cotinine	Mean cigs/day among smoking mothers: Tx: 38% reduction Ctrl: 8% increase (*p* <.01) Only 296 infants had complete urine data Both groups increased and did not differ from each other
Wall et al. 1995[60]	49 offices, 128 pediatricians, 2901 families	Offices randomly assigned Tx: N = 26 Ctrl: N = 23 Ctrl: advice and education Tx: same plus advice at 2, 4, and 6 months	Pediatrician letter to smoking mothers advising to quit and not to expose kids to ETS Oral/written advice given at well-baby visits at 2 weeks and 2, 4, and 6 months	Mailed questionnaires at BL and 6-month follow-up about smoking patterns Measured smoking cessation and relapse rates	Cessation rates: Ctrl: 2.7% Tx: 5.9% (*p* <.05) Sustained quits: Ctrl: 45% of quitters Tx: 55% of quitters (*p* <.01)
Eriksen et al. 1996[57]	443 families attending a well-child medical exam	Randomly assigned Tx: N = 221 Ctrl: N = 222 80 dropped out or did not complete measures 1-month follow-up questionnaire	Three brochures on health effects of ETS and how to reduce exposure, and smoking cessation at a well-baby visit, plus 5 min extra counseling on ETS exposure and cessation	Mailed questionnaire Indoor smoking, and % who smoked near child	No significant group differences Indoor cessation: Tx n=4 Ctrl n=4 No smoking near child: Tx: 34.5% Ctrl: 32.1%

TABLE 9.1 (*continued*)
Clinical Interventions for ETS Control in Residences

Author and date	Sample	Design	Intervention	Primary measures	Results
Wahlgren et al. 1997[73] (follow-up to Hovell et al. 1994[69])	91 families with an asthmatic child exposed to at least 1 cig/day	Randomly assigned No Tx control Monitoring control Counseling Interviews at baseline, 1 and 2 months; follow-up at 6, 9, and 12 months; extended follow-up through 30 months	Printed self-help ETS counseling advice plus debriefing at the conclusion of the initial trial (subsequent funding allowed for this extended follow-up)	# cigs to which child was exposed in the home	Maintenance of reduced ETS exposure in counseled group through 30 months (24 months post-Tx) Debriefing and self-help materials resulted in significantly reduced exposure among control groups: Monitoring 66% Usual care 25% Counseling no further change
Irvine et al. 1999[59]	73 M.D. offices, 501 asthmatic child/families	Self-administered questionnaire at BL and 1 year follow-up	Brief counseling on ETS and smoking cessation; plus leaflets mailed with a letter of M.D. recommendations. at 4 and 8 months	Smoking habits, child's overall ETS exposure, child's saliva cotinine	Mean decrease in cotinine not signif. different: Tx = 0.70 ng/ml Ctrl = 0.88 ng/ml Number who reportedly smoked less near child not signif. different: Tx = 28% Ctrl = 22%

Note: Abbreviations: cig = cigarette; Tx = treatment/treatment group; Ctrl = control group; M.D. = physician; RN = nurse; BL = baseline; signif. = significant; and Mdn = median.

measured explicitly in either cohort. Thus, the investigators did not have objective information that controls did not provide advice or that the "experimental" group had provided advice.

Eriksen et al. conducted a controlled trial of ETS exposure reduction for healthy children.[57] During well-child clinical visits, parents were counseled and provided brochures explaining ETS effects, how to protect children from ETS exposure, and how to quit smoking. No significant differences were found for smoking indoors or smoking while the child was present.

McIntosh and associates provided a more extensive, but minimal intervention for families with an asthmatic child.[58] Patients were assigned at random to conditions. Controls were provided with physician advice and a booklet about ETS exposure. The treatment group obtained the same plus 1 month later received cotinine feedback, a letter from the physician recommending parents not to expose their child to ETS, and a self-help manual. Significant group differences were found for "trying" to smoke outside, but differences for actually smoking outside did not reach significance. Irvine and associates conducted a similar study of asthmatic children.[59] Patients were assigned at random to control and experimental conditions. Parents were provided with education/counseling and leaflets regarding ETS exposure and health effects, exposure reduction, and smoking cessation. Leaflets were mailed with a letter at 4 and 8 months. At 1 year, no significant differences were found for child cotinine or parents smoking in the same room with the child.

Wall and co-investigators[60] conducted a controlled trial of newborns. Offices were assigned at random, and analyses were adjusted for cluster effects as well as other covariates. Over 2000 mothers completed the study. Control patients were provided with an ETS information packet and a letter from their physician advising tobacco control. The "extended" intervention group received the same packet along with 2 min of physician counseling and written materials at 2-, 4-, and 6-month well-baby visits. Physician training and quality assurance procedures were employed to maximize physician counseling. Although ETS exposure was not measured, results showed significantly higher rates of smoking cessation (2.7 vs. 5.9%) and sustained non-smoking status (45 vs. 55%) at 6 months in the extended treatment group. This study showed that under precise supervision minimal interventions carried out over repeated contacts can reduce smoking rates and by implication ETS exposure for newborns. The authors also noted that the presence of another smoker in the home (often a husband) was associated with failure to quit among smokers or with relapse among quitters. The presence of another smoker also raises doubt about the degree of ETS exposure that can be assumed from one smoker quitting. Thus, though this study demonstrated very strong methods, it remains only an implied effect for ETS exposure.

9.2.7 Minimal Interventions and SLT

All of the forgoing studies employed various physician or assistant administered counseling and "feedback" to parents to influence their children's exposure to ETS. From a practical standpoint, all investigated interventions could be employed easily in most medical facilities. However, all studies had important shortcomings, and only one found significant differences between control and experimental groups.

None of the studies employed procedures that followed the principles of behavior based on SLT or the Behavioral Ecological Model. All provided remarkably small amounts of "instruction" to parents about tobacco (i.e., limited dose interventions). This could prompt some change in smoking practices, especially for those concerned with newborns or asthmatic children. However, given the addictive properties of tobacco use, even small changes in smoking patterns (i.e., smoking away from an infant) would be unlikely to continue in the absence of some type of feedback or reinforcement. None of these studies provided feedback or physician encouragement soon enough to reinforce possible early changes due to initial advice. Three studies[54,58,59] provided physician follow-up contacts. Chilmonezyk and McIntosh provided parents with their child's cotinine feedback. However, this intervention was provided too late and too infrequently to serve as likely reinforcement for decreased ETS exposure practices. Worse, the use of such feedback based on one measure runs the very real risk that the cotinine level and parent smoking practices were not concordant. If the cotinine was low and this followed parent exposure practices that were no less than prior to intervention, it could encourage parents to continue ETS exposure or even increase it on the grounds that their child is not obtaining much nicotine. Similarly, some parents might have reduced their child's exposure substantially, but the cotinine level might imply no change or even an increase. This could lead parents to be very discouraged with their attempts to protect the child. They might react by discontinuing their tobacco control efforts. Biological assays are quite variable and have imperfect relationships with parent exposure practices as defined by smoking when the child is present. The use of biological feedback should not be based on one-time samples or without considering the relationship to reported ETS exposure practices. They could backfire.

Irvine et al.[59] provided repeated interventions and, in this sense, more closely approximated principles of behavior and shaping techniques of behavior change. However, here too it was a near miss. The physician follow-up contacts were not individualized and did not provide specific instructions to parents based on their recent history of change, if any. Moreover, waiting 4 months for the first follow-up is too long a delay from initial counseling and possible early changes in ETS exposure practices. In short, these studies were designed without consideration of behavior science theory or techniques for behavior change, and they were highly unlikely to result in significant changes, accordingly.

One[56] of the two studies that reported a significant exposure effect did so in the absence of a defined physician intervention or even evidence that it was provided to the experimental cohort and was not provided to the control cohort. Reported measures were employed and interviewers were not blind to conditions. Most importantly, for changes in asthma severity, there is a very real possibility that improvements were due to other asthma treatments (e.g., new medications) or to reduced triggers other than tobacco smoke. Thus, this study did not rule out many alternative explanations and can be viewed as an effect in search of a cause.

A few tentative conclusions seem warranted from these studies. The first is that minimal clinical interventions for ETS exposure reduction have not produced effects about which we can be confident. Most have failed to show reduction in smoking or ETS exposure. However, this may be a bit of an overgeneralization. Most of the studies were conducted with relatively small samples and may simply have not had the statistical power to find expected small effects. The relative success of the Wall et al. study,[60] with a large sample size, implies that minimal interventions may be more efficacious than other studies suggest. Second, even minimal interventions may require some level of repeated contact and contact repeated within a "modest" period of time. Again, even though Wall et al. provided only brief counseling (e.g., approximately 2 min) and written materials, they did so on a regular basis at 2, 4, and 6 months. These contacts may have prompted and reinforced greater change. Finally, even when minimal interventions designed to evoke cessation are effective (e.g., Wall et al.), they do not decrease ETS exposure for the majority of families to whom the intervention is provided because quit rates are small (i.e., < 6%). This means that the main value of minimal interventions is their relative ease of administration and low cost for benefit. However, if the goal is to make a substantial difference in children's ETS exposure in their home, these "minimal" clinical interventions are not sufficient. These studies suggest that brief, usually one time, counseling sessions are not sufficient to reduce ETS exposure in asthmatic or healthy children. It is not surprising to find that a powerful addiction such as smoking is difficult or impossible to change by brief counseling. The reinforcing contingencies that maintain smoking and exposure practices must be changed or new contingencies must be added to counter them. One-time interventions are unlikely to change contingencies. However, some parents may be undergoing changes in social contingencies — friends and family, media, and worksite influences to stop smoking or avoid exposing others. The addition of physician advice may yield the cumulative level of combined interventions to evoke reductions in ETS exposure practices for a small proportion of all such families. To detect the few individuals for whom the confluence of events might lead to ETS exposure changes requires very large samples and very accurate measures. Since these interventions are relatively inexpensive, future studies should be conducted with sufficient sample size and accurate measures to test minimal interventions more thoroughly.[61,62] Future studies also should be designed based on behavioral principles. This should include measures of context (e.g., friends encouragement and media exposure) variables unique to individual families that might enhance the efficacy of the minimal clinical intervention. Use of behavioral principles also would include more frequent counseling/feedback with contacts based on individual families' performance, not simply the provision of standardized instructions or brochures. With individualization, very brief contacts from clinicians might influence ETS exposure, especially for newborn infants or ill children. Refined minimal interventions for ETS exposure reduction remain important to test experimentally. If minimal interventions are found to be efficacious for a small, but important proportion of the population, the remaining majority of "non-responders" may be appropriate candidates for repeated and more extensive counseling services designed to shape ETS exposure reduction practices. This seems especially important for children with asthma or other respiratory diseases.

9.2.8 CLINICIAN COMPLIANCE

Even if these interventions are efficacious for a small proportion of patients, getting clinicians to provide these services is not automatic. Hymowitz[63] surveyed pediatricians and found that only 45% distributed smoking control materials to smoking parents, only 28% counseled parents, and only 11% provided follow-up sessions. Narce-Valente and Kligman[64] provided physicians with training to conduct screening and counseling for passive smoking. Fewer than 50% attended the training, and chart reviews showed no more than 6% provided screening or counseling. These studies suggest that more will be needed to get clinicians to perform even minimal interventions for tobacco control. Our investigations suggest that SLT applies to clinician prevention services as they do to patients' practices.[65,66]

9.2.9 SLT-Based Counseling

Similar to and predating the Wall et al.[60] study, two investigative teams conducted experimental studies of repeated counseling provided to families with children who had lower respiratory disease and were exposed to ETS. Table 9.1 summarizes these investigations as well. Meltzer and colleagues conducted a quasi-experimental study where five 30-min counseling sessions were provided over 1 month.[67] Counseling included information about asthma, ETS exposure effects, and means of reducing exposure without having to quit smoking. Counselors used problem solving and shaping to guide parents to smoke outdoors or away from children and provided praise and corrective feedback as appropriate. Counseling was associated with 40 to 80% reductions in reported ETS exposure. Additional reductions were obtained for four of five families at a 1-month follow-up.

Greenberg and co-investigators conducted a controlled trial of newborns using multivariate analyses of trends in ETS exposure.[68] A nurse provided four 45-min in-home counseling sessions over 6 months using procedures similar to those used by Meltzer et al.[67] Refinements were negotiated at follow-up visits. Measures were obtained at baseline and 7 and 12 months. For babies with mothers who smoked, exposure in the experimental group decreased significantly more than for controls. Among non-smoking mothers, the overall level of babies' exposure was lower than for smokers and trends showed an increase in both experimental and control groups. However, the increase was smaller for the experimental group, and this difference reached statistical significance. Most of the reported intervention effect was attributable to change in smoking mothers. Patterns of reported ETS exposure reduction among smoking mothers showed 6-month maintenance from end of counseling to 12 months. Cotinine results showed increases over time in both the control and experimental groups, and these differences did not reach statistical significance.

Hovell and co-workers[69] replicated the Meltzer[67] intervention with asthmatic children at about the same time that Greenberg et al.[68] conducted their trial. Families of asthmatic children (n = 91) were assigned at random to a control, monitoring, or counseling condition. Six counseling sessions (about 30 min each) in the clinic or home were provided over 6 months. Results for reported ETS exposure showed decreases in all groups with monitoring alone. Exposure in the control group decreased the least, followed by the monitoring group and then the experimental group. Group by time interactions showed that the counseling group obtained greater decreases in ETS exposure than either the control or monitoring condition. Lowered ETS exposure was sustained for 6 months. Correlations with nicotine dosimeter and symptoms validated reported ETS exposure.[70-72] Wahlgren et al. showed that decreases in ETS exposure were sustained for up to 2 years, and decreases in symptom reports were obtained for the counseling group.[73] Wahlgren also showed that with debriefing (i.e., one counseling session) and provision of a brochure that explained the counseling procedures, the control and monitoring groups obtained significant decreases in ETS exposure. These results suggest that a history of monitoring ETS, a one-time counseling, and written instructions may yield an approximation of the same degree of ETS exposure reduction obtained from the original counseling procedure.

9.2.10 EFFECTS SIZE, CLINICAL, AND PUBLIC HEALTH SIGNIFICANCE

These trials were conducted by two separate teams of investigators who employed the same theoretical foundation and intervention procedures. The use of repeated counseling sessions and shaping procedures was considerably more intensive than the brief clinical interventions noted earlier. However, even these counseling procedures did not involve extensive time or highly trained counselors. Compared to diabetic care or other chronic conditions, these remain relatively efficient interventions. Considering the addictive properties of tobacco and the difficulty most smokers have avoiding smoking as the "urge" arises, these interventions were remarkably powerful. Meltzer et al.[67] reported decreases in ETS exposure from 40 to 80% compared to baseline. Greenberg et al.[68] reported about 50% mean decrease, and Hovell et al.[69] reported 79% mean decrease. These are huge effects for clinical interventions, and it is most likely that these effects were a function of following SLT. At the behavioral level, these studies showed marked clinical significance. Though asthma symptoms were associated with ETS reduction, it remains to be seen the degree of clinical effectiveness achievable as measured by health outcomes. These results also suggest that more needs to be done to fully control ETS exposure and to maximize the public health significance of these procedures.

The trials that showed significant decreases in ETS exposure provided only limited treatment over 1 to 6 months. Most families remained well above zero level exposure at the end of counseling. This suggests that the length of intervention to bring most children to near zero exposure has not yet been determined. It is possible that extending the length of counseling could yield both more substantive decreases in ETS exposure and more parents attempting to quit. Often, smokers who quit return to smoking after counseling is completed, and it can be expected that some smokers who have successfully decreased their child's ETS exposure will return to practices that re-expose the child once counseling is completed. Sometimes, the only reinforcement that might sustain changes in patient behavior is social reinforcement delivered by clinicians; therefore, it is not surprising that patient behavior often returns to pretreatment levels after discontinuing tested programs. Thus, it seems important to test longer term counseling programs in order to reduce ETS exposure for a larger proportion of the population and possibly to avoid (possibly prevent) recidivism once counseling is discontinued. Families with asthmatic or otherwise illness-compromised children for whom ETS exposure is a serious exacerbating influence might warrant continuous counseling services in order to sustain as near to zero exposure as possible.

Clinical treatment programs should adopt ETS exposure reduction counseling following traditional service guidelines and continue services as long as they seem to be working for a given family and as long as the child continues to be exposed to ETS.

These trials suggest that counseling may be efficacious, and the two controlled trials provided evidence of valid reported measures. However, Greenberg et al.[68] were not able to confirm effects using cotinine as an outcome measure, and Hovell et al.[69] did not include cotinine outcome measures. The degree to which cotinine measures are appropriate remains controversial, but confirmation of counseling effects remains to be verified by changes in biological markers or health outcomes.

Two of these trials concerned children with respiratory disease and, thus, raise questions about the degree to which similar counseling procedures might work with families who have healthy children. The range of studies also raises questions about the degree to which similar procedures can be effective for children who are exposed predominantly by individuals other than their mothers. This is especially likely among certain ethnic groups (e.g., Latino families). Since Matt et al.[71] have shown that non-smoking mothers were less accurate in reporting their children's ETS exposures from other sources (e.g., husband), measures available for assessing trials for such families may not be sensitive (or sufficiently reliable) for detecting intervention effects.

Hovell and associates recently completed a similar study of healthy children exposed to ETS in their home, and preliminary results suggest similar success from counseling, including therapeutic differences in urine cotinine.[74] Hovell and associates are currently completing a study of Latino asthmatic children's exposure to ETS, where about 67% of the mothers are not smokers. Again, preliminary analyses suggest that significant decreases in ETS exposure can be obtained even by mothers who are not the primary source of ETS exposure.[75] Even with these more recent studies showing promising outcomes for healthy children, and for asthmatic Latino children whose mothers do not smoke, additional studies are needed to verify the degree to which such intervention effects are generalizable. Future studies should be conducted with samples representing varying populations and should include greater use of environmental and biological measures from which changes can be confirmed. If results are replicated, research should begin determining the components needed to make these procedures most efficient and to explore health benefits.

9.3 MEASURES AS PREREQUISITE FOR BOTH DETECTING AND EFFECTING ETS EXPOSURE

As noted previously, measures of ETS are not yet fully developed, and the degree of accuracy can be a concern for both accurate studies of ETS exposure and for effecting control. Hovell et al. recently summarized some of the more salient concerns about measures of ETS exposure.[76] They suggest that available measures emphasize reported exposure, urine or blood cotinine, air samples from which nicotine is assayed, and less frequently air particle dosimeter estimates. The relationships between reported ETS exposure and either cotinine or nicotine assays tend to be positive and significant. Mean correlations ranged from 0.43 for lower bounds to 0.63 for higher bounds among studies reviewed. These represented relationships where the shared variance in one measure of ETS exposure and the other ranged from about 19 to 40%. These results suggest considerable variance yet to be explained, but also suggest that reported measures correspond in the right direction with that of environmental and biological measures. Perhaps most reassuring for reported measures (albeit most disconcerting for environmental and biological measures) is that the relationships between environmental and biological measures are about the same magnitude as the relationships between reported measures and cotinine.[72] This suggests that none of these measures is uniquely superior to another. Even if we assume relationships in the neighborhood of 0.60 between two measures, considerable unexplained variance is implied. We think this means that no available measure is assessing the same thing and that the most complete assessment of ETS exposure will require multiple measures, obtained repeatedly, if not continuously.

9.3.1 REACTIVITY

The validity of reported measures is qualified by the fact that correspondence between reported ETS exposure measures and environmental or biological markers has been under conditions where respondents were aware of possible confirmation of their reports. The use of nicotine or cotinine assays with reported measures is likely to sensitize respondents to be more careful observers and more honest reporters. This reactivity has been described as a "bogus pipeline" when the confirmation process is not actually used.[77] Since investigators can determine only general reporting patterns and general patterns for objective measures, the use of nicotine/cotinine can be considered a partial bogus pipeline or partial validity test. This issue is highlighted by striking non-concordance reported by Pattishall and associates: 20% of children living in homes with a smoker did not show evidence of cotinine, and 26% of those living in homes without a smoker showed evidence of cotinine.[78] Emmons et al. reported that 29 to 47% of adults reported exposure, but no saliva cotinine was detected.[79] These discrepancies also suggest that cotinine may not be sensitive to all ETS exposure and that reported measures may include overestimates of true exposure. Matt et al. showed that mothers' reports of exposure to their own smoking were more accurate than reports of children's exposure to others' smoking.[71] Non-smoking mothers tended to overestimate exposure from others. These findings suggest that current evidence of the validity of reported measures of ETS exposure may not represent estimates in the absence of environmental or biological confirmation measures and that substantial sources of unexplained variance remain in both the reported, environmental, and biological measures. It is important to note, however, that reported measures that overestimate the degree of exposure may be "tolerable" in intervention studies where the family member, such as a parent, implicitly is judging greater amounts of ETS exposure than is likely to be true under some circumstances. This is likely to lead to greater efforts to reduce exposure than might be true with more accurate measures. Thus, mothers who are not smokers themselves may be more vigilant in their efforts to protect their child from a husband's or other family members' smoking when overestimating the degree of exposure.

9.3.2 No "Gold Standard"

While there is general correspondence between parent reported exposure and children's cotinine (i.e., 8 to 50% shared variance) and environmental nicotine (5 to 56% shared variance), these measures are far from providing identical information (i.e., 50 to 92% of the variance is not shared for cotinine, and 44 to 95% is not shared for nicotine). The failure to obtain greater correspondence is likely due to error in both reported and biological or environmental measures and because these measures do not assess the same things. Future research should address inter-individual differences affecting the air physics taking into account toxic concentrations, type of cigarettes, ventilation, dispersion, and other factors that may influence the degree of exposure in microenvironments. Similarly, greater understanding of breathing patterns, lung function and size, and metabolism are needed to understand the sources of variance in biomarkers and to better understand possible toxic mechanisms. Reported measures warrant investigation as well. Matt et al. have shown that "fuzzy set" measures, reporting high and low ranges of exposure, can account for an additional 8.7% of variance in babies' cotinine compared to point estimates alone.[72] This suggests that the type of question and computed estimate of exposure may influence validity. The increase in coefficients over time suggests that respondents learn to report ETS exposure more accurately with experience.[70,72] While this is problematic when comparing groups of unequal experience, it suggests that research should determine the experience that equips participants with accurate observation and reporting skills.

9.3.3 Acute vs. Cumulative Exposure

Reported, environmental, or biological measures of ETS exposure provide information about acute exposure. Cumulative exposure estimates require continuous or frequently repeated measures. Hair samples offer "longer term" (i.e., 1 to 2 months of exposure per centimeter of hair) estimates of nicotine.[80,81] However, estimates of cotinine are less well studied, sources of contamination are many, and hair sampling procedures have yet to be fully standardized. Hair sampling is a promising area to be developed as an additional measure of ETS exposure. Thus, common measures serve well as estimates for settings that do not change often and as satisfactory measures of interventions for reducing ETS exposure in the short term. Studies of cumulative toxic exposure or long-term behavior patterns require more frequent use of common measure or new continuous measures.

9.3.4 Real-Time and Continuous Measures

Dosimeter and biomarkers are collected intermittently in field studies, and samples are returned to laboratories for delayed analyses. Reported measures are usually taken intermittently, and it may take weeks to summarize the data. Only diaries approximate continuous and real-time indicators of ETS exposure. The failure to obtain real-time and continuous (or close approximation) measures of ETS exposure limits our understanding of true exposure levels, cumulative exposure levels, and the variability of exposure over time. This failure compromises our understanding of disease risks and severely limits our understanding of the causes of exposure-related behavior and our ability to change it. Passive and active dosimeters offer promise for both real-time and continuous measurement.[82-84] However, these remain to be miniaturized, made sufficiently versatile, and useful for feedback purposes.

9.3.5 FEEDBACK

A profound means of influencing human behavior is immediate feedback. Real-time measures of ETS exposure, as feedback, could influence ETS exposure practices. Most of the technology needed to develop fine particle dosimeters that can be convenient, nearly continuous, and from which immediate feedback can be derived is already available.[85] To our knowledge, no real-time measures of biomarkers are available, but by analogy, it should be possible to develop tools similar to those used by diabetic patients to obtain frequent blood (or other biological sample) glucose samples. Research and development should be directed toward more accurate, real-time, and continuous indicators of exposure. The development of these measures and their use for public health studies of ETS exposure and its control will be invaluable. Some of these measurement systems also may be powerful interventions in their own right.

9.4   DISCUSSION AND NEW RESEARCH

9.4.1   REDUCING ETS EXPOSURE

Policy interventions in public buildings/places have resulted in lower level ETS exposure in public and might influence ETS exposure in homes. More research is needed to determine the degree of generalization to private residences from policies directed to public settings. Policies and bans for residential exposure have been suggested, but not yet implemented, and many problems remain to be overcome in determining how such bans could be effected without compromising rights to privacy. Such bans also need to be tested to determine how they affect ETS exposure. The possible use of policies/bans to regulate residential exposure to ETS requires much more investigation.

Minimal clinical interventions, such as one-time physician advice, appear ineffective, but the quality and scope of the research to date lacks satisfactory rigor from which to conclude null effects. These studies also have not been designed following SLT and, hence, may have been less powerful than might be possible for minimal interventions. Additional research of minimal interventions designed following SLT should be tested in larger scale controlled trials.

Repeated counseling interventions based on SLT have been few in number, but well-designed community trials. Results show large reductions in reported ETS exposure; reported measures have been significantly correlated with environmental and/or biological markers of ETS, suggesting satisfactory validity; and asthma symptoms have been associated with reduction in ETS exposure. The rigor and effects observed in these studies strongly suggest that EST exposure can be reduced successfully by employing SLT-based counseling procedures. However, these interventions, too, might be improved, and numerous areas of investigation remain to be completed, from which improvements in the power of these interventions can be expected. If media programs (including entertainment programs), public bans, minimal clinical interventions, and SLT-based counseling were provided routinely, the combination would approximate the Behavioral Ecological Model of interlocking contingencies, and they might be sufficient to reduce and sustain low rates of tobacco initiation, smoking, and ETS exposure. Sustained and integrated community-wide programs for tobacco control might accelerate the political pressure needed to discontinue the legal production and marketing of tobacco. Additional research is needed to set the stage for more comprehensive and powerful tobacco control programs, including programs aimed at ETS control in residences.

9.4.2   IMPROVED MEASURES AND INCREASED INTERVENTION EFFECTS

Efficacious counseling to control ETS exposure has been dependent on "clinical measures" (e.g., physician assessed symptoms) as a basis for directing behavior change. These have been supplemented by interviews about frequency, timing, and conditions under which a child might have been exposed to the parents' smoking. These "clinical" measures have proved adequate for lowering ETS exposure in spite of their relatively imprecise nature.

Improved measures might increase the power of minimal interventions previously found ineffective.[53-55] Clinicians are in a unique position to employ both biomarkers (e.g., cotinine) and interview measures repeatedly over office visits. If results were used for both advice and feedback, the efficacy of clinician counseling might increase. Even though ideal measures are not yet available, clinicians could use existing cotinine and interview measures. The combination of cotinine and interview information might be sufficient to make repeated clinician advice (or counseling) efficacious. For example, physicians could routinely ask about smoking and ETS exposure and collect urine cotinine assays. With the combination of interview information and cotinine feedback (to the physician and the family), the physician (or assistant) is in an excellent position to discuss apparent exposure from either reported or urine assay measures. This counseling would be relatively unique, because families have not yet obtained such counseling in the context of biological (or environmental) feedback. This process also would be recorded in medical charts as part of the medical diagnostic process, making clinicians more accountable and more capable of addressing ETS exposure issues over time. This could lead to clinical associations with illnesses and other complications that could direct ongoing medical advice, treatment, and efforts to control tobacco use.

Perhaps most intriguing in this imagined possibility is the likelihood that routine cotinine assays could become reimbursable procedures. This would provide a discrete incentive for physicians to assess ETS exposure, and once the information is available, it might promote physician counseling to control ETS exposure. Thus, employing existing measurement technology with routine medical care may be a means of integrating counseling and enhancing it with the use of more complete feedback to families. The combination of payment to physicians should enhance such services. The combination might be a sustainable service. No such system for ETS exposure control has been tested. However, related biological feedback and counseling have been effective for smoking cessation.[86] Extending such a test to ETS exposure control seems a logical next step.

As soon as feedback systems are adequately developed and possibly automated or made available to the individual family, it may be possible for feedback per se to influence ETS exposure. For instance, if smoke and carbon monoxide alarms, now standard in most homes, were to be supplemented by similar particle or ETS exposure detectors in key rooms or the whole house, it might be possible to program automatic feedback to families when ETS is present. Such instrumentation could be mandated by public policy. This might be sufficient to control many families' ETS in homes. However, this model relies on implied, if not explicitly, aversive control and would require considerable development effort to refine to the point that homes could be equipped with a machine that could not be circumvented by determined smokers.

Personal monitors to be worn by a child or placed in the child's living space might be more acceptable to most families. These could be used to titrate when, where, and how much exposure actually occurs. They also could provide parents with cumulative dose feedback and time trends. Portable particle dosimeters that approximate these capabilities might serve this function.

If feedback systems were equipped with telemetry capabilities, physicians might be able to monitor progress as well. This could integrate medical management for children already under medical care (e.g., asthma). Feedback systems such as this also could set the stage for use of formal incentives for reducing ETS exposure.

9.4.3   NEW INTERVENTION APPROACHES

While the use of incentives for health behavior change is still controversial, their use is essentially impossible in the absence of reliable and valid measures of exposure on which to base some type of "payment." A recent study of smoking cessation showed that the use of incentives increased quit rates.[87] This suggests that the use of incentives could increase the reduction in ETS exposure similarly. The use of incentives to decrease ETS exposure will require the development of measures independent of manipulation by the family (except by reducing exposure), but may be a valuable means of increasing the power of an intervention for families with the greatest difficulty changing otherwise.

To our knowledge, no one has investigated the combination of ETS exposure counseling and tobacco cessation counseling. This remains an exciting direction for ultimate tobacco control. As most studies of tobacco cessation have been limited to adults who have agreed to join a cessation trial, current interventions for cessation have not be tested adequately for "non-volunteers." Ethically, this is difficult. However, there may be a means of recruiting the usually reluctant smoker to cessation via ETS exposure counseling. In our experience,[69,74] many smokers became interested in quitting even though they had joined the project with the understanding that cessation was neither asked of them nor was cessation counseling provided. This suggests that experience with ETS exposure counseling brings some smoking parents to be interested in cessation assistance. We have just started a study on the combination of ETS exposure reduction counseling and cessation counseling for those parents who become interested.[88] This study will provide initial evidence of the effects of combined ETS and cessation counseling, and we expect it will greatly decrease ETS exposure.

Two groups of tobacco users and/or exposed populations tend to get ignored in the tobacco control research effort. The first and most important are youths. Preteens and teens are rarely entered into cessation trials and, except for school education programs, are not usually recruited for prevention studies. Thus, the initiation of new smokers and the early treatment of teen addicts have received very little research attention. To our knowledge, no one has systematically assessed the degree of ETS exposure among preteens and teens or directly worked with preteens or teens to decrease their own ETS exposure. Family studies to date have counseled parents to alter smoking practices to protect their children. Counseling has not been formally directed to children. This seems to be another area for important investigations. From anecdotal reports from selected families in which a preteen or teen was the target child to be protected from the parents' smoking, we have learned that the preteen/teen youth has been very instrumental in effecting reduced ETS exposure. Individual teens may take an active role and participate in counseling with parents. The teens have also reminded parents to smoke outside or have left the room while the parent was smoking. These observations suggest that older youths could be the targets for "self-management" counseling to avoid ETS exposure. Such an intervention might be extended for youths to avoid not only exposure from family members, but also from other teens. This might have two beneficial consequences. It could protect the non-smoking preteen from influences that could promote smoking, making ETS counseling a means of preventing tobacco initiation. It also might have some subtle effects on peers who smoke. As the preteen withdraws to avoid ETS exposure, he/she also withdraws social support for the teen's smoking behavior. If this could be engineered on a large enough scale, it might be a means of effecting cessation among teens. This is consistent with the Behavioral Ecological Model and social influence thought to be important for adolescent practices.[27,28] School-based education or pediatric-based counseling services might enable children to avoid ETS exposure even if the smoking parent is not willing to participate. Since such schools and physicians already provide this type of "instruction" as part of quality health education and medical care, it should be possible for public education and clinical care systems to incorporate counseling/instruction to youths in order to assist them in avoiding ETS exposure. This direction of development should be researched with sensitivity to sustaining positive relations among family members.

The second population that deserves attention regarding ETS is the elderly. Many seniors still smoke, and many expose their adult children and grandchildren. Often, they are insensitive to the risks, and adult children may be uncomfortable asking a senior member of the family to quit or not smoke in the home. These conditions may lead to exemptions for grandparents, where the grandparent(s) is allowed to smoke and expose a grandchild even in homes where smoking is not allowed. Perhaps directing ETS exposure counseling to smoking seniors for whom these circumstances exist could help them reduce their smoking around grandchildren, if not all others. Since this might also contribute to senior quitting, and since quitting still conveys health benefits for seniors, this direction of research seems warranted, also.

9.5   CONCLUSION

The study and control of ETS is critical to reduce children's risk of morbidity and mortality. This is especially true for children with pulmonary or other chronic diseases likely to be exacerbated by ETS exposure. These relationships more than justify greater research and program development to reduce ETS exposure.

With increasing proportions of· the population from low-income and minority groups, or internationally, with an increased proportion of developing countries targeted by the tobacco industry, we can expect higher rates of smoking and ETS exposure among disenfranchised populations. These will be relatively poor, racial/ethnic minorities, including women and children. This raises the possibility that financial or race/ethnicity-based social prejudices and related cultural complications may define susceptibility to smoking, to ETS exposure, and to tobacco-related morbidity/mortality. If so, attention to these disparities in the overall social structure of the community may be a prerequisite to complete control of tobacco and ETS exposure of children in homes.

These tobacco industry trends in marketing also justify more attention to ETS exposure. A logical argument used by the tobacco industry to promote tobacco production, marketing, and use is an individual's right to choose to smoke or not. This "right" is theoretically debatable, but has been used successfully by the tobacco industry for decades to defend the production of a known health compromising and addictive substance. However, regardless of the accuracy of an individual's ability to choose freely, children do not have the privilege to avoid ETS exposure. Removing the free choice argument as a defense for the industry to continue production and marketing of tobacco may lead to greater controls of the industry than likely by concentrating public health efforts to delimit the use of tobacco on adult smokers and their health consequences. Efforts to control ETS may reduce both the prevalence of smoking and ETS exposure. Thus, ETS research and program development may be critical for the larger agenda of tobacco control for preventive medicine and health promoting objectives.

This chapter summarizes the policy and clinical research concerned with children's ETS exposure in their homes. Though much more research is needed to determine the effects of policies, minimal physician advice, and counseling programs, evidence to date for reducing children's exposure to ETS is encouraging. Counseling based on SLT appears to be efficacious for lowering ETS exposure among children. Refinements in these procedures should enhance their efficacy. Future studies should include refined measures of exposure, effectiveness trials, and longer term interventions from which health benefits can be determined. Following the Behavioral Ecological Model, the effects of multiple, concurrent, and community-wide programs for ETS and general tobacco control remain to be tested, but offer promise for countering the effects of the tobacco industry.

ACKNOWLEDGMENTS

This work was conducted through the Center for Behavioral Epidemiology and Community Health (C-BEACH), in the Graduate School of Public Health at San Diego State University. It was supported in part by, and based on the work funded through, grants to Dr. Hovell by the National Heart, Lung, and Blood Institute, NIH, #1 R18 HL52835-01A2; the Tobacco Related Disease Research Program (TRDRP), #1RT509 and #4RT0092; the Maternal and Child Health Bureau, #5 R40MC00093 and #R40MC00185; the Robert Wood Johnson Foundation, #027946; and through grants by the TRDRP to Dr. C. Richard Hofstetter, #9RT-0073, and to Dr. Joachim Reimann, #6KT0117. The authors would also like to thank Susan B. Meltzer, M.P.H., C. Richard Hofstetter, Ph.D., and Jennifer A. Jones, M.P.H. for their invaluable contributions throughout the course of these research projects.

REFERENCES

1. U.S. Department of Health and Human Services, Reducing the Health Consequences of Smoking: 25 Years of Progress. A Report of the Surgeon General, DHHS Publication No. (CDC) 89-8411, U.S. Department of Health and Human Services, Public Health Service, Centers for Disease Control, Center for Chronic Disease Prevention and Health Promotion, Office on Smoking and Health, 1989.
2. U.S. Department of Health and Human Services, The Health Consequences of Involuntary Smoking. A Report of the Surgeon General, DHHS Publication No. (CDC) 87-8398, U.S. Department of Health and Human Services, Public Health Service, Centers for Disease Control, Center for Health Promotion and Education, Office on Smoking and Health, 1986.
3. U.S. Environmental Protection Agency, Respiratory Health Effects of Passive Smoking: Lung Cancer and Other Disorders, EPA Publication No. EPA/600/6-90/006F, U.S. Environmental Protection Agency, Office of Research and Development, Office of Health and Environmental Assessment, 1992.
4. California Environmental Protection Agency, Health Effects of Exposure to Environmental Tobacco Smoke, California Environmental Protection Agency, Office of Environmental Health Hazard Assessment, September 1997.
5. Australian National Health and Medical Research Council, The Health Effects of Passive Smoking, Australian Government Publishing Service, 1997, ISBN 0-642-27270-0.
6. Scientific Committee on Tobacco and Health, Report of the Scientific Committee on Tobacco and Health, 1998. Available online at http://www.official-documents.co.uk/document/doh/tobacco/contents.htm.
7. Cook, D.G. and Strachan, D.P., Summary of effects of parental smoking on the respiratory health of children and implications for research, *Thorax*, 54, 357–366, 1999.
8. World Health Organization, International Consultation on Environmental Tobacco Smoke and Child Health: Consultation Report, World Health Organization, Geneva, 1999. Available online at http://www.who.int/toh/TFI/consult.htm.
9. Wahlgren, D.R., Hovell, M.F., Meltzer, E.O., et al., Involuntary smoking and asthma, *Curr. Opin. Pulm. Med.*, 6, 31–36, 2000.
10. Glantz, S.A. and Parmley, W.W., Passive smoking and heart disease: epidemiology, physiology, and biochemistry, *Circulation*, 83, 1–12, 1991.

11. Centers for Disease Control, State-specific prevalence of cigarette smoking among adults, and children's and adolescents' exposure to environmental tobacco smoke — United States, 1996. *MMWR*, 46, 1038–1043, 1997.

12. U.S. Department of Health and Human Services, Preventing Tobacco Use Among Young People: A Report of the Surgeon General, U.S. Department of Health and Human Services, Public Health Service, Centers for Disease Control and Prevention, National Center for Chronic Disease Prevention and Health Promotion, Office on Smoking and Health, Atlanta, GA, 1994.

13. Pierce, J.P. and Gilpin, E., How long will today's new adolescent smoker be addicted to cigarettes?, *Am. J. Pub. Health*, 86, 253–256, 1996.

14. Pirkle, J.L., Flegal, K.M., Bernert, J.T., et al., Exposure of the U.S. population to environmental tobacco smoke: the third National Health and Nutrition Examination Survey, 1988 to 1991, *J. Am. Med. Assoc.*, 275, 233–240, 1996.

15. Gergen, P.J., Fowler, J.A., Maurer, K.R., et al., The burden of environmental tobacco smoke exposure on the respiratory health of children 2 months through 5 years of age in the United States: Third National Health and Nutrition Examination Survey, 1988 to 1994, *Pediatrics*, 101, E8, 1998.

16. Collaborative Group S.I.D.R.I.A (Italian Studies on Respiratory Disorders in Childhood and the Environment), Parental smoking, asthma and wheezing in children and adolescents. Results of S.I.D.R.I.A., *Epidemiol. Prev.*, 22, 146–154, 1998.

17. Lister, S.M. and Jorm, L.R., Parental smoking and respiratory illnesses in Australian children aged 0-4 years: ABS 1989-90 National Health Survey results, *Aust. N.Z. J. Public Health*, 22, 781–786, 1998.

18. Huss, K., Rand, C., Butz, A., et al., Home environmental risk factors in urban minority asthmatic children, *Ann. Allergy*, 72, 173–177, 1994.

19. Jaakkola, M.S. and Jaakkola, J.J.K., Assessment of exposure to environmental tobacco smoke, *Eur. Respir. J.*, 10, 2384–2897, 1997.

20. Repace, J., Risk management of passive smoking at work and at home, *St. Louis Univ. Public Law Rev.*, 13, 763–785, 1994.

21. Löfroth, G., Environmental tobacco smoke: multicomponent analysis and room-to-room distribution in homes, *Tob. Control*, 2, 222–225, 1993.

22. Borland, R., Background Paper: Theories of Behavior Change in Relation to Environmental Tobacco Control to Protect Children, International Consultation on Environmental Tobacco Smoke and Child Health, World Health Organization, Geneva, 1999. Available online at http://www.who.int/toh/TFI/consult.htm.

23. Wahlgren, D.R., Meltzer, S.B., Zakarian, J.M., et al., Predictors of physician counseling to reduce tobacco smoke exposure in children with asthma, *Ann. Behav. Med.*, 15(S1 Suppl.), 36, 1993.

24. Soulier-Parmeggiani, L., Griscom, S., Bongard, O., Avvanzino, R., and Bounameaux, H., One-year results of a smoking-cessation programme, *Schweiz. Med. Wochenschr.*, 129, 395–398, 1999.

25. McBride, C.M., Curry, S.J., Lando, H.A., Pirie, P.L., Grothaus, L.C., and Nelson, J.C., Prevention of relapse in women who quit smoking during pregnancy, *Am. J. Public Health*, 89, 706–711, 1999.

26. Daughton, D.M., Fortmann, S.P., Glover, E.D., et al., The smoking cessation efficacy of varying doses of nicotine patch delivery systems 4 to 5 years post-quit day, *Prev. Med.*, 28, 113–118, 1999.

27. Hovell, M., Hillman, E., Blumberg, E., et al., A behavioral-ecological model of adolescent sexual development: a template for AIDS prevention, *J. Sex Res.*, 31, 267–289, 1994.

28. Wahlgren, D., Hovell, M.F., Slymen, D., et al., Predictors of tobacco use initiation in adolescents: a two-year prospective study and theoretical discussion, *Tob. Control*, 6, 95–103, 1997.

29. Glenn, S.S., Contingencies and metacontingencies: relations among behavioral, cultural, and biological evolution, in *Behavioral Analysis of Societies and Cultural Practices*, Lamal, P.A., Ed., Hemisphere, New York, 1991.

30. Traynor, M.P. and Glantz, S.A., California's tobacco tax initiative: the development and passage of Proposition 99, *J. Health Politics Policy Law*, 21, 543–585, 1996.

31. Elder, J.P., Edwards, C.C., Conway, T.L., Kenney, E., Johnson, C.A., and Bennett, E.D., Independent evaluation of the California Tobacco Education Program, *Public Health Rep.*, 111, 353–358, 1996.

32. Pechmann, C. and Goldberg, M., Evaluation of Ad Strategies for Preventing Youth Tobacco Use, Final report submitted to the Tobacco Related Disease Research Program for Grant #6RT-0038, 1998.

33. "Florida Online Tobacco Education Resources, " Florida's Tobacco Control Program. Available online at http://www.state.fl.us/tobacco/.

34. "Tobacco Related Disease Research Program," California's Tobacco Control Program. Available online at http://www.ucop.edu/srphome/trdrp/.

35. "Massachusetts Tobacco Control Program," Massachusetts' Tobacco Control Program. Available online at http://www.state.ma.us/dph/mtcp/home.htm.

36. Jacobson, P.D. and Wasserman, J., The implementation and enforcement of tobacco control laws: policy implications for activists and the industry, *J. Health Politics Policy Law*, 24, 567–598, 1999.

37. Thompson, B., Emmons, K., Abrams, D., et al., ETS exposure in the workplace, *J. Occup. Environ. Med.*, 37, 1086–1092, 1995.

38. Moskowitz, J., Lin, Z., and Hudes, E., The impact of California's smoking ordinances on worksite smoking policy and exposure to environmental tobacco smoke, *Am. J. Health Prom.*, 13, 278–281, 1999.

39. Eisner, M., Smith, A., and Blanc, P., Bartenders' respiratory health after establishment of smoke-free bars and taverns, *J. Am. Med. Assoc.*, 280, 1909–1914, 1998.

40. Winkelstein, M., Tarzian, A., and Wood, R., Parental smoking behavior and passive smoke exposure in children with asthma, *Ann. Allergy Asthma Immunol.*, 78, 419–423, 1997.

41. Baker, R.R. and Proctor, C.J., The origins and properties of environmental tobacco smoke, *Environ. Int.*, 16, 231–245, 1990.

42. Eatough, D.J., Hansen, L.D., and Lewis, E.A., The chemical characterization of environmental tobacco smoke, *Environ. Technol.*, 11, 1071–1085, 1990.

43. Nelson, P.R., Heavner, D.L., Collie, B.B., Maiolo, K.C., and Ogden, M.W., Effect of ventilation and sampling time on environmental tobacco smoke component ratios, *Environ. Sci. Technol.*, 26, 1909–1915, 1992.

44. Apte, M., Gundel, L., Klepeis, N., and Sextro, R., Indoor measurements of environmental tobacco smoke. Poster presented at AIM '98, TRDRP Annual Investigator's Meeting, Los Angeles, CA, *Annu. Invest. Meet. Proc.*, p. 58, 1998.

45. Apte, M., Gundel, L., Chang, G., and Sextro, R., Indoor measurements of environmental tobacco smoke, Poster presented at AIM '99, TRDRP Annual Investigator's Meeting, Los Angeles, CA, *Annu. Invest. Meet. Proc.*, p. 52, 1999.

46. Abernathy, T., Grady, B., and Dukeshire, S., Changes in ETS following anti-smoking legislation, *Can. J. Public Health*, 89, 33–34, 1998.

47. Farrelly, M., Evans, W., and Sfekas, A., The impact of workplace smoking bans: results from a national survey, *Tob. Control*, 8, 272–277, 1999.

48. Farkas, A., Gilpin, E., Distefan, J., et al., The effects of household and workplace smoking restrictions on quitting behaviours, *Tob. Control*, 8, 261–265, 1999.

49. Borland, R., Mullins, R., Trotter, L., et al., Trends in environmental tobacco smoke restrictions in the home in Victoria, Australia, *Tob. Control*, 8, 266–271, 1999.

50. Ezra, D., Sticks and stones can break my bones, but tobacco smoke can kill me: can we protect children from parents that smoke?, *St. Louis Univ. Public Law Rev.*, 13, 547–590, 1994.

51. Ashley, M. and Ferrence, R., Reducing children's exposure to environmental tobacco smoke in homes: issues and strategies, *Tob. Control*, 7, 61–65, 1998.

52. Sweda, E., Protecting children from exposure to environmental tobacco smoke, *Tob. Control*, 7, 1–2, 1998.

53. Woodward, A., Owen, N., Grgurinovich, N., et al., Trial of an intervention to reduce passive smoking in infancy, *Pediatr. Pulmonol.*, 3, 173–178, 1987.

54. Chilmonczyk, B., Palomaki, G., Knight, G., et al., An unsuccessful cotinine-assisted intervention strategy to reduce environmental tobacco smoke exposure during infancy, *Am. J. Dis. Child.*, 146, 357–360, 1992.

55. Vineis, P., Ronco, G., Ciccone, G., et al., Prevention of exposure of young children to parental tobacco smoke: effectiveness of an educational program, *Tumor*, 79, 183–186, 1993.

56. Murray, A. and Morrison, B., The decrease in severity of asthma in children of parents who smoke since the parents have been exposing them to less cigarette smoke, *J. Allergy Clin. Immunol.*, 91, 102–110, 1993.

57. Eriksen, K., Sorum, K., and Bruusgaard, D., Effects of information on smoking behaviour in families with school children, *Acta Pediatr.*, 85, 209–212, 1996.

58. McIntosh, N., Clark, N., and Howatt, W., Reducing tobacco smoke in the environment of the child with asthma: a cotinine-assisted minimal-contact intervention, *J. Asthma*, 31, 453–462, 1994.

59. Irvine, L., Crombie, I.K., Clark, R.A., et al., Advising parents of asthmatic children on passive smoking: randomised controlled trial, *Br. Med. J.*, 318, 1456–1459, 1999.

60. Wall, M., Severson, H., Andrews, J., et al., Pediatric office-based smoking intervention: impact on maternal smoking and relapse, *Pediatr.*, 96, 622–628, 1995.

61. Slymen, D. and Hovell, M., Cluster versus individual randomization in adolescent tobacco and alcohol studies: illustrations for design decisions, *Int. J. Epidemiol.*, 26, 765–771, 1997.

62. Hovell, M.F., Elder, J., Blanchard, J., et al., Behavior analysis and public health perspectives: combining paradigms to effect prevention, *Educ. Treat. Child.*, 9, 287–306, 1986.

63. Hymowitz, N., A survey of pediatric office-based interventions on smoking, *N.J. Med.*, 92, 657–660, 1995.

64. Narce-Valente, S. and Kligman, E., Increasing physician screening and counseling for passive smoking, *J. Fam. Pract.*, 34, 722–728, 1992.

65. Russos, S., Keating, K., Hovell, M.F., et al., Counseling youth for tobacco use prevention: determinants of clinician compliance, *Prev. Med.*, 29, 13–21, 1999.

66. Hovell, M.F., Wahlgren, D., and Russos, S., Preventive Medicine and Cultural Contingencies: A natural experimentm in *Cultural Contingencies: Behavior Analytic Perspectives on Cultural Practices*, Lamal, P.A., Ed., Praeger, New York, 1997.

67. Meltzer, S.B., Hovell, M.F., Meltzer, E.O., et al., Reduction of secondary smoke exposure in asthmatic children: parent counseling, *J. Asthma*, 30, 391–400, 1993.

68. Greenberg, R.A., Strecher, V.J., Bauman, K.E., et al., Evaluation of a home-based intervention program to reduce infant passive smoking and lower respiratory illness, *J. Behav. Med.*, 17, 273–290, 1994.

69. Hovell, M.F., Meltzer, S.B., Zakarian, J.M., et al., Reduction of environmental tobacco smoke exposure among asthmatic children: a controlled trial, *Chest*, 106, 440–446, 1994.

70. Emerson, J.A., Hovell, M.F., Meltzer, S.B., et al., The accuracy of environmental tobacco smoke exposure measures among asthmatic children, *J. Clin. Epidemiol.*, 48, 1251–1259, 1995.

71. Matt, G.E., Wahlgren, D.R., Hovell, M.F., et al., Measuring ETS exposure in infants and young children through urine cotinine and memory-based parental reports: empirical findings and discussion, *Tob. Control*, 8, 282–289, 1999.

72. Matt, G.E., Hovell, M.F., Zakarian, J.M., et al., Parent-reports of environmental tobacco smoke exposure in infants: reliability and validity in a sample of low-income families, *Health Psych.*, 19, 232–241, 2000.

73. Wahlgren, D.R., Hovell, M.F., Meltzer, S.B., et al., Reduction of environmental tobacco smoke exposure in asthmatic children: a two-year follow-up, *Chest*, 111, 81–88, 1997.

74. Hovell, M.F., Zakarian, J.M., Matt, G.E., et al., Effect of counseling mothers on their children's exposure to environmental tobacco smoke: randomised controlled trial, *BMJ*, 321, 337–342, 2000.

75. Hovell, M.F., Risk Reduction and Adherence Among High Risk Adults and Other Populations, Presented to the University of Pittsburgh Center for Research in Chronic Disorders, Nov. 1, 1999.

76. Hovell, M.F., Zakarian, J.M., Wahlgren, D.R., Matt, G.E., and Emmons, K., Measurement of environmental tobacco smoke exposure: trials and tribulations, *Tob. Control*, 9, 0–6, 2000.

77. Murray, D.M., O'Connell, C.M., Schmid, L.A., and Perry, C.L., Validation of smoking self-report by adolescents: a re-examination of bogus pipeline procedures, *Addict. Behav.*, 12, 7–15, 1987.

78. Pattishall, E., Stope, G., Etzel, R., et al., Serum cotinine as a measure of tobacco smoke exposure in children, *Am. J. Dis. Child.*, 139, 1101–1104, 1985.

79. Emmons, K.M., Abrams, D.B., Marshall, R., et al., An evaluation of the relationship between self-report and biochemical measures of environmental tobacco smoke exposure, *Prev. Med.*, 23, 35–39, 1994.

80. Zahlsen, K. and Nilsen, O.G., Nicotine in hair of smokers and non-smokers: sampling procedure and gas chromatograph/mass spectrometric analysis, *Pharmacol. Toxicol.*, 75, 143–149, 1994.

81. Kintz, P., Gas chromatographic analysis of nicotine and cotinine in hair, *J. Chromatogr.*, 580, 347–353, 1992.

82. Turner, W., Spengler, J., Dockery, D.I., et al., Design and performance of a reliable personal monitoring system of respirable particulates, *J. Air Pollut. Control Assoc.*, 29, 747–748, 1979.

83. Spengler, J., Treltman, R., Tosteson, D.I., et al., Personal exposures to respirable particulates and implications for air pollution epidemiology, *Environ. Sci. Technol.*, 19, 700–707, 1985.

84. Muramatsu, M., Umemura, S., Okada, T., et al., Estimation of personal exposure to tobacco smoke with a newly developed nicotine personal monitor, *Environ. Res.*, 35, 218–227, 1984.

85. Brauer, M., Hirtle, R.D., Hall, A.C., et al., Monitoring personal fine particle exposure with a particle counter, *J. Exp. Anal. Environ. Epidemiol.*, 9, 228–236, 1999.

86. Hoffman, D., Use of Repeated Cotinine Determinations as a Motivational and Educational Tool in Smoking Cessation Counseling for Pregnant Women, Grant #027941 SFP, Robert Wood Johnson Smoke-Free Families Program, 11/95–12/97.

87. Prows, S., Significant-Other Supporter (SOS) Program, Grant #027945 SFP, Robert Wood Johnson Smoke-Free Families Program, 11/95–12/97.
88. Hovell, M.F., WIC Families Who Smoke: A Behavioral Counseling Study. Maternal and Child Health Bureau (Title V, Social Security Act), Grant #R40 MC 00185, Health Resources and Services Administration, Department of Health and Human Services, 1/00–12/03.

10 Passive Cigarette Smoking and Breast Cancer

*Alfredo Morabia, Timothy Lash, and
Ann Aschengrau*

CONTENTS

10.1 INTRODUCTION

The well-established risk factors for breast cancer offer few opportunities for intervention and account for less than half of all breast cancer cases.[1] Most of these factors involve aspects of a woman's reproductive course that are intimately related to her lifestyle and culture — so are difficult to predict and therefore to change — or are currently immutable (e.g., genotype). While pharmaceutical intervention (e.g., tamoxifen) could be considered as a prophylactic strategy in high risk women,[2] recommendations for lifestyle changes to reduce the risk of breast cancer remain elusive. If active smoking or environmental tobacco smoke were found to be a cause of breast cancer, they would offer opportunities for lifestyle changes that would

0-8493-0311-7/00/$0.00+$.50
© 2001 by CRC Press LLC

reduce the risk of breast cancer in some women. A history of tobacco smoke exposure would also help to identify women who are at high risk and who may wish to consider prophylactic pharmaceutical intervention.

Exposure to tobacco smoke has long been thought to not cause breast cancer.[3] Most studies of active smoking indicate either a weakly positive or a null effect.[3–8] Four studies have measured a protective effect.[9–12] Recent studies consistently show a direct association between exposure to environmental tobacco smoke and the occurrence of breast cancer.[13–16] Other reports suggest that genetic susceptibility (e.g., having a slow or a fast genotype for N-acetyltransferase 2) may result in heterogeneity of risk in populations and for the inconsistent pattern of effects observed across studies.[17–20]

This chapter begins with a review of the biologic rationale to suspect that tobacco smoke may cause breast cancer. A brief review of the association between active cigarette smoking and breast cancer risk follows, since this association informs the studies of the association between exposure to environmental tobacco smoke and breast cancer risk, which is the focus of the remainder of the chapter.

10.2 BIOLOGIC RATIONALE TO SUSPECT TOBACCO SMOKE MAY CAUSE BREAST CANCER

Different models of breast carcinogenesis have been proposed. One of them suggests that it is a two-stage process.[21] In the first stage, susceptible breast tissue is initiated by a mutagenic lesion of an oncogene in a cell's DNA. This initiated cell can then divide, die, or progress to the second stage. In the second stage, another mutagenic event in an oncogene leads to the clonal carcinogenic cell. This cell can then divide, die, or be promoted through growth and progress to a tumor. Sex hormones (mainly estrogen and progesterone) are responsible for promoting breast cancer growth.[22] Few breast cell mutagens have been identified, but estrogen-induced mitogenesis has been well characterized. Most of the established breast cancer risk factors reflect the cumulative exposure of breast tissue to sex hormones. These factors include lower age at menarche, later age at menopause, increasing parity, use of hormone replacement therapy, and higher postmenopausal body mass index.

One potential explanation for the dearth of breast cancer initiators that have been identified is that these initiators are likely to act when women are very young, perhaps even before the onset of puberty.[23] Few cohort studies have enrolled women at young enough ages to measure the effect of such early age exposures. Retrospective studies may not be able to assess early age exposures accurately enough to measure their effect, given errors of recall by older participants who must remember their early age exposure histories.

10.2.1 BREAST TISSUE SUSCEPTIBILITY TO CARCINOGENIC INITIATION

To initiate carcinogenesis, a mutagen must act while breast tissue is susceptible and before mitogenesis. According to Russo and Russo,[24] breast tissue development and differentiation determines its susceptibility to mutagenesis. Breast tissue development is a function of a woman's reproductive milestones. From birth to puberty, the breast contains macrostructures called lobules 1. At puberty, these lobules differentiate to structures called lobules 2, which predominate until a woman's first pregnancy. With the first term pregnancy, these lobules 2 further differentiate to structures called lobules 3. With the onset of lactation, the lobules 3 differentiate to lobules 4, the terminal differentiation state. After lactation, the lobules 4 dedifferentiate to lobules 3. Although the majority of breast tissue follows this progression, some tissue will remain in early differentiation stages even after a first term pregnancy. With a second pregnancy, the proportion of fully differentiated breast tissue increases and so on with each additional pregnancy.[23] In nulliparous women, there is some extrahormonal differentiation of lobules 2 to lobules 3, but the breast tissue contains primarily lobules 2.[24] At menopause, lobules 3 involute to lobules 2 and then to lobules 1, but these lobules do not have the susceptibility to carcinogenesis of prepregnancy lobules 1 and 2 described later.

In vitro studies show that lobule type strongly influences three surrogates for breast tissue susceptibility to carcinogenesis.[24] Tissues were collected from women at breast reduction surgery. First, the DNA labeling index demonstrates that cells from lobules 1 grow more rapidly than cells from lobules 2, which in turn grow more rapidly than cells from lobules 3. Second, the binding index of dimethyl benzanthracene (an artificial polycyclic aromatic hydrocarbon that is a mammary carcinogen in rats) shows that cells from lobules 1 bind this carcinogen more efficiently than cells from lobules 2, which in turn bind it more efficiently than cells from lobules 3. Third, cells from lobules 1 and 2 can be chemically transformed, whereas cells from lobules 3 are immune to transformation by the chemicals tested thus far.[24]

This model suggests that the time of exposure with respect to reproductive milestones to breast carcinogens ought to determine susceptibility to carcinogenesis. Breast tissue exposed premenopausally to mutagenic events while the proportion of lobules 1 and 2 is high should be susceptible. Promotion of the exposed and susceptible tissue by mitogenic estrogen compounds ought to further increase the risk of tumor development,[25] whereas inhibition of mitogenesis ought to reduce the risk.[22]

Studies of the modification of the relative risk of breast cancer by age at exposure to radiation support this model of breast tissue susceptibility to carcinogenesis.[26] In a study of survivors of the atomic bomb, women exposed before age 9 had the highest relative risk, women exposed between ages 10 and 19 had intermediate relative risk, and women exposed after age 20 but before age 40 had the lowest relative risk.[27] Similar modification of the relative risk by age at exposure to medical radiation has been observed in studies of women treated with radiation for scoliosis,[28] tinea capitis,[29] and Hodgkin's disease.[30] Radiation is an established initiator of breast carcinogenesis.[23] Early age of exposure strongly modifies the relative risk associated with exposure to radiation, which is consistent with the foregoing model of age-dependent breast tissue susceptibility to carcinogenic initiation.

In addition, an early age at first pregnancy and a higher number of term pregnancies both reduce a woman's risk of breast cancer. This epidemiologic evidence further supports the model for breast tissue susceptibility to carcinogenesis. An early age at first pregnancy reduces the time period during which mutagenic insults can accumulate in susceptible tissue. Multiple term pregnancies increase the proportion of the breast tissue that has differentiated from the susceptible lobules 1 and 2 to the immune (or less susceptible) lobules 3.

10.2.2 Tobacco Smoke and Breast Carcinogenesis

Tobacco smoke is well known to cause cancer at sites with which it comes in direct contact, such as the lung, larynx, and oral mucosa. Tobacco smoke also causes cancer at sites with which it has no direct contact, such as the pancreas, renal pelvis, and bladder.[31] Carcinogenesis at these sites may be initiated by one of the 400 to 500 gaseous components of tobacco smoke or 3500 chemicals in the particulate phase. These include at least 43 chemicals that have been demonstrated to be animal carcinogens and several that are known human carcinogens.[32] When smoke is collected directly at the tip of the cigarette, the identified carcinogens are usually found at higher concentrations in sidestream smoke than in mainstream smoke.[31] Studies of breast tumor tissue[32] and adjacent normal breast tissue[32,33] of smokers found the diagonal radiation zone (DRZ) that is characteristic of the DNA-adduct pattern observed in other human tissues that are susceptible to tobacco smoke carcinogenesis. Breast tissues taken from non-smokers did not display the DRZ.[32] Tobacco smoke condensate provoked DNA repair in an assay of secondary cultures of human mammary epithelial cells collected from reduction mammoplasty specimens from five women ages 19 to 22 years.[34] Assays of excised mouse mammary glands treated with arylamines[35] and an *in vitro* assay of human mammary epithelial cells treated with benzo[a]pyrene[36] showed that mammary cells possess the metabolic activity required to activate chemical carcinogens included among the constituents of tobacco smoke. Taken together, this evidence suggests that constituents of tobacco smoke could initiate carcinogenesis in susceptible human breast tissue.

Among the constituents of tobacco smoke that have been identified as carcinogens are the aromatic and heterocyclic amines.[32] These chemicals require host-mediated metabolic activation to electrophiles, which readily bind nucleophilic DNA, in order to induce a heritable mutation and ultimately cancer.[37] There are two pathways by which aromatic amines can be metabolized.[38] First, aromatic amines can be N-acetylated by N-acetyltransferase 2 (NAT2) or NAT1 in the liver. This is a detoxifying pathway. Second, aromatic amines and heterocyclic amines can be N-oxidized by P450 enzymes in the liver or in extrahepatic tissues such as the breast. This oxidation competes with the hepatic N-acetylation for aromatic amines, but not heterocyclic amines. The product of the oxidation is then either O-acetylated by NAT2 or NAT1, a reaction that yields the activated electrophile, or detoxified by competing enzymatic pathways. Therefore, NAT2 has a dual role: it detoxifies aromatic amines hepatically, but may play a role in the activation of aromatic amines and heterocyclic amines in extrahepatic tissues such as the breast.

NAT2 is a polymorphic enzyme in humans. Those who possess homozygous wildtype alleles are called fast acetylators, and those who possess the mutant polymorphisms are called slow acetylators. Depending on which metabolic pathway predominates at critical junctures of exposure and tissue susceptibility, fast acetylators may be at higher or lower risk of smoking-induced breast carcinogenesis than slow acetylators.

10.2.3 TOBACCO SMOKE AND ESTROGENS

An antiestrogenic effect of tobacco smoking in women has been proposed by Baron et al.[39] Women who smoke are at higher risk than non-smokers for conditions of estrogen deficiency, such as early menopause and osteoporotic fracture. Smokers are at lower risk than non-smokers for conditions of estrogen excess, such as endometrial cancer, fibrocystic disease, and vomiting of pregnancy. While these epidemiologic investigations suggest an antiestrogenic effect, few studies have observed a difference in serum estrogen concentrations between smokers and non-smokers. The only consistently observed difference is an increase among smokers in the concentration of the adrenal androgens (androstenedione and dehydroepiandrosterone sulfate [DHEAS]), which are precursors of estrone particularly in postmenopausal women.[40] Several alternative mechanisms have been proposed to explain the apparent antiestrogenic effect of smoking,[40] but the current status of the science has not resolved the paradox of consistent epidemiologic findings of an antiestrogenic effect in the face of the unaffected hormonal milieu.

The ratio of 2-hydroxylated catechol estrogen to 16-hydroxylated estrogen (estriol) is higher among women who smoke than among women who do not smoke.[41,42] The 2-hydroxylated catechol estrogen competes with estrogen for the estrogen receptor protein, but is inactive in the receptor, whereas estriol is a weak estrogen agonist.[40] The higher ratio of 2-hydroxylated estradiol to 16-hydroxylated estradiol in smokers may explain part of the antiestrogenic effect of smoking. The 2- and 4-hydroxylated catechol estrogens are metabolized by catechol-O-methyltransferase (COMT) to non-reactive species.

There are two COMT alleles, one encoding high activity (COMT-H) and the second encoding approximately fourfold lower activity (COMT-L). Women who are homozygous for low activity (COMT-LL) and who smoke should have an enriched concentration of the 2-hydroxylated catechol estrogen. It can be hypothesized that these women have a lower smoking-related risk of breast cancer than women who are heterozygous (COMT-LH) or homozygous for the high activity allele (COMT-HH).

10.2.4 ESTROGEN RECEPTORS

Whether smoking is specifically associated with estrogen receptor (ER) negative breast cancer is a key question to reveal the biologic bases of the carcinogenic effect of tobacco smoke on the mammary gland. Constituents of tobacco smoke may bind competitively to ERs and permanently change the binding properties of the malignant cells.[43] Also, if ER negative tumors are likely to be more aggressive and to have a worse prognosis,[44] an association with smoking could explain the association between smoking and the risk of fatal breast cancer.[7]

The available evidence is inconclusive because of the contradictory results in the literature about a possible interaction between a tumor's ER status, smoking history, and breast cancer.[43,45–47] In the U.S. Nurses Study,[45] the relative risk of breast cancer for women smoking ≥25 cigarettes per day (vs. never smokers) was 1.4 (95% confidence interval [CI]: 1.0 to 1.8) among women with ER positive tumors and 1.05 (95% CI: 0.7 to 1.6) for ER negative tumors. An investigation among

Japanese women reported that the relative risk of breast cancer among ever smokers was of similar magnitude in ER positive (odds ratio [OR] = 1.4; 95% CI:1.04 to 1.9) and ER negative (OR = 1.3; 95% CI: 0.9 to 1.5) tumors.[47] Australian[43] and Swiss[46] case-control studies showed somewhat stronger relative risks for ER negative tumors, with a twofold difference between ER groups, but the interaction was statistically significant only in the Australian study (p = 0.036). In the former study,[43] there was no effect of ever-active smoking in ER positive tumors; while in the Swiss study, the effect of ever-active smoking was seen in both ER negative and ER positive tumors. This inconsistency may be related to the grouping of passive smokers and non-exposed women in the same reference category in the Australian study. The grouping may result in an underestimate of the relative risk in both groups and, thus, mask a weaker effect of active smoking in ER positive tumors.

10.3   ACTIVE SMOKING AND BREAST CANCER RISK

10.3.1   NON-ACTIVE SMOKERS AS THE REFERENCE CATEGORY

Epidemiologic investigations of the association between tobacco smoke and breast cancer occurrence can be separated into three generations.[48] The first generation of studies — reviewed by McMahon and colleagues,[49] Palmer and Rosenberg,[3] and Morabia and Bernstein[48] — compared active cigarette smokers with never-active cigarette smokers. Subgroups of smokers with different cumulative smoking history, age at smoking initiation, or age at smoking cessation were often analyzed. Both cohort and case-control studies were conducted. The accumulated evidence from these studies indicated that smoking had, at most, a small effect on breast cancer risk.[48] One report of an association limited to heavy smokers who began smoking at young age generated considerable interest,[50] but was not confirmed by subsequent studies.[5,51,52]

10.3.2   NON-ACTIVE, NON-PASSIVE SMOKERS AS
THE REFERENCE CATEGORY

The second generation of studies compared active cigarette smokers with women who never smoked themselves and never lived with a smoker. This design, which was first suggested by Sandler and colleagues in 1986,[53] differed from the first generation studies in that the reference group included no passive smokers. Second generation studies also compared the breast cancer risk of passive smokers with the risk of never-active, never-passive smokers (discussed later). The five second generation studies have shown a consistent effect of active smoking and passive smoking on breast cancer risk. A meta-analysis of the five incidence studies show an overall relative risk of 2.2 (95% CI: 1.6 to 2.9) for active smoking compared with never-active, never-passive smoking.[48]

10.3.3   GENE-ACTIVE SMOKING INTERACTION

The third generation of studies measured the effects of tobacco smoke within categories of certain genetic polymorphisms. The two best-studied genetic polymorphisms are NAT2 and COMT.

10.3.3.1 NAT2

Postmenopausal women who smoke and who have a reduced ability to detoxify byproducts of tobacco smoke, as measured by their NAT2 genotype, have an excess risk of breast cancer occurrence.[17,19,20] In a similar study, it was the postmenopausal women who detoxified byproducts of tobacco smoke most quickly who were at highest risk.[18] A fourth study observed little dependence of smoking risk on acetylation rate, although women who smoked only before their first pregnancy had an elevated breast cancer risk.[54] Morabia and colleagues argued that an alternative analysis would have demonstrated an interaction between smoking and slow acetylation,[55] but Hunter and colleagues disagreed.[56]

A study of the interaction between smoking, NAT2, and breast cancer — which included measurement of passive smoking — had an intriguing finding.[20] When active smokers were compared with never-active smokers, a reference category that includes both unexposed women and passive smokers, the results suggested that postmenopausal *slow* acetylators were at higher risk of breast cancer if they smoked. This conclusion was consistent with the study of Ambrosone et al.[17] On the other hand, when non-active, non-passive smokers were used as the reference category, analyses led to the opposite conclusion. That is, postmenopausal *fast* acetylators were at higher risk of breast cancer if they smoked. We consider the latter results as being more valid, because it is logical to use the lowest level of exposure to tobacco as the reference category, especially in view of the epidemiological evidence of a specific effect of passive smoking described later.

10.3.3.2 COMT

In the first study of the interaction between COMT polymorphisms and tobacco smoke, COMT-LL postmenopausal women were at reduced risk of breast cancer occurrence, and the protective effect was confined to women who smoked.[57] However, COMT-LL premenopausal women were at excess risk of breast cancer occurrence, a finding that was also confined to women who smoked.[57] Thompson et al.[58] found an overall protective effect of COMT-LL in postmenopausal women, primarily among those who were most lean. In premenopausal women, whose circulating estrogen is carefully regulated, the COMT-LL allele conferred a higher risk, primarily among those who were heaviest. The authors suggested that in the low estrogen environment of postmenopausal women, the antiestrogenic effect of COMT-LL dominated. In the high estrogen environment of premenopausal women, DNA damage secondary to formation of oxidative metabolites of the catechol estrogens[59] dominated. Lavigne et al. found the opposite pattern of effects, a protective effect of COMT-LL in premenopausal women and a causal effect in postmenopausal women,[60] although the number of participants in this study was very low. Finally, Millikan et al. found no effect of COMT-LL or interaction between COMT and menopausal status, body weight, or smoking history.[61]

10.3.3.3 Other Genes

Studies have also demonstrated interaction between tobacco smoke and NAT1, BRCA1/BRCA2 (the "breast cancer genes"),[63] and cytochrome P450 1A1 gene polymorphisms.[64]

10.4   EXPOSURE TO ENVIRONMENTAL TOBACCO SMOKE AND BREAST CANCER RISK

None of the first generation studies discussed earlier measured the effect of active smoking relative to women who were neither active nor passive smokers. A large proportion of the non-smokers in the referent groups in these studies could have been substantially exposed to passive smoking.

10.4.1   NON-ACTIVE AND NON-PASSIVE SUBJECTS AS THE REFERENCE CATEGORY

Suspicions that breast cancer may be linked to passive smoking began in 1985 when Sandler et al.[65] observed a twofold risk of breast cancer associated with smoking by a spouse. They were the first to postulate that in studies comparing smokers to non-smokers, "non smokers may have included women who were passively exposed, limiting the possibility of observing any effect that might result from exposure to sidestream smoke." They suggested that "Future studies of this question should collect data that will allow for stratification into at least three smoking categories: true non exposed, those with passive exposure only, and active smokers."[53]

There are now seven published studies of the association between exposure to environmental tobacco smoke and breast cancer risk (see Table 10.1), four of which were recently reviewed by Wells.[66] These were conducted[13,15,67,68] or reanalyzed[15,65,69] according to the foregoing methodological recommendation. The endpoint of the Japanese cohort study conducted by Hirayama was mortality rather than incidence, so it is not included in the literature synthesis. Synthetic ORs and CIs for the six remaining studies have been computed using the methodology described by Wells.[66]

These six studies consistently show that women exposed to passive smoking are at increased risk of breast cancer relative to women never exposed to either active or environmental tobacco smoke. The synthetic OR of these five studies is 1.7 (95% CI: 1.3 to 2.3) for the association of breast cancer with exposure to environmental tobacco smoke. The consistency of the findings (1) across seven different study populations from widely varying geographic regions; (2) in studies conducted by both case-control designs and follow-up designs; and (3) in populations that were primarily premenopausal, primarily postmenopausal, or of mixed menopausal status adds credence to the plausibility of the association. The most recent of these seven investigations, a prospective cohort study of Korean women, found a relative risk of 1.3 (95% CI: 0.9 to 1.8) for women whose husbands were current smokers, a relative risk of 1.2 (95% CI: 0.8 to 1.8) for women whose husbands were former smokers, and a relative risk 1.7 (95% CI: 1.0 to 2.8) for women whose husbands were current smokers and had smoked for more than 30 years.

10.4.2   GENE-PASSIVE SMOKING INTERACTION

Relative risks for passive smoking are reportedly higher for women with the fast than with the slow NAT2 genotype, especially among postmenopausal women.[20]

TABLE 10.1
Epidemiologic Studies of the Association Between Exposure to Environmental Tobacco Smoke and Breast Cancer Risk

Ref.	Description	Adjusted relative effect (95% confidence interval)
Sandler, D., personal communication with A. Judson Wells; published in Wells, A.J., Breast cancer, cigarette smoking, and passive smoking (Letter), *Am. J. Epidemiol.*, 133, 208–210, 1991.	Case-control study of incident pre- and post-menopausal breast cancer patients diagnosed in North Carolina. Relative risk compares never-smoking wives of men who smoked with wives of men who never smoked.	1.62 (0.76–3.44)
Hirayama, T., personal communication with A. Judson Wells; published in Wells, A.J., Breast cancer, cigarette smoking, and passive smoking (Letter), *Am. J. Epidemiol.*, 133, 208–210, 1991.	Prospective follow-up study of fatal breast cancer cases diagnosed in Japan. Relative risk compares never-smoking wives of men who smoked with wives of men who never smoked.	1.32 (0.83–2.09)
Smith, S.J., Deacon, J.M., Chilvers, C.E.D., et al., Alcohol, smoking, passive smoking and caffeine in relation to breast cancer risk in young women, *Br. J. Cancer*, 70, 112–119, 1994.	Case-control study of incident breast cancer patients less than 36 years old at diagnosis in the U.K. Relative risk compares never-smoking wives of men who smoked with wives of men who never smoked.	1.58 (0.81–3.10)
Morabia, A., Bernstein, M., Heritier, S., et al., Relation of breast cancer with passive and active exposure to tobacco smoke, *Am. J. Epidemiol.*, 143, 918–928, 1996.	Case-control study of incident breast cancer patients less than 75 years old at diagnosis in Geneva, Switzerland. Relative risk compares never-smoking women who had more than 1 h/day exposure to passive smoke for 12 consecutive months at home, workplace, or leisure environment with women who had never smoked and had less passive smoke exposure.	2.3 (1.5–3.7)
Lash, T.L. and Aschengrau, A., Active and passive cigarette smoking and the occurrence of breast cancer, *Am. J. Epidemiol.*, 149, 5–12, 1999.	Case-control study of incident breast cancer patients, 90% postmenopausal, diagnosed on Cape Cod, MA. Relative risk compares never-smoking women who lived with a smoker to never-smoking women who never lived with a smoker.	2.0 (1.1–3.7)

TABLE 10.1 (*continued*)
Epidemiologic Studies of the Association Between Exposure to Environmental Tobacco Smoke and Breast Cancer Risk

Ref.	Description	Adjusted relative effect (95% confidence interval)
Jee, S.H., Ohrr, H., and Kim, I.S., Effects of husbands' smoking on the incidence of lung cancer in Korean women, *Int. J. Epidemiol.*, 28, 824–828, 1999.	Prospective follow-up study of incident breast cancer among Korean women aged 40–88 years. The relative rate compares never-smoking wives of current smokers with wives of never smokers.	1.3 (0.9–1.8)
Johnson, K.C., Semenciw, R., Hu, J., et al., Active and passive smoking and breast cancer, *Cancer Causes and Control*, 11, 211–221, 2000.	Case-control study of incident breast cancer among premenopausal women diagnosed in Canada. The relative risk compares never-smoking women who had a history of exposure to passive smoke in the home or workplace with never-smoking women who had no such history.	2.3 (1.2–4.6)

Because fast acetylators are at increased risk, these results suggest that a genetic susceptibility to tobacco heterocyclic amines, such as PhIP (pyridine), rather than bicyclic aromatic amines, such as beta-naphthylamine and 4-aminobiphenyl, is involved in ETS-induced breast carcinogenesis. The concentration of heterocyclic amines in cigarette smoke is greater than the concentration of bicyclic aromatic amines, and heterocyclic amines are at least ten times more concentrated in the sidestream smoke than in the mainstream smoke. An analogy can be made here with colon cancer, which seems to be caused by dietary heterocyclic amines and for which a higher risk in fast acetylators has been consistently reported.[70] A relation of breast cancer to dietary heterocyclic amines is not well established,[71] but is supported by the results of the Iowa Woman Health Study.[72]

10.4.3 THE PARADOX

In most of these studies, the association with breast cancer is of similar magnitude for passive and active smoking. This observation is paradoxical, since active smokers are apparently more exposed to tobacco smoke than passive smokers. It is usually believed that the risk of breast cancer should be larger for active than for passive smoking because exposure to passive smoking is considered to be equivalent to a weak exposure to active smoke. It has, for example, been stated that "Passive smoking is on average perhaps equivalent to smoking one quarter to a half cigarette per day."[73] Thus, one might expect exposure to environmental tobacco smoke to be proportionally less carcinogenic.

However, it is well documented that sidestream smoke (an important component of environmental tobacco smoke) contains several polycyclic aromatic hydrocarbons (PAHs), nitrosamines, and aromatic amines, which are, at the tip of the cigarette, in much greater concentrations than in direct smoke.[31] This is due to the incomplete combustion that occurs when the tobacco burns at a lower temperature and because the sidestream smoke is unfiltered.[74] Some chemicals are up to 100 times more concentrated in the sidestream smoke (SS) than in the mainstream smoke (MS) of a non-filtered cigarette.[74] This SS/MS ratio is about 4 for the concentration of benzo(a)pyrene (a PAH and a known respiratory carcinogen) and about 30 for the concentration of another known carcinogen of the bladder, 4-aminobiphenyl.[75] However, passive smokers breathe diluted SS, and their exposure depends on the number of smokers, the size of the room, and the duration of exposure. Once inhaled into the lungs, carcinogens eventually pass into the bloodstream,[76] are transformed, and their metabolites chemically bond with, for example, hemoglobin or DNA to form adducts. These adducts have been measured in the blood and urine of smokers and non-smokers exposed to environmental tobacco smoke.[73] Vineis et al.[77] have found that women exposed to environmental tobacco smoke had more DNA adducts than active smokers. Thus, passive smoking may not be a weak equivalent of active smoking, but a different type of exposure.

An analogy with tobacco-induced lung cancer also leads to an expectation of a larger relative risk differential between active and passive smokers. The lung cancer model may not be relevant for the mammary gland because it is not directly exposed to tobacco smoke. Carcinogens need to circulate in the blood before reaching the gland and seem to be able to remain for a very long time in the mammary epithelial cells.[78,79] Environmental tobacco smoke particles are small and, therefore, poorly filtrated by the lung. The risk differential between passive and active smokers may be smaller for breast cancer than for lung cancer.

In addition, the relative risk observed for passive smokers may be magnified by "the low dose effect." Vineis and McMichael formulated this hypothesis for colon cancer:[70] the modifying effect of a genotype or phenotype can be more evident at low dose. It is reasonable, for example, to conceive that high levels of exposure to tobacco smoke saturate the NAT2 enzyme activity, resulting in formation of abundant DNA adducts from O-acetylation of hydroxylated heterocyclic amines in both slow and fast acetylators. In contrast, at low doses of exposure to tobacco, DNA adducts may accumulate only among fast acetylators. Among slow acetylators, the N-acetoxy derivatives are either produced in too small a quantity or detoxified at the same rate they are produced.

Furthermore, women may be exposed to environmental tobacco smoke before the onset of sexual maturity, when breast tissue is most susceptible to carcinogenic initiation, whereas exposure to active smoking usually begins near the time of sexual maturity, when susceptibility to carcinogenic initiation is reduced. Horton[80] correlated breast cancer incidence with cigarette sales and hypothesized that indoor tobacco smoke initiates breast cancer in a process that leads to tumor diagnosis 15 to 30 or more years later.

Thus, the paradox of equivalent relative risks for breast cancer associated with active smoking and exposure to environmental tobacco smoke may stem from some combination of the different chemistries of the two types of smoke and different age-dependent exposure patterns and their relationship with the breast tissue susceptibility to carcinogenesis.

10.5 CONCLUSIONS

The association between exposure to tobacco smoke, actively or through environmental tobacco smoke, is complex. The exposures may both cause the disease — through initiation of cancer while the breast tissue is susceptible — and prevent the disease through an apparent antiestrogenic effect. Because of the differences in exposure patterns by age, which are strongly correlated with reproductive milestones and, therefore, breast tissue susceptibility to carcinogenesis, both active smoking and passive smoking should be included in definitions of the exposed and unexposed conditions in epidemiologic studies. In addition to its dependence on breast tissue susceptibility, tobacco smoke's effect on breast cancer risk has shown some dependence on genetic polymorphisms.

No study has examined the effect of genetic polymorphisms within groups of ever-active smokers; passive-only smokers; and never-active, never-passive smokers while simultaneously taking account of breast tissue susceptibility at different reproductive milestones. Studies of the interaction between the genetic polymorphisms and biologically based categories of cigarette exposure would help to resolve the effect of cigarette smoking on breast cancer risk. A new generation of case-control studies are now needed to evaluate these complex associations. The next generation of studies should incorporate the following design features:

- Smoking groups divided into active, passive, and neither with information on lifetime exposure to environmental tobacco smoke in the home, workplace, and leisure environment
- These crude exposure groups further divided into exposure periods that correspond to reproductive milestones
- Analyses of the interaction between these well-defined exposures and genetic polymorphisms
- Sufficient sample size to allow investigations of the gene–environment interactions within the biologically relevant categories of exposure to active smoke and environmental tobacco smoke
- Recall enhancements to reduce the potential for disease-dependent recall or reporting of exposure to tobacco smoke and to reduce non-differential misclassification

While these studies would be difficult and resource intensive, the undertaking is worthwhile, given the dearth of lifestyle changes that can be recommended to reduce breast cancer risk. Furthermore, understanding how the risk of breast cancer associated with tobacco smoke varies with genotype and exposure relative to reproductive milestones will help to define biologically relevant exposure groups for other candidate causes of breast cancer.

REFERENCES

1. Madigan, M.P., Ziegler, R.G., Benichou, J., et al., Proportion of breast cancer cases in the United States explained by well-established risk factors, *J. Natl. Cancer Inst.*, 87, 1681–1685, 1995.
2. Fisher, B., Costantino, J.P., Wickerham, D.L., Redmond, C.K., Kavanah, M., Cronin, W.M., Vogel, V., Robidoux, A., Dimitrov, N., Atkins, J., Daly, M., Wieand, S., Tan-Chiu, E., Ford, L., Wolmark, N., and Other National Surgical Adjuvant Breast and Bowel Project Investigators, Tamoxifen for prevention of breast cancer: report of the National Surgical Adjuvant Breast and Bowel Project P-1 Study, *J. Natl. Cancer Inst.*, 90, 1371–1388, 1998.
3. Palmer, J.R. and Rosenberg, L., Cigarette smoking and the risk of breast cancer, *Epidemiol. Rev.*, 15, 145–156, 1993.
4. Braga, C., Negri, E., La Vecchia, C., et al., Cigarette smoking and the risk of breast cancer, *Eur. J. Cancer Prev.*, 5, 159–164, 1996.
5. Baron, J.A., Newcomb, P.A., Longnecker, M.P., et al., Cigarette smoking and breast cancer, *Cancer Epidemiol. Biomark. Prev.*, 5, 399–403, 1996.
6. Ranstam, J. and Olsson, H., Alcohol, cigarette smoking, and the risk of breast cancer, *Cancer Detect. Prev.*, 19, 487–493, 1995.
7. Calle, E.E., Miracle-McMahill, H.L., Thun, M.J., et al., Cigarette smoking and the risk of fatal breast cancer, *Am. J. Epidemiol.*, 139, 1001–1007, 1994.
8. Bennicke, K., Conrad, C., Sabroe, S., et al., Cigarette smoking and breast cancer, *Br. Med. J.*, 310, 1431–1433, 1995.
9. Vessey, M., Baron, J., Doll, R., et al., Oral contraceptives and breast cancer: final report of an epidemiological study, *Br. J. Cancer*, 47, 455–462, 1983.
10. O'Connell, D.L., Hulka, B.S., Chambless, L.E., et al., Cigarette smoking, alcohol consumption, and breast cancer risk, *J. Natl. Cancer Inst.*, 78, 229–234, 1987.
11. Gammon, M.D., Schoenberg, J.B., Teitelbaum, S.L., Brinton, L.A., Potischman, N., Swanson, C.A., Brogan, D.J., Coates, R.J., Malone, K.E., and Stanford, J.L., Cigarette smoking and breast cancer risk among young women (United States), *Cancer Causes Control*, 9, 583–90, 1998.
12. Ghadirian, P., Lacroix, A., Perret, C., Masionneuve, P., and Boyle, P., Sociodemographic characteristics, smoking, medical and family history, and breast cancer, *Cancer Detect. Prev.*, 22, 485–494, 1998.
13. Morabia, A., Bernstein, M., Heritier, S., et al., Relation of breast cancer with passive and active exposure to tobacco smoke, *Am. J. Epidemiol.*, 143, 918–928, 1996.
14. Wells, A.J., Breast cancer, cigarette smoking, and passive smoking. Letter to the Editor, *Am. J. Epidemiol.*, 133, 208–210, 1991.
15. Smith, S.J., Deacon, J.M., Chilvers, C.E.D., et al., Alcohol, smoking, passive smoking, and caffeine in relation to breast cancer risk in young women, *Br. J. Cancer*, 70, 112–119, 1994.
16. Jee, S.H., Ohrr, H., and Kim, I.S., Effects of husbands' smoking on the incidence of lung cancer in Korean women, *Int. J. Epidemiol.*, 28, 824–828, 1999.
17. Ambrosone, C.B., Freudenheim, J.L., Graham, S., et al., Cigarette smoking, N-acetyltransferase 2 genetic polymorphisms, and breast cancer risk, *J. Am. Med. Assoc.*, 276, 1494–1501, 1996.
18. Millikan, R.C., Pittman, G.S., Newman, B., et al., Cigarette smoking, N-Acetyltransferases 1 and 2, and breast cancer risk, *Cancer Epidemiol. Biomark. Prev.*, 7, 371–378, 1998.

19. Ambrosone, C.B. and Shields, P.G., Molecular epidemiology of breast cancer, *Prog. Clin. Biol. Res.*, 396, 83–99, 1997.

20. Morabia, A., Bernstein, M.S., Bouchardy, I., Kurtz, J., and Morris, M.A., Breast cancer, active and passive smoking: the role of the *N*-Acetyltransferase 2 genotype, *Am. J. Epidemiol.*, 152, 226–232, 2000.

21. Moolgavkar, S.H., Day, N.E., and Stevens, R.G., Two-stage model for carcinogenesis: epidemiology of breast cancer in females, *J. Natl. Cancer Inst.*, 65, 559–569, 1980.

22. Spicer, D.V. and Pike, M.C., Sex steroids and breast cancer prevention, *Monogr. Natl. Cancer Inst.*, 16, 139–147, 1994.

23. Colditz, G.A. and Frazier, A.L., Models of breast cancer show that risk is set by events of early life: prevention efforts must shift focus, *Cancer Epidemiol. Biomark. Prev.*, 4, 567–571, 1995.

24. Russo, J. and Russo, I.H., Toward a physiological approach to breast cancer prevention, *Cancer Epidemiol. Biomark. Prev.*, 3, 353–364, 1994.

25. Pike, M.C., Spicer, D.V., Dahmoush, L., et al., Estrogens, progestogens, normal breast cell proliferation, and breast cancer risk, *Epidemiol. Rev.*, 15, 17–35, 1993.

26. Miller, R.W., Special susceptibility of the child to certain radiation-induced cancers, *Environ. Health Perspect.*, 103(Suppl. 6), 41–44, 1995.

27. Tokunaga, M., Land, C.E., Tokuoka, S., Nishimori, I., Soda, M., and Akiba, S., Incidence of female breast cancer among atomic bomb survivors, 1950–1985, *Radiat. Res.*, 138, 209–223, 1994.

28. Hoffman, D.A., Lonstein, J.E., Morin, M.M., Visscher, W., Harris, B.S.H., and Boice, J.B., Breast cancer in women with scoliosis exposed to multiple diagnostic x-rays, *J. Natl. Cancer Inst.*, 81, 1307–1312, 1989.

29. Modan, B., Chetrit, A., Alfandary, E., and Katz, L., Increased risk of breast cancer after low-dose irradiation, *Lancet*, 1, 629–631, 1989.

30. Aisenberg, A.C., Finkelstein, D.M., Doppke, K.P., Koerner, F.C., Boivin, J., and Willet, C.G., High risk of breast carcinoma after irradiation of young women with Hodgkin's disease, *Cancer*, 79, 1203–1210, 1997.

31. IARC, Tobacco smoking, in *IARC Monographs on the Evaluation of the Carcinogenic Risk of Chemicals to Humans, Volume 38*, International Agency for Research on Cancer, Lyon, 1986.

32. Phillips, D.H., DNA adducts in human tissues: biomarkers of exposure to carcinogens in tobacco smoke, *Environ. Health Perspect.*, 104(Suppl. 3), 453–458, 1996.

33. Li, D., Wang, M., Dhingra, K., and Hittelman, W.N., Aromatic DNA adducts in adjacent tissues of breast cancer patients: clues to breast cancer etiology, *Cancer Res.*, 56, 287–293, 1996.

34. Eldridge, S.R., Gould, M.N., and Butterworth, B.E., Genotoxicity of environmental agents in human mammary epithelial cells, *Cancer Res.*, 52, 5617–5621, 1992.

35. Tonelli, Q.J., Custer, R.P., and Sorof, S., Transformation of cultured mouse mammary glands by aromatic amines and amides and their derivatives, *Cancer Res.*, 39, 1784–1792, 1979.

36. Stampfer, M.R., Bartholomew, J.C., Smith, H.S., and Bartley, J.C., Metabolism of benzo[a]pyrene by human mammary epithelial cells: toxicity and DNA adduct formation, *Proc. Natl. Acad. Sci. U.S.A.*, 78, 6251–6255, 1981.

37. Hein, D.W., Acetylator genotype and arylamine-induced carcinogenesis, *Biochim. Biophys. Acta*, 948, 37–66, 1988.

38. Presentation by Christine Ambrosone, Workshop on Tobacco, Breast Cancer, and Genetics, Geneva, Switzerland, August 1999.

39. Baron, J.A., La Vecchia, C., and Levi, F., The antiestrogenic effect of cigarette smoking in women, *Am. J. Obstet. Gynecol.*, 162, 502–514, 1990.

40. Presentation by Professor John Baron, Workshop on Tobacco, Breast Cancer, and Genetics, Geneva, Switzerland, August 1999.

41. Michnovicz, J.J., Hershcopf, R.J., Naganuma, H., et al., Increased 2-hydroxylation of estradiol as a possible mechanism for the antiestrogenic effect of cigarette smoking, *N. Engl. J. Med.*, 315, 1305–1309, 1986.

42. Michnovicz, J.J., Naganuma, H., Hershcopf, R.J., et al., Increased urinary catechol estrogen excretion in female smokers, *Steroids*, July-August, 69–83, 1988.

43. Cooper, J.A., Rohan, T.E., Cant, E.L., Horsfall, D.J., and Tilley, W.D., Risk factors for breast cancer by oestrogen receptor status: a population-based case-control study, *Br. J. Cancer*, 59, 119–123, 1989.

44. Habel, L.A. and Stanford, J.L., Hormone receptors and breast cancer, *Epidemiol. Rev.*, 15, 209–219, 1993.

45. London, S.J., Colditz, G.A., Stampfer, M.J., Willett, W.C., Rosner, B.A., and Speizer, F.E., Prospective study of smoking and the risk of breast cancer, *J. Natl. Cancer Inst.*, 81, 1625–1631, 1989.

46. Morabia, A., Bernstein, M., Ruiz, J., Heritier, S., Berger, S.D., and Borisch, B., Relation of smoking to breast cancer by estrogen receptor status, *Int. J. Cancer*, 75, 339–342, 1998.

47. Yoo, K.Y., Tajima, K., Miura, S., Takeuchi, T., et al., Breast cancer risk factors according to combined estrogen and progesterone receptor status: a case-control analysis, *Am. J. Epidemiol.*, 146, 307–314, 1997.

48. Morabia, A. and Bernstein, M., A review of the relation of smoking (active and passive) to breast cancer, *J. Women's Cancer*, 2, 1–9, 2000.

49. MacMahon, B., Wald, N., and Baron, J., Eds., Cigarette smoking and cancer of the breast, *Smoking and Hormone-Related Disorders*, Oxford University Press, London, 1990, 15.

50. Palmer, J.R., Rosenberg, L., Clarke, E.A., Stolley, P.D., Warshauer, M.E., Zauber, A.G., and Shapiro, S., Breast cancer and cigarette smoking: a hypothesis, *Am. J. Epidemiol.*, 134, 1–13, 1991.

51. Field, N.A., Baptiste, M.S., Nasca, P.C., and Metzger, B.B., Cigarette smoking and breast cancer, *Int. J. Epidemiol.*, 21, 842–848, 1992.

52. Ewertz, M., Breast cancer and cigarette smoking: a hypothesis [letter], *Am. J. Epidemiol.*, 135, 1185, 1992.

53. Sandler, D.P., Everson, R.B., and Wilcox, A.J., Cigarette smoking and breast cancer [letter], *Am. J. Epidemiol.*, 123, 370–371, 1986.

54. Hunter, D.J., Hankinson, S.E., Hough, H., Gertig, D.M., Garcia-Closas, M., Spiegelman, D., Manson, J.E., Colditz, G.A., Willett, W., Speizer, F.E., and Kelsey, K., A prospective study of NAT2 acetylation genotype, cigarette smoking, and risk of breast cancer, *Carcinogenesis*, 18, 2127–2132, 1997.

55. Morabia, A., Bernstein, M., and Heritier, S., Re: A prospective study of NAT2 acetylation genotype, cigarette smoking and risk of breast cancer, *Carcinogenesis*, 19, 1705, 1998.

56. Hunter, D., Gertig, D., and Spiegelman, D., A prospective study of NAT2 acetylation genotype, cigarette smoking and risk of breast cancer, *Carcinogenesis*, 19, 1705, 1998.

57. Ambrosone, C.B., Thompson, P.A., Shields, P.G., et al., Catechol-O-methyltransferase (COMT) genetic polymorphisms, smoking, and breast cancer risk, Abstract #259, *Am. J. Epidemiol.*, 147(Suppl. 1), S65, 1998.

58. Thompson, P.A., Shields, P.G., Freudenheim, J.L., Stone, A., Vena, J.E., Marshall, J.R., Graham, S., Laughlin, R., Nemoto, T., Kadlubar, F.F., and Ambrosone, C.B., Genetic polymorphisms in catechol-O-methyltransferase, menopausal status, and breast cancer risk, *Cancer Res.*, 58, 2107–2110, 1998.

59. Yager, J.D. and Liehr, J.G., Molecular mechanisms of estrogen carcinogenesis, *Annu. Rev. Pharmacol. Toxicol.*, 36, 230–232, 1996.

60. Lavigne, J.A., Helzlsouer, K.J., Huang, H., Strickland, P.T., Bell, D.A., Selmin, O., Watson, M.A., Hoffman, S., Comstock, G.W., and Yager, J.D., An association between the allele coding for a low activity variant of catechol-O-methyltransferase and the risk for breast cancer, *Cancer Res.*, 57, 5493–5497, 1997.

61. Millikan, R.C., Pittman, G.S., Tse, C.J., Duell, E., Newman, B., Savitz, D., Moorman, P.G., Boissy, R.J., and Bell, D.A., Catechol-O-methyltransferase and breast cancer risk, *Carcinogenesis*, 19, 1943–1947, 1998.

62. Zheng, W., Deitz, A.C., Campbell, D.R., Wen, W., Cerhan, J.R., Sellers, T.A., Folsom, A.R., and Hein, D.W., N-acetyltransferase 1 genetic polymorphism, cigarette smoking, well-done meat intake, and breast cancer risk, *Cancer Epidemiol. Biomark. Prev.*, 8, 233–239, 1999.

63. Brunet, J.S., Ghadirian, P., Rebbeck, T.R., Lerman, C., Garber, J.E., Tonin, P.N., Abrahamson, J., Foulkes, W.D., Daly, M., Wagner-Costalas, J., Godwin, A., Olopade, O.I., Moslehi, R., Liede, A., Futreal, P.A., Weber, B.L., Lenoir, G.M., Lynch, H.T., and Narod, S.A., Effect of smoking on breast cancer in carriers of mutant BRCA1 or BRCA2 genes, *J. Natl. Cancer Inst.*, 90, 761–766, 1998.

64. Ishibe, N., Hankinson, S.E., Colditz, G.A., Spiegelman, D., Willett, W.C., Speizer, F.E., Kelsey, K., and Hunter, D.J., Cigarette smoking, cytochrome P4501A1 polymorphisms, and breast cancer risk in the Nurses' Health Study, *Cancer Res.*, 58, 667–671, 1998.

65. Sandler, D.P., Everson, R.B., and Wilcox, A.J., Passive smoking in adulthood and cancer risk, *Am. J. Epidemiol.*, 121, 37–48, 1985.

66. Wells, A.J., Breast cancer, cigarette smoking, and passive smoking [letter], *Am. J. Epidemiol.*, 147, 991–992, 1998.

67. Johnson, K.C., Semenciw, R., Hu, J., et al., Active and passive smoking and premenopausal breast cancer [Abstract], *Am. J. Epidemiol.*, 47, S68, 1998.

68. Lash, T.L. and Aschengrau, A., Active and passive cigarette smoking and the occurrence of breast cancer, *Am. J. Epidemiol.*, 149, 5–12, 1999.

69. Hirayama, T., Epidemiología y factores de riesgo del cáncer de mama. Cancer de mama, *Avances en Diagnóstico y Tratamiento*, Diaz-Faes, J., Eds., Santiago García, León, 1990, 21–38.

70. Vineis, P. and McMichael, A., Interplay between heterocyclic amines in cooked meat and metabolic phenotype in the etiology of colon cancer, *Cancer Causes Control*, 7, 479–486, 1996.

71. Snyderwine, E.G., Dietary heterocyclic amines and breast cancer, *Women Cancer*, 1, 21–29, 1999.

72. Zheng, W., Gustafson, D.R., Sinha, R., et al., Well-done meat intake and the risk of breast cancer, *J. Natl. Cancer Inst.*, 90, 1724–1729, 1998.

73. Hackshaw, A.K., Lung cancer and passive smoking, *Stat. Methods Med. Res.*, 7, 119–136, 1998.

74. U.S. Environmental Protection Agency, Respiratory Healths Effects of Passive Smoking: Lung Cancer and Other Disorders, EPA/600//6-90/006F, U.S. EPA, Office of Research and Development RD-689, Washington, D.C., 1992.

75. Guerin, M.R., Jenkins, R.A., and Tomkins, B.A., The Chemistry of ETS: Composition and Measurement, Lewis Publishers, Chelsea, MI, 1992.

76. Hecht, S.S., Carmella, S.G., Murphy, S.E., Akerkar, S., Brunnemann, K.D., and Hoffmann, D., A tobacco-specific lung carcinogen in the urine of men exposed to cigarette smoke, *N. Engl. J. Med.*, 329, 1543–1546, 1993.
77. Vineis, P., Bartsch, H., Caporaso, N., Harrington, A.M., et al., Genetically based n-acetyltransferase metabolic polymorphism and low-level environmental exposure to carcinogens, *Nature*, 369, 154–156, 1994.
78. Petrakis, N.L., Gruenke, L.D., Beelen, T.C., et al., Nicotine in breast fluid of nonlactating women, *Science*, 199, 303–305, 1978.
79. Petrakis, N.L., Nipple aspirate fluid in epidemiologic studies of breast disease, *Epidemiol. Rev.*, 15, 188–195, 1993.
80. Horton, A.W., Epidemiologic evidence for the role of indoor tobacco smoke as an initiator of human breast carcinogenesis, *Cancer Detect. Prev.*, 16, 119–127, 1992.

11 Environmental Tobacco Smoke: Potential Effects on Cardiovascular Function in Children and Adolescents

Samuel S. Gidding

CONTENTS

11.1 INTRODUCTION

This chapter reviews known cardiovascular effects of environmental tobacco smoke exposure on children and adolescents, as well as young adults. Though it can easily be postulated based on the relevant literature that adverse effects occur, it has not yet been demonstrated that past smoke exposure in childhood produces cardiovascular morbidity in adulthood. It seems likely that passive smoke exposure is synergistic with other known risk factors for coronary artery disease risk. Further studies are needed to determine if passive smoke exposure is related to vascular disease in childhood.

It has been concluded that there is a significant relationship between environmental tobacco smoke and coronary disease, with most studies showing a 20 to 40% increase in cardiovascular morbidity and mortality in non-smokers who are consistently exposed to environmental tobacco smoke. These studies have been conducted both at home and in the workplace.[1-9] This is in contrast to an approximate tripling of cardiovascular risk from tobacco use itself. These conclusions are derived from a careful review of scientific evidence both by individuals and by many scientific

0-8493-0311-7/00/$0.00+$.50
© 2001 by CRC Press LLC

advisory panels.[1-9] There are many known adverse cardiovascular effects of passive smoke exposure. One final common pathway is thought to be vascular endothelial injury, which results in endothelial dysfunction and the production of atherosclerotic lesions. Mechanisms along the pathway to endothelial dysfunction include an increased oxidation potential of low density lipoprotein (LDL) after exposure to smoke, diminished high density lipoprotein (HDL) cholesterol levels, an increase in blood coaguability, and an overall diminished systemic oxygen transport.

The U.S. Surgeon General has reported on the all-cause morbidity and mortality related to tobacco smoke exposure in children and adolescents.[10] Support for this document has come from the American Heart Association and the American Academy of Pediatrics.[11,12] This morbidity is related to both environmental smoke exposure and to tobacco use in adolescence. However, there has only been limited research on the cardiovascular effects of passive smoke exposure in children. There have been limitations to conduct such research: the absence of easily measured cardiovascular disease endpoints, the non-routine availability of good measures of environmental tobacco smoke exposure other than the medical history of familial smoking, and ethical constraints on chronically exposing children to environmental tobacco smoke for research purposes. Also, there are no non-invasive measures of early athero-sclerosis sufficiently sensitive to test the hypothesis that environmental tobacco smoke exposure can cause premature atherogenesis in children.

This chapter will review the cardiovascular effects of environmental tobacco smoke from the perspective of the limited literature on children and adolescents and will also rely on animal and adult studies for further support where plausible. Endpoints in these studies are generally intermediate as coronary artery disease is chronic, with origins in childhood and endpoints in adulthood.

11.2   INITIATION OF ATHEROSCLEROSIS

That smoking can initiate atherosclerosis is well substantiated in the literature.[1-4,6,8] Since the earliest atherosclerotic lesions (fibrous plaques) occur in the second decade, and these plaque formations are associated with tobacco use, it can safely be presumed that smoking initiates atherosclerosis in adolescence.[13]

Passive smoking by itself can initiate endothelial injury.[14,15] Studies have shown increased blood endothelial cell counts in individuals exposed to environmental tobacco smoke and have also shown enhanced platelet aggregability. Thus, environ-mental tobacco smoke can initiate the earliest steps in plaque formation, endothelial injury. Many chemicals in tobacco smoke have been thought to be responsible; 1,3-butadiene and possibly nicotine may have major roles.[16-20] Progression of atherosclerosis with LDL accumulation in the carotid arterial wall has also been associated with environmental tobacco smoke.[20] Tobacco smoke exposure may enhance this process by increasing the binding of oxidized LDL to the carotid vessel wall.[20,21] Animal studies in cockerels and the cholesterol-fed rabbit have both shown that smoke exposure increases the development of atherosclerotic plaque.[22-24] In the vascular tissue of rats, glycosaminoglycans and glycoproteins accumulate in response to cigarette smoke exposure, another early event in atherosclerosis.[25] It does not appear that the sympathetic nervous system is involved in this process.[26]

Clinical correlation of the basic research finding that environmental smoke exposure enhances atherosclerosis has been established using carotid wall thickness, an accepted non-invasive measure of atherosclerotic burden with good correlation to coronary atherosclerosis.[27–29] These studies have consistently shown that non-smokers with environmental tobacco smoke exposure have greater carotid wall thickness than non-smokers without environmental tobacco smoke exposure. Further, in follow-up studies, progression of atherosclerosis has been associated with environmental tobacco smoke exposure.[29] These findings are independent of other cardiovascular risk factors. Interestingly, for individuals with diabetes mellitus, the rates of atherosclerosis progression from passive smoke exposure are accelerated. Though the environmental smoke exposure effect was smaller than that seen in regular smokers, it remained highly statistically significant.

It is known that children exposed to environmental tobacco smoke have levels of thiocyanate in their blood sufficiently high to cause alterations in oxygen transport.[30] From the Pathobiological Determinants of Atherosclerosis in Youth (PDAY) Study, it is also known that the amount of atherosclerosis in youth is directly related to thiocyanate levels in the blood.[13,21] It has further been shown in this study that the incorporation of oxidized LDL into early atherosclerotic lesions is also associated with thiocyanate levels. Though the PDAY Study findings are most relevant to regular smoking, the role of passive smoke exposure and other causes of thiocyanate in the blood cannot be excluded.

11.3 OXYGEN TRANSPORT

Carbon monoxide is an important component of tobacco smoke and binds tightly to hemoglobin, displacing oxygen in the blood. This results in a reduction of oxygen carrying capacity. The physiologic effects of passive smoking on oxygen transport have been well reviewed.[4,11] In children, the concentration of red cell 2,3-diphosphoglycerate, a compound which regulates oxygen delivery at the tissue level, is known to be higher in children of smokers and is also proportional to the serum thiocyanate levels of these children.[30,31] Thus, passive smoking can be related to other chronic causes of tissue hypoxia in childhood, such as anemia, cyanotic heart disease, chronic lung disease, and prolonged exposure to high altitude. Adult studies have shown that exercise performance and time to recovery of the baseline heart rate are impaired in both healthy adult males and male survivors of myocardial infarction. All these findings are associated with elevated carbon monoxide levels.[32,33]

There is a substantial body of literature from the animal laboratory on smoke exposure, oxygen transport, and myocardial metabolism. These have included studies of experimental myocardial infarction in both adult and immature animals. They have consistently shown that the effects on cardiac metabolism in adults are also true in younger animals.[34,35] In particular, extension of infarct size after exposure to environmental tobacco smoke is similar in young and adult rats. Explanations for this extension of infarct size after environmental tobacco smoke exposure could include increased myocardial oxygen needs, increased carboxyhemoglobin levels, abnormal mitochondrial cytochrome oxidase activity, and sensitivity to ischemia reperfusion injury.[36–39] L-Arginine attenuates this effect, suggesting environmental tobacco smoke impairs endothelial dependent vasodilation.[40] One study has also suggested that early environmental tobacco smoke exposure may adversely impact ischemic events later in life.[34]

However, the clinical significance of environmental tobacco exposure in childhood to acute vascular events in childhood remains uncertain because vascular ischemic events are rare at this age. Further study of ischemic events in childhood should be undertaken to determine if smoke exposure is in fact a risk factor.

11.4 LIPOPROTEINS

Lipoproteins are well known to mediate cardiovascular risk. Adverse effects from environmental tobacco smoke on lipoproteins have recently been observed. Lower HDL cholesterol levels, a known coronary risk factor, have recently been demonstrated in children with environmental smoke exposure.[30,41] The magnitude of change was in the range of 5 to 10%. It has recently been confirmed that children with severe hypercholesterolemia and family members who smoke have lower HDL cholesterol.[42] Studies in Japan and Turkey have not shown a similar reduction in HDL cholesterol, but they have shown changes in the LDL to HDL cholesterol ratio.[43,44] Collectively, all pediatric studies on environmental smoke exposure and lipoproteins show changes similar to those observed in adolescent regular smokers.[45] The mechanism for this effect may be inhibition of plasma lecithin/cholesterol acyl transferase activity and also altered chylomicron remnant clearance by the liver.[46,47]

Another effect on lipoproteins is the association of LDL oxidation with environmental smoke exposure. This is particularly important because of the relationship of oxidized LDL to early atherogenesis and the progression of atherosclerotic lesions. Already mentioned in this chapter is the interrelationship of oxidized LDL and thiocyanate in the blood to accelerated atherosclerosis progression.[21] Several adult studies have shown a positive relationship between environmental tobacco smoke and oxidative stress.[48–51] One study showed lower serum levels of ascorbic acid, an important anti-oxidant, with smoke exposure.[48] Another study, conducted in a work environment, demonstrated higher levels of oxidative DNA mutagen and catalase and glutatione peroxidase.[49] There are also studies demonstrating limited serum anti-oxidant defense and LDL capacity to resist oxidation.[50] In smoke-exposed subjects, increased components of lipid peroxidation have appeared in serum and their LDL is more readily taken up by cultured macrophages. In hypercholesterolemic rabbits, exposure to anti-oxidant vitamins attenuates the effects of environmental tobacco smoke exposure.[51] Human plasma from smoke-exposed individuals has been used in animal studies to show similar anti-oxidant effects.[47]

Lower HDL cholesterol may be associated with ease of LDL oxidation. HDL is lower in smoke-exposed individuals. Since HDL is the major lipoprotein carrier of cholesterol ester hydroperoxides, *in vitro* HDL may decrease lipid peroxides generated on LDL during oxidation.[52,53]

11.5 VASCULAR FUNCTION

As a final common pathway to the previous observations, it would be useful to know that vascular dysfunction after environmental smoke exposure can be demonstrated.

In fact, this has occurred in two studies of individuals who smoke, those who have environmental smoke exposure and those who do not smoke.[54,55] The response of the endothelium to smoke exposure is dose dependent, being most abnormal in those with the highest smoke exposure, and suggests the presence of vascular injury. Also, this endothelial dysfunction appears to be a manifestation of the presence of autoantibodies to oxidixed LDL.[56] These effects have been observed not only in smokers, but also in adolescents and young adults who have been exposed to environmental tobacco smoke.[57] Again, these effects were dose dependent, and at the high exposure end, adverse effects were similar to those individuals who use tobacco chronically. In a recent publication, these effects have been shown to be partially reversible in those whose tobacco smoke exposure is more distant.[58] Vascular dysfunction has been shown in large vessels in cardiac catheterization studies of cardiac elasticity in adult males with chest pain. Here, comparisons were made between smokers and non-smokers before and after passive smoke exposure. The adverse effect of passive smoke exposure was similar to the effect of smoking chronically.[59] However, not all studies of vascular dysfunction related to smoke exposure have shown statistically significant effects. In a study which used cotinine as a surrogate for smoke exposure as part of a larger study of risk factors on the vascular reactivity of 9 to 11 year olds, no relationship was found.[60] Effects on endothelial dysfunction may be expressed differently in different races as Chinese do not appear to respond in the same way as Caucasians.[61] Laboratory studies of endothelial dysfunction after environmental smoke exposure are consistently positive.[62,63] Testosterone may exacerbate these effects.[64]

To conclude, there is some evidence that endothelial dysfunction after exposure to environmental tobacco smoke exists in adolescents and young adults. However, the magnitude of the effect and the severity of the injury remain unknown. It does appear that this effect is reversible when it can be demonstrated.

11.6 THROMBOSIS

To have a myocardial infarction, one must first have an atherosclerotic plaque which, after a particular stress, initiates an acute thrombotic event. A critical step in the evolution of understanding the relationship between smoking and myocardial infarction has been the demonstration of enhanced potential for thrombosis after exposure to environmental tobacco smoke. This is true whether the smoke exposure is active or passive.[1,7] Increased thrombotic potential may be a critical explanation of the observed increase in ischemic events in adults who smoke or have passive smoke exposure.[7] Animal studies have consistently observed prothrombotic effects, including decreased bleeding time and enhanced platelet aggregability, after passive smoke exposure.[23,35] Several studies in humans have shown this finding, and also, it has been shown that the non-smoker acutely exposed to ETS may be more sensitive to thrombotic and platelet aggregating effects.[14,15,65] The large body of literature on this topic has recently been reviewed.[66] Tobacco smoke directly interferes with platelet activating factor acetylhydrolase in neutralizing platelet activating factor activity.[66] An interesting observation is that platelet activating factor is developmentally regulated and does not reach its full expression until 6 weeks of age. Therefore, maternal smoke exposure of the fetus and newborn may increase neonatal morbidity from thrombosis.[67]

11.7   FETAL AND INFANT EXPOSURE

That fetal exposure to tobacco smoke has adverse effects is well known. Metabolites of tobacco smoke have been identified at birth in the hair of infants born to mothers who smoke.[68] A number of adverse outcomes in the neonate could be attributed to cardiovascular effects on the fetus of maternal smoking. It is known that maternal smoking leads to increases in carboxyhemoglobin in maternal and fetal blood, with higher levels in the fetus than the mother because of increased fetal hemoglobin affinity for carbon monoxide.[69] The fetus and neonate will have reduced oxygen carrying capacity and also reduction in the partial pressure of oxygen where oxygen is delivered to the fetal tissues. There are also ultrastructural changes consistent with diminished placental blood flow in maternal placentas.[70] Microcirculatory function during the first few days of life has been shown to be impaired by maternal smoking.[71] This finding may be explained by reduced prostacyclin biosynthesis in the blood vessels of smokers and the umbilical arteries of infants of smoking mothers.

An association with higher cotinine levels in cord blood has been shown in infants born with persistent pulmonary hypertension.[72] Maternal smoke exposure can either be active or passive. It is of interest that animal studies of passive smoke exposure in cholesterol-fed rabbits have shown both systemic and pulmonary arterial atherosclerosis.[24]

Another association with maternal smoking relates to the diet of low-income children. If their parents smoke, the diet quality of these low-income children is worse than those whose parents do not smoke.[73] Other health behaviors such as decreased participation in physical activity, degree of overweight, and increased hours of television watched are seen in children of smokers.[74] Most of the differences observed in these studies would increase cardiovascular risk.

Finally, a recent report suggests maternal smoking may influence infant heart rate variability and the autonomic nervous system.[75]

11.8   ADVERTISING AND THE MEDIA

As a final consideration, though not "smoke" exposure, the passive and involuntary exposure of children and adolescents to the promotional activities of tobacco companies should be considered a form of environmental tobacco exposure. The usefulness of advertising and promotional activities to entice children to become regular smokers has been recognized by behavioral research on tobacco advertising, interviews with children, and in recently released memoranda from tobacco companies.[10,76,77] In legal challenges in the U.S. to bans on billboard advertising of tobacco products, the courts have accepted as evidence of efficacy the large investment in advertising and promotion of tobacco products without requiring scientific evidence.[76] The goal of promotional activities is to make tobacco use seem an acceptable and pleasurable part of everyday life; children's exposure to these activities is involuntary whether it be through exposure to advertising (billboards, magazine ads), by solicitation to use merchandise, or in the media (product placement in movies and television, prominent entertainers and sports figures shown enjoying tobacco).

11.9 SUMMARY AND CONCLUSIONS

Environmental smoke exposure in children has measurable cardiovascular effects. What remains unknown is the clinical significance of these effects. The distance between childhood and coronary events in adulthood can be measured in decades; it is unlikely that any study will ever prove that environmental smoke exposure directly contributes to vascular disease later in life. However, there is a wealth of evidence that suggests environmental smoke exposure is synergistic with other known coronary risk factors and exacerbates the atherosclerotic process.[78] Most important, some data are evolving which suggest that endothelial dysfunction may be present, at least to a small degree, in relationship to passive smoke exposure. Given the impressive weight of basic scientific evidence on adverse effects of environmental smoke exposure, it will be important to conduct pediatric studies to determine if environmental tobacco smoke exposure and/or fetal smoke exposure contribute to the risk for ischemic events or other vascular diseases in neonates and children.

REFERENCES

1. Glantz, S.A. and Parmley, W.W., Passive smoking and heart disease. Epidemiology, physiology, and biochemistry, *Circulation*, 83, 1, 1991.
2. Taylor, A.E., Johnson, D.C., and Kazemi, H., Environmental tobacco smoke and cardiovascular disease (a position paper from the Council on Cardiopulmonary and Critical Care, American Heart Association), *Circulation*, 86, 699, 1992.
3. Wells, A.J., Passive smoking as a cause of heart disease, *J. Am. Coll. Cardiol.*, 24, 546, 1994.
4. Glantz, S.A. and Parmley, W.W., Passive smoking and heart disease. Mechanisms and risk, *J. Am. Med. Assoc.*, 273, 1047, 1995.
5. California Environmental Protection Agency, Health effects of exposure to environmental tobacco smoke, 1997. (http://www.calepa.cahwnet.gov/oehha/docs/finalets.htm)
6. Thomas, D., Passive smoking and cardiovascular disease (article in French), *Bull. Acad. Natl. Med.*, 181, 743, 1997.
7. Law, M.R., Morris, J.K., and Wald, N.J., Environmental tobacco smoke exposure and ischaemic heart disease: an evaluation of the evidence, *Br. Med. J.*, 315, 973, 1997.
8. Wells, A.J., Heart disease from passive smoking in the workplace, *J. Am. Coll. Cardiol.*, 31, 1, 1998.
9. United Kingdom Department of Health, Report of the Scientific Committee on Tobacco and Health, 1998. (http://www.open.gov.uk/doh/public/scoth.htm)
10. U.S. Department of Health and Human Services, Preventing tobacco use among young people: a report of the Surgeon General. Executive Summary, *Morb. Mortal Wkly. Rep. Mar. 11*, 43(RR-4), 1, 1994.
11. American Heart Association, Active and passive tobacco exposure: a serious pediatric health problem, A Statement from the Committee on Atherosclerosis and Hypertension in Children, Council on Cardiovascular Disease in the Young, *Circulation*, 90, 2581, 1994.
12. Gidding, S.S. and Schydlower, M., Active and passive tobacco exposure: a serious pediatric health problem (Commentary), *Pediatrics*, 94, 750, 1994.

13. PDAY Research Group, Relationship of athersclerosis in young men to serum lipo-protein cholesterol concentrations and smoking, A preliminary report from the Patho-biological Determinants of Atherosclerosis in Youth (PDAY) Research Group, *J. Am. Med. Assoc.*, 264, 3018, 1990.

14. Davis, J.W., Shelton, L., Watanabe, I.S., and Arnold, J., Passive smoking affects endothelium and platelets, *Arch. Intern. Med.*, 149, 386, 1989.

15. Davis, J.W., Some effects of smoking on endothelial cells and platelets: pathogenesis and cellular mechanisms, in *Tobacco Smoking and Atherosclerosis*, Diana, J.N., Ed., Plenum Press, New York, 1990; *Adv. Exp. Med. Biol.*, 273, 1990.

16. Zimmerman, M. and McGeachie, J., The effect of nicotine on aortic endothelium. A quantitative ultrastructural study, *Atherosclerosis*, 63, 33, 1987.

17. Lin, S.J., Hong, C.Y., Chang, M.S., Chiang, B.N., and Chien, S., Long-term nicotine exposure increases aortic endothelial cell death and enhances transendothelial mac-romolecular transport in rats, *Arterioscler. Thromb.*, 12, 1305, 1992.

18. Penn, A. and Snyder, C.A., Butadiene inhalation accelerates arteriosclerotic plaque development in cockerels, *Toxicology*, 113, 351, 1996.

19. Penn, A. and Snyder, C.A., 1,3 Butadiene, a vapor phase component of environmental tobacco smoke, accelerates arteriosclerotic plaque development, *Circulation*, 93, 552, 1996.

20. Roberts, K.A., Rezai, A.A., Pinkerton, K.E., and Rutledge, J.C., Effect of environ-mental tobacco smoke on LDL accumulation in the artery wall, *Circulation*, 94, 2248, 1996.

21. Scanlon, C.E., Berger, B., Malcom, G., and Wissler, R.W., Evidence for more extensive deposits of epitopes of oxidized low density lipoprotein in aortas of young people with elevated serum thiocyanate levels, *Atherosclerosis*, 121, 23, 1996.

22. Penn, A. and Snyder, C.A., Inhalation of sidestream cigarette smoke accelerates development of arteriosclerotic plaques, *Circulation*, 88(4 Pt. 1), 1820, 1993.

23. Penn, A., Chen, L.C., and Snyder, C.A., Inhalation of steady-state sidestream smoke from one cigarette promotes arteriosclerotic plaque development, *Circulation*, 90, 1363, 1994.

24. Zhu, B.Q., Sun, Y.P., Sievers, R.E., Isenberg, W.M., Glantz, S.A., and Parmley, W.W., Passive smoking increases experimental atherosclerosis in cholesterol-fed rabbits, *J. Am. Coll. Cardiol.*, 21, 225, 1993.

25. Latha, M.S., Vijayammal, P.L., and Krup, P.A., Changes in the glycosaminoglycans and glycoproteins in the tissues in rats exposed to cigarette smoke, *Atherosclerosis*, 31, 49, 1991.

26. Sun, Y.P., Zhu, B.Q., Sievers, R.E., Glantz, S.A., and Parmley, W.W., Metroprolol does not attenuate atherosclerosis in lipid-fed rabbits exposed to environmental tobacco smoke, *Circulation*, 89, 2260, 1994.

27. Howard, G., Wagenknecht, L.E., Burke, G.L., Diez-Roux, A., Evans, G.W., McGovern, P., Nieto, F.J., and Tell, G.S., Active and passive smoking are associated with increased carotid wall thickness, *Arch. Intern. Med.*, 154, 1277, 1994.

28. Diez-Roux, A.V., Nieto, F.J., Comstock, G.W., Howard, G., and Szklo, M., The relationship of active and passive smoking to carotid atherosclerosis 12-14 years later, *Prev. Med.*, 24, 48, 1995.

29. Howard, G., Wagenknecht, L.E., Burke, G.L., Diez-Roux, A., Evans, G.W., McGovern, P., Nieto, F.J., and Tell, G.S., Cigarette smoking and progression of atherosclerosis: the Atherosclerosis Risk in Communities (ARIC) Study, *J. Am. Med. Assoc.*, 279, 119, 1998.

30. Moskowitz, W.B., Mosteller, M., Schieken, R.M., Bossano, R., Hewitt, J.K., Bodurtha, J.N., and Segrest, J.P., Lipoprotein and oxygen transport alterations in passive smoking preadolescent children. The MCV Twin Study, *Circulation*, 81, 586, 1990.

31. Moskowitz, W.B., Mosteller, M., Hewitt, J.K., Eaves, L.J., Nance, W.E., and Schieken, R.M., Univariate genetic analysis of oxygen transport regulation in children: the Medical College of Virginia Twin Study, *Pediatr. Res.*, 33, 645, 1993.

32. McMurray, R.G., Hicks, L.L., and Thompson, D.L., The effects of passive inhalation of cigarette smoke on exercise performance, *Eur. J. Appl. Physiol.*, 54, 196, 1985.

33. Leone, A., Mori, L., Bertanelli, F., Fabiano, P., and Filippelli, M., Indoor passive smoking: its effect on cardiac performance, *Int. J. Cardiol.*, 33, 247, 1991.

34. Zhu, B.Q., Sun, Y.P., Sudhir, K., Sievers, R.E., Browne, A.E., Gao, L., Hutchison, S.J., Chou, T.M., Deedwania, P.C., Chatterjee, K., Glantz, S.A., and Parmley, W.W., Effects of second-hand smoke and gender on infarct size of young rats exposed in utero and in the neonatal to adolescent period, *J. Am. Coll. Cardiol.*, 30, 1878, 1997.

35. Zhu, B.Q., Sun, Y.P., Sievers, R.E., Glantz, S.A., Parmley, W.W., and Wolfe, C.L., Exposure to environmental tobacco smoke increases myocardial infarct size in rats, *Circulation*, 89, 1282, 1994.

36. Gvozdjakova, A., Bada, V., Sany, L., Kucharska, J., Kruty, F., Bozek, P., Trstansky, L., and Gvozdjak, J., Smoke cardiomyopathy: disturbance of oxidative processes in myocardial mitochondria, *Cardiovasc. Res.*, 18, 229, 1984.

37. Gvozdjakova, A., Kucharska, J., and Gvozdjak, J., Effect of smoking on the oxidative processes of cardiomyocytes, *Cardiology*, 81, 81, 1992.

38. Aronow, W.S., Effect of passive smoking on angina pectoris, *N. Engl. J. Med.*, 299, 21, 1978.

39. Van Jaarsveld, H., Kuyl, J.M., and Alberts, D.W., Antioxidant vitamin supplementation of smoke-exposed rats partially protects against myocardial ischaemic/reperfusion injury, *Free Rad. Res. Commun.*, 17, 263, 1992.

40. Zhu, B., Sun, Y., Sievers, R.E., Shuman, J.L., Glantz, S.A., Chatterjee, K., Parmley, W.W., and Wolfe, C.L., L-Arginine decreases infarct size in rats exposed to environmental tobacco smoke, *Am. Heart J.*, 132, 91, 1996.

41. Feldman, J., Shenker, I.R., Etzel, R.A., Spierto, F.W., Lilienfield, D.E., Nussbaum, M., and Jacobson, M.S., Passive smoking alters lipid profiles in adolescents, *Pediatrics*, 88, 259, 1991.

42. Neufeld, E.J., Mietus-Snyder, M., Beiser, A.S., Baker, A.L., and Newburger, J.W., Passive cigarette smoking and reduced HDL cholesterol levels in children with high-risk lipid profiles, *Circulation*, 96, 1403, 1997.

43. Iscan, A., Uyanik, B.S., Vurgun, N., Ece, A., and Yigitoglu, M.R., Effects of passive exposure to tobacco, socioeconomic status and a family history of essential hypertension on lipid profiles in children, *Jpn. Heart J.*, 37, 917, 1996.

44. Misawa, K., Matsuki, H., Kasuga, H., Yokoyama, H., and Hinohara, S., An epidemiological study on the relationships among HDL-cholesterol, smoking and obesity, *Jpn. J. Hyg.*, 44, 725, 1989.

45. Craig, W.Y., Palomaki, G.E., Johnson, A.M., and Haddow, J.E., Cigarette smoking-associated changes in blood lipid lipoprotein levels in the 8-19 year-old age group: a meta-analysis, *Pediatrics*, 85, 155, 1990.

46. Bielicki, J.K., McCall, M.R., van den Berg, J.J., Kuypers, F.A., and Forte, T.M., Copper and gas-phase cigarette smoke inhibit plasma lecithin: cholesterol acyltransferase activity by different mechanisms, *J. Lipid Res.*, 36, 322, 1995.

47. Pan, X.M., Staprans, I., Hardman, D.A., and Rapp, J.H., Exposure to cigarette smoke delays in plasma clearance of chylomicrons and chylomicron remnants in rats, *Am. J. Physiol.*, 273(1 Pt. 1), G158, 1997.

48. Tribble, D.L., Giuliano, L.J., and Fortmann, S.P., Reduced plasma ascorbic acid concentrations in nonsmokers regularly exposed to environmental tobacco in smoke, *Am. J. Clin. Nutr.*, 58, 886, 1993.

49. Howard, D.J., Ota, R.B., Briggs, L.A., Hampton, M., and Pritsos, C.A., Environmental tobacco smoke in the workplace induces oxidative stress in employees, including increased production of 8-hydroxy-2′deoxyguanosine, *Cancer Epidemiol. Biomark. Prev.*, 7, 141, 1998.

50. Valkonen, M. and Kuusi, T., Passive smoking induces atherogenic changes in low-density lipoprotein, *Circulation*, 97, 2012, 1998.

51. Schwarzacher, S.P., Hutchison, S., Chou, T.M., Sun, Y.P., Zhu, B.Q., Chatterjee, K., Glantz, S.A., Deedwania, P.C., Parmley, W.W., and Sudhir, K., Antioxidant diet preserves endothelium-dependent vasodilatation in resistance arteries of hyper-cholesterolemic rabbits exposed to environmental tobacco, *J. Cardiovasc. Pharmacol.*, 31, 649, 1998.

52. Mackness, M.I., Arrol, S., Abbott, C., and Durrington, P.N., Protection of low-density lipoprotein against oxidative modification by HDL associated paraoxanase, *Atherosclerosis*, 104, 129, 1993.

53. Mackness, M.I. and Durrington, P.N., HDL, its enzymes and its potential to influence lipid peroxidation, *Atherosclerosis*, 115, 243, 1995.

54. Celemajer, D.S., Sorensen, K.E., Georgakopoulos, D., Bull, C., Thomas, O., Robinson, J., and Deanfield, J.E., Cigarette smoking is associated with dose-related and potentially reversible impairment of endothelium-dependent dilation in health young adults, *Circulation*, 88, 2149, 1993.

55. Rangemark, C., et al., Endothelium-dependent and independent vasodilation and reactive hyperemia in healthy smokers, *J. Cardiovasc. Pharmacol.*, 20(Suppl. 2), S198, 1992.

56. Heitzer, T., Yla-Herttuala, S., Luoma, J., Kurz, S., Munzel, T., Just, H., Olschewski, M., and Drexler, H., Cigarette smoking potentiates endothelial dysfunction of forearm resistance vessels in patients with hypercholesterolemia. Role of oxidized LDL, *Circulation*, 93, 1346, 1996.

57. Celermajer, D.S., Adams, M.R., Clarkson, P., Robinson, J., McCredie, R., Donald, A., and Deanfield, J.E., Passive smoking and impaired endothelium-dependent arterial dilatation in healthy young adults, *N. Engl. J. Med.*, 334, 150, 1996.

58. Raitakari, O.T., Adams, M.R., McCredie, R.J., Griffiths, K.A., and Celermajer, D.S., Arterial endothelial dysfunction related to passive smoking is potentially reversible in healthy young adults, *Ann. Intern. Med.*, 130, 578, 1999.

59. Stefanadis, C., Vlachopoulos, C., Tsiamis, E., Diamantopoulos, L., Toutouzas, K., Giatrakos, N., Vaina, S., Tsekoura, D., and Toutouzas, P., Unfavorable effects of passive smoking on aortic function in men, *Ann. Intern. Med.*, 128, 426, 1998.

60. Leeson, C.P.M., Whincup, P.H., Cook, D.G., Donald, A.E., Papacosta, O., Lucas, A., and Deanfield, J.E., Flow-mediated dilation in 9-to-11-year-old children: the influence of intrauterine and childhood factors, *Circulation*, 96, 2233, 1997.

61. Woo, K.S., Robinson, J.T., Chok, P., Adams, M.R., Yip, G., Mai, Z.J., Lam, C.W., Sorensen, K.E., Deanfield, J.E., and Celermajer, D.S., Differences in the effect of cigarette smoking on endothelial function in chinese and white adults, *Ann. Intern. Med.*, 127, 372, 1997.

62. Jorge, P.A., Ozaki, M.R., and Almeida, E.A., Endothelial dysfunction in coronary vessels and thoracic aorta of rats exposed to cigarette smoke, *Clin. Exp. Pharmacol. Physiol.*, 22, 410, 1995.

63. Hutchinson, S.J., Reitz, M.S., Sudhir, K., Sievers, R.E., Zhu, B.Q., Sun, Y.P., Chou, T.M., Deedwania, P.C., Chatterjee, K., Glantz, S.A., and Parmley, W.W., Chronic dietary l-arginine prevents endothelial dysfunction secondary to environmental tobacco smoke in normocholesterolemic rabbits, *Hypertension*, 29, 1186, 1997.

64. Hutchison, S.J., Sudhir, K., Chou, T.M., Sievers, R.E., Zhu, B.Q., Sun, Y.P., Deedwania, P.C., Glantz, S.A., Parmley, W.W., and Chatterjee, K., Testosterone worsens endothelial dysfunction associated with hypercholesterolemia and environmental tobacco smoke exposure in male rabbit aorta, *J. Am. Coll. Cardiol.*, 19, 800, 1997.

65. Burghuber, O.C., Punzengruber, C., Sinzinger, H., Haber, P., and Silberbauer, K., Platelet sensitivity to prostacyclin in smokers and non-smokers, *Chest*, 90, 34, 1986.

66. Kritz, H. and Sinzinger, H., Passive smoking, platelet function and atherosclerosis, *Wien Klin. Wochenschr.*, 108, 582, 1996.

67. Caplan, M., Hsueh, W., Kelly, A., and Donovan, M., Serum PAF acetylhydrolase increases during neonatal maturation, *Prostaglandins*, 39, 705, 1990.

68. Eliopoulos, C., Kein, J., Phan, M.K., Knie, B., Greenwald, M., Chitayat, D., and Koren, G., Hair concentrations of nicotine and cotinine in women and their newborn infants, *J. Am. Med. Assoc.*, 271, 621, 1994.

69. Cole, P.V., Hawkins, L.H., and Roberts, D., Smoking during pregnancy and its effects on the fetus, *J. Obstet. Gynaecol. Br. Commonw.*, 79, 782, 1972.

70. Asmussen, I., Ultrastructure of the villi and fetal capillaries in placentas from smoking and nonsmoking mothers, *Br. J. Obstet. Gynaecol.*, 87, 239, 1980.

71. Ahlsten, G., Ewald, U., and Tuvemo, T., Maternal smoking reduces prostacyclin formation in human umbilical arteries: a study on strictly selected pregnancies, *Acta Obstet. Gynecol. Scand.*, 65, 645, 1986.

72. Bearer, C., Emerson, R.K., O'Riordan, M.A., Roitman, E., and Shackleton, C., Maternal tobacco smoke exposure and persistent pulmonary hypertension, *Environ. Health Perspect.*, 105, 202, 1997.

73. Johnson, R.K., Wang, M.Q., Smith, M.J., and Connolly, G., The association between parental smoking and the diet quality of low-income children, *Pediatrics*, 97, 312, 1996.

74. Burke, V., Gracey, M.P., Milligan, R.A., Thompson, C., Taggart, A.C., and Beilin, L.J., Parental smoking and risk factors for cardiovascular disease in 10-12 year old children, *J. Pediatr.*, 133, 206, 1998.

75. Franco, P., Chabanski, S., Szliwowski, H., Dramaix, M., and Kahn, A., Influence of maternal smoking on autonomic nervous system in healthy infants, *Pediatr. Res.*, 47, 215, 2000.

76. Garner, D.W. and Whitney, R.J., Protecting children from Joe Camel and his friends: a new first amendment and federal preemption analysis of tobacco billboard regulation, *Emory Law J.*, 46, 482, 1997.

77. Kluger, R., *Ashes to Ashes: America's Hundred-Year Cigarette War, the Public Health, and the Unabashed Triumph of Philip Morris*, Alfred A. Knopf, New York, 1996.

78. Moskowitz, W.B., Schwartz, P.F., and Schieken, R.M., Childhood passive smoking, race, and coronary artery disease risk: the MCV Twin Study. Medical College of Virginia, *Arch. Pediatr. Adolesc. Med.*, 153, 446, 1999.

12 Smoking Effects on the Estrogen and Progesterone Balance in Women

Murray D. Meek

CONTENTS

12.1 BACKGROUND

Cigarette smoke is an aerosol composed of volatile agents in the vapor phase and semivolatiles in the particulate phase. Approximately 400 to 500 individual gaseous compounds are contained in the vapor phase, including ammonia, hydrogen cyanide, toluene, benzene, hydrogen sulfide, and formaldehyde. More than 3500 compounds have been identified in the particulate phase, including naphthalene, pyrenes, phenols, benzofurans, aniline, toluidines, *N*-nitrosoamines, and the addictive substance nicotine. Additionally, the particulate phase contains inorganic compounds including hydrazine, arsenic, nickel, chromium, cadmium, lead, and polonium.[1] This list of known chemicals is incomplete because the identities of flavor and other additives are trade secrets.

Considering that cigarette smoke contains approximately 4000 known compounds, it is not surprising that smoking exhibits pleiotropic effects. Researchers have known for a long time that cigarette smoking is associated with lung cancer and certain other cancers and have identified certain compounds in cigarette smoke as carcinogens. For example, polyaromatic hydrocarbons (PAHs) have been shown to induce lung tumors in mice. Oral cancer has been shown to develop with 4-(methylnitrosamine)-1-(3-pyridyl)-1-butanone (NNK) treatment in rodents.[1] More recently, scientists have identified some of the molecular effects of cigarette smoke, including effects on gene transcription and enzymes that affect hormone balance.

0-8493-0311-7/00/$0.00+$.50
© 2001 by CRC Press LLC

12.2 EPIDEMIOLOGY

Cigarette smoking affects the hormonal balance of humans, especially females. Therefore, cigarette smoke is considered to be an endocrine disruptor because it disrupts the normal homeostatic mechanisms that control hormone levels in the body. The endocrine disrupting properties were discovered in epidemiological studies that showed that smoking exerts an important anti-estrogenic effect in women. For example, epidemiological studies have shown that smoking reduces the risk of endometrial cancer, an estrogen-responsive cancer, by as much as 50%.[2-5] The reduction in risk is greatest in women who are multiparous, obese, or not using estrogen replacement therapy. Among postmenopausal women, current smokers showed the greatest reduction in risk (relative risk = 0.4). Former smokers, including those who had recently stopped, were less affected (relative risk = 0.8), suggesting that smoking is protective.[6]

Other studies have shown that women who smoke experience menopause 2 to 3 years earlier than non-smoking women.[7] This process normally occurs with cessation of estrogen production by the ovaries.[5,8] Rosenberg et al.[9] have shown that women who smoke exhibit a significantly greater frequency of irregular bleeding during their menstrual cycle. Epidemiological data have also linked smoking to increased rates of osteoporosis, a condition associated with decreased serum estrogen levels.[5]

12.3 MECHANISMS

The mechanism(s) of the anti-estrogenic effect of smoking is unclear. Several hypotheses have been proposed. Catabolic inactivation of endogenous estrogen has been proposed as a mechanism by which smoking could elicit an anti-estrogenic effect.[10] This hypothesis is based on the observation that the rate of 2-hydroxylation of 17β-estradiol (E2) is significantly increased in premenopausal female smokers who smoked at least 15 cigarettes per day compared to non-smokers. Biologically, 2-hydroxyestrogens are significantly less potent estrogen receptor (ER) agonists than E2 and are rapidly cleared from the circulation.[11,12] Thus, C2 hydroxylation effectively terminates the peripheral activity of estradiol.

In humans, CYP1A1 and CYP1A2 are the primary enzymes catalyzing the 2-hydroxylation of estradiol.[13] The regulation of CYP1A1/1A2 activation occurs via an aryl hydrocarbon receptor (Ah receptor)-mediated process.[14,15] Mechanistically, ligand binding to the Ah receptor and release of heat shock protein 90 allows heterodimerization of the liganded Ah receptor with the aryl hydrocarbon receptor nuclear translocator (ARNT) protein. The activated receptor–ligand complex translocates from the cytoplasm to the nucleus where it can interact with DNA response elements, termed dioxin or xenobiotic response elements (DRE or XRE, respectively), in the vicinity of target genes, resulting in gene expression.[14]

12.4 EXPERIMENTAL EVIDENCE

Direct experimental evidence indicates that cigarette smoke binds to and activates the Ah receptor. Further, the activated receptor is able to stimulate CYP1A1/1A2, the enzymes responsible for the catabolic inactivation by estrogen by 2-hydroxylation.

The Ah receptor complex formation assay was performed by incubating guinea pig liver cytosol with cigarette smoke and a radiolabeled DRE oligonucleotide. The resultant product was resolved on SDS-PAGE and visualized by autoradiography. The analysis showed an inducible protein–DNA complex that became absent if no DRE was present or was competitively inhibited by incubation with an excess of unlabeled mutated DRE sequence.[16,17]

Further evidence that cigarette smoke can bind to and functionally activate the Ah receptor came from reporter gene studies. In this series of experiments, induction of Ah receptor-regulated luciferase reporter gene activity in liver cells occurred in a dose-dependent and saturable manner. Cigarette smoke exhibits approximately 50% of the potency of the prototypical Ah receptor ligand, 2,3,7,7-tetrachloro-p-dibenzodioxin (TCDD). In contrast, there was no luciferase activity in variant liver cells that did not contain either the Ah receptor or the ARNT protein.[16] This demonstrates the requirement for a functional Ah receptor and ARNT protein for DRE-regulated luciferase expression.

Finally, when liver cells were exposed to cigarette smoke, CYP1A1/1A2 activity increased in a dose-dependent and saturable manner. The maximal activity was approximately 50% of that of TCDD. The lack of any detectable increase in CYP1A1/1A2 activity in Ah receptor-deficient or ARNT-deficient variant cells is consistent with the requirement of these proteins to activate CYP1A1/1A2 activity.[18,19]

The role of the ER in mediating the anti-estrogenic effect of smoking is another mechanism by which smoking could elicit an anti-estrogenic effect. The proposed mechanism of action of the ER involves ligand binding to the inactive oligomer complexed to several heat shock proteins.[20] Upon ligand binding, the complex dissociates to allow dimerization of ligand-bound receptor monomers.[21] The dimerized complex then binds to palindromic DNA binding sites termed estrogen response elements in the vicinity of estrogen-responsive genes to elicit transcriptional activation.[22] In order for cigarette smoke to elicit an anti-estrogenic effect via the ER, components in the smoke would be expected to bind to but not transcriptionally activate the ER. Such an antagonistic effect would be consistent with the observed anti-estrogenic effect of smoking. However, this is not observed experimentally.

Ligand binding experiments using the hydroxyapatite method[23] have demonstrated that cigarette smoke binds to the human ER in a concentration-dependent and saturable manner. The IC_{50} value for cigarette smoke is about six orders of magnitude less potent than for the IC_{50} of E2 (1 nM), the endogenous ER ligand.[16] Using a firefly luciferase reporter gene system that can only be activated by the agonist-occupied human ER, cigarette smoke was able to induce luciferase activity in human breast cancer cells transfected with the reporter gene components.[16] Taken together, this evidence indicates that cigarette smoke not only binds the ER, but activates it functionally to induce gene transcription. Extrapolating, it is reasonable to expect that estrogen-responsive genes would be activated *in vivo*. This has been confirmed experimentally in rodents. In bilaterally ovariectomized 3-week-old female rats, uterine wet was increased slightly upon exposure to cigarette smoke.[24] The slight increase may not be representative of gene transcriptional events because this assay is not very sensitive since it assesses gene transcriptional events at an integrated organ level.

12.5 IMPROVING *IN VIVO* GENE TRANSCRIPTION ACTIVATION ASSAYS

A more sensitive assay would involve the assessment of specific estrogen-responsive genes. The lactoferrin gene would be a reasonable candidate for *in vivo* studies because is contains estrogen-responsive elements in the upstream region and has been shown to be ER responsive.[25] It is expressed in rat uterus, making it a convenient *in vivo* model to assess estrogen action. Ovariectomized and non-ovariectomized female rats could be exposed to cigarette smoke in smoking chambers for various periods of time and their lactoferrin gene transcription levels assessed using quantitative reverse transcriptase-polymerase chain reaction. If the exposed animals showed higher levels of lactoferrin than non-exposed animals, it would suggest that cigarette smoke is modulating ER-responsive genes *in vivo*. This method of assessing the *in vivo* ER-mediated gene transcription would be direct and would be much more sensitive than the current "gold standard" uterotropic or vaginatropic assays. These methods rely on the increase in uterine wet weight or vaginal cornification (Allen-Doisy assay), respectively, as indexes of estrogen action.

Another mechanism by which cigarette smoke might elicit an anti-estrogenic effect via the Ah receptor is by ER downregulation. PAH congeners which bind to the Ah receptor, including benzo[a]pyrene, benz[a]anthracene, and 7,12-dimethylbenz[a]anthracene, caused a decrease in nuclear ER levels. In contrast, benzo[ghi]perylene, a congener which did not bind to the Ah receptor, did not affect nuclear ER levels in human breast cancer cells.[26] The mechanism is not clear, but the effect is Ah receptor-dependent.

Experimental evidence that cigarette smoke is able to bind to and activate the ER and induce transcription is unexpected given the obvious anti-estrogenic effects of smoking and the absence of reported estrogenic effects. Considering the pleiotropic effects of smoking and the sensitive mechanisms of hormone regulation, however, this finding may be physiologically significant. It is conceivable that the two pathways are indirectly competing against one another and that the observed anti-estrogenic effect of smoking is actually the net result of this competition. It is equally conceivable that the ER pathway may be exerting subtle estrogenic effects that make it difficult to detect an association between cause and effect. For example, several studies have shown a weak to moderate link between smoking and breast cancer,[27–30] while other studies show no link.[31–34] This apparent contradiction would be consistent with the idea that cigarette smoke exerts a subtle estrogenic effect in females. Alternatively, others have proposed a temporal relationship between cigarette smoke and its estrogen agonist/antagonistic effects. The evidence for this is as follows: cigarette smoke is estrogenic in young female rats (3 weeks old) because it increases uterine wet weight; by comparison, in 3-month-old animals, cigarette smoke inhibits the ability of estrogen to induce progesterone receptors in uterine tissue.[24] The study did not address the effect of cigarette smoke on uterine wet weight in the older animals as would have been both appropriate and expected. Induction of the progesterone receptors occurs via the ER and is permissive for progesterone activity. This suggests a window of opportunity in which cigarette smoke may act as an estrogen and a later window when it acts as an anti-estrogen.

12.6   PREGNANCY

In the U.S., approximately 25% of women smoke during pregnancy. Smoking during pregnancy is associated with an increased incidence of spontaneous abortion, thought to be due to corpus luteum insufficiency. This hypothesis is supported by experimental findings demonstrating that incubation of human granulosa cells with cotinine; anabasine; the combination of nicotine, cotinine; and anabasine; or with an aqueous extract of cigarette smoke resulted in inhibition of progesterone synthesis.[35]

It is equally conceivable that compounds present in cigarette smoke may disrupt the effects of progesterone by preventing the upregulation of progesterone receptors. This might occur by two distinct mechanisms. First, the Ah receptor-mediated anti-estrogenic effect could biologically inactivate estradiol levels directly through 2-hydroxylation. Second, PAHs in cigarette smoke may cause a decrease in ER levels as shown *in vitro*. Both of these mechanisms would lead to the same biological effect — decreased estrogen activity leading to decreased progesterone activity through lack of receptors. These mechanisms may also contribute to the fact that cigarette smoking disrupts menses by increasing the frequency of spot bleeding and altering the normal menstrual frequency. On the other hand, the estrogenic activity of cigarette smoke should upregulate progesterone receptors to enhance progesterone activity. However, cigarette smoke inhibits progesterone synthesis so the estrogenic effect may be blunted.[36]

It has been known for decades that women who smoke have poorer outcomes in pregnancy. One study showed that nicotine augmented estradiol secretion and inhibited progesterone secretion by human granulosa cells in a dose-dependent manner. Nicotine had little effect on steroid secretion. If however, granulosa cells were stimulated with luteinizing hormone, nicotine suppressed estradiol secretion and progesterone secretion. The results suggest that cigarette smoking specifically affects the control mechanisms of intraovarian processes which are responsible for normal luteal function.[37] However, studies that examine isolated effects of a single compound are not reflective of the *in vivo* situation. Cigarette smoke is a complex mixture of aerosols and semivolatiles that contains endocrine disrupting chemicals. These chemicals may be sequestered by binding to plasma proteins. Furthermore, the half-lives of these chemicals in humans are unknown in most instances and the effect on liver detoxification enzymes is not known. Further studies on human granulosa cells are required to assess the effect of whole cigarette smoke on progesterone secretion and progesterone receptor levels.

12.7   SUMMARY

In summary, although we have long known that smoking causes lung cancer, it is only relatively recently that cigarette smoke has been shown to be an endocrine disruptor. It has complex estrogenic and anti-estrogenic effects mediated by the estrogen and Ah receptors, respectively. These effects, in turn, have consequences for progesterone receptor up- and downregulation. In addition, cigarette smoke inhibits the synthesis of progesterone. The effect of cigarette smoke on estrogen and progesterone action is not fully known. Given the increasing importance of endocrine disruptors in environmental research and the societal movement towards anti-smoking, research in this area is expected to intensify.

REFERENCES

1. Hoffmann, D. and Hoffmann, I., The changing cigarette, 1950–1995, *J. Toxicol. Environ. Health*, 50, 307–364, 1997.
2. Leskio, S.M., Rosenburg, L., and Kaufman, D.W., Cigarette smoking and the risk of endometrial cancer, *N. Engl. J. Med.*, 313, 593–596, 1985.
3. Kelsey, J.L., LiVolsi, V.A., and Holford, T.R., A case-control study of cancer of the endometrium, *Am. J. Epidemiol.*, 116, 333–342, 1982.
4. Smith, E.M., Sowers, M.F., and Burns, T.L., Effects of smoking on the development of female reproductive cancers, *J. Natl. Cancer Inst.*, 73, 371–376, 1984.
5. Baron, J.A., Smoking and estrogen-related disease, *Am. J. Epidemiol.*, 119, 9–22, 1984.
6. Brinton, L.A., Barrett, R.J., Berman, M.L., Mortel, R., Twiggs, L.B., and Wilbanks, G.D., Cigarette smoking and the risk of endometrial cancer, *Am. J. Epidemiol.*, 137, 281–291, 1993.
7. Schmeiser-Rieder, A., Schoberberger, R., and Kunze, M., Women and smoking, *Wiener Med. Wochenschr.*, 145, 73–76, 1995.
8. MacMahon, B., Trichopoulos, D., Cole, P., and Brown, J., Cigarette smoking and urinary estrogens, *N. Engl. J. Med.*, 307, 1062–1065, 1982.
9. Rosenberg, M.J., Waugh, M.S., and Stevens, C.M., Smoking and cycle control among oral contraceptive users, *Am. J. Obstet. Gynecol.*, 174, 628–632, 1996.
10. Michnovicz, J.J., Hershcopf, R.J., Naganuma, H., Bradlow, H.L., and Fishman, J., Increased 2-hydroxylation of estradiol as a possible mechanism for the anti-estrogenic effect of cigarette smoking, *N. Engl. J. Med.*, 315, 1305–1309, 1986.
11. Martucci, C.P. and Fishman, J., Impact of continuously administered catechol estrogens in uterine growth and luteinizing hormone secretion, *Endocrinology*, 105, 1288–1292, 1979.
12. Jellinck, P.H., Krey, L., and Davis, P.G., Central and peripheral action of estradiol and catecholestrogens administered at low concentrations by constant infusion, *Endocrinology*, 108, 1848–1854, 1981.
13. Ball, P. and Knuppen, R., Formation, metabolism and physiologic importance of catecholestrogens, *Am. J. Obstet. Gynecol.*, 163, 2163–2170, 1990.
14. Hankison, O., The aryl hydrocarbon receptor complex, *Annu. Rev. Pharmacol. Toxicol.*, 35, 307–340, 1995.
15. Quattrochi, L.C., Vu, T., and Tukey, R.H., The human CYP1A2 gene and induction by 3-methylchloranthrene: a region of DNA that supports AH-receptor binding and promoter specific binding, *J. Biol. Chem.*, 269, 6949–6954, 1994.
16. Meek, M.D. and Finch, G.L., Diluted mainstream cigarette smoke condensates activate estrogen receptor and aryl hydrocarbon receptor-mediated gene transcription, *Environ. Res.*, 80, 9–17, 1999.
17. Denison, M.S., Fischer, J.M., and Whitlock, J.P., The DNA recognition site for the dioxin-Ah receptor complex. Nucleotide sequence and functional analysis, *J. Biol. Chem.*, 263, 17221–17224, 1988.
18. Hankison, O., Brooks, B.A., Weir-Brown, K.I., Hoffman, E.C., Johnson, B.S., Nanthur, J., Reyes, H., and Watson, A.J., *Biochemie*, 73, 61–65, 1991.
19. Zhang, J., Watson, A.J., Probst, M.R., Minehart, E., and Hankinson, O., Basis for the loss of aryl hydrocarbon receptor gene expression in clones of a mouse hepatoma cell line, *Mol. Pharmacol.*, 50, 1454–1462, 1996.
20. Landel, C.C., Kushner, P.J., and Greene, G.L., The interaction of human estrogen receptor with DNA is modulated by receptor-associated proteins, *Mol. Endocrinol.*, 8, 1407–1419, 1994.

21. Parker, M.G., Arbuckle, N., Dauvois, S., Danielian, P., and White, R., Structure and function of the estrogen receptor, *Ann. N.Y. Acad. Sci.*, 684, 119–126, 1993.
22. Ruh, M.F., Cox, L.K., and Ruh, T.S., Estrogen receptor interaction with specific histones. Binding to genomic DNA and an estrogen response element, *Biochem. Pharmacol.*, 52, 869–878, 1996.
23. Laws, S.C., Carey, S.A., Kelch, W.R., Cooper, R.L., and Gray, L.E., Vinclozolin does not alter progesterone receptor (PR) function in vivo despite inhibition of PR binding by its metabolites in vitro, *Toxicology*, 110, 1–11, 1996.
24. Berstein, L.M., Tsyrlina, E.V., Krjukova, O.G., and Dzhumasultanova, S.V., Influence of tobacco smoke on DNA unwinding and uterotrophic effect of estrogens in rats, *Cancer Lett.*, 127, 95–98, 1998.
25. Teng, C., Mouse lactoferrin gene — a marker for estrogen and epidermal growth factor, *Environ. Health Perspect.*, 103, 17–20, 1995.
26. Chaloupka, K., Krishnan, V., and Safe, S., Polynuclear aromatic hydrocarbon carcinogens as antiestrogens in MCF-7 human breast cancer cells: role of the Ah receptor, *Carcinogenesis*, 13, 2233–2239, 1992.
27. Bennicke, K., Conrad, C.G., Sabroe, S., and Sorensen, H.T., Smoking and breast cancer, *Ugeskr. Laeg.*, 158, 4909–4911, 1996.
28. Morabia, A., Bernstein, M., Heritier, S., and Khatchatrian, N., Relation of breast cancer with passive and active exposure to tobacco smoke, *Am. J. Epidemiol.*, 143, 918–928, 1996.
29. Bennicke, K., Conrad, C., Sabroe, S., and Sorensen, H.T., Cigarette smoking and breast cancer, *Br. Med. J.*, 310, 1431–1433, 1995.
30. Calle, E.E., Miracle-McMahill, H.L., Thun, M.J., and Heath, C.J., Cigarette smoking and risk of fatal breast cancer, *Am. J. Epidemiol.*, 139, 1001–1007, 1994.
31. Baron, J.A., Newcomb, P.A., Longnecker, M.P., Mittendorf, R., Storer, B.E., Clapp, R.W., Bogdan, G., and Yuen, J., Cigarette smoking and breast cancer, *Cancer Epidemiol. Biomark. Prev.*, 5, 399–403, 1996.
32. Braga, C., Negri, E., La, V.C., Filiberti, R., and Franceschi, S., Cigarette smoking and the risk of breast cancer, *Eur. J. Cancer Prev.*, 5, 159–164, 1996.
33. Ranstam, J. and Olsson, H., Alcohol, cigarette smoking, and the risk of breast cancer, *Cancer Detect. Prev.*, 19, 487–493, 1995.
34. Smith, S.J., Deacon, J.M., and Chilvers, C.E., Alcohol, smoking, passive smoking and caffeine in relation to breast cancer risk in young women. UK National Case-Control Study Group, *Br. J. Cancer*, 70, 112–119, 1994.
35. Gocze, P.M., Szabo, I., and Freeman, D.A., Influence of nicotine, cotinine, anabasine and cigarette smoke extract on human granulosa cell progesterone and estradiol synthesis, *Gynecol. Endocrinol.*, 13, 266–272, 1999.
36. Gocze, P.M., Porpaczy, Z., and Freeman, D.A., Effect of alkaloids in cigarette smoke on human granulosa cell progesterone synthesis and cell viability, *Gynecol. Endocrinol.*, 10, 223–228, 1996.
37. Bodis, J., Hanf, V., Torok, A., Tinneberg, H.R., Borsay, P., and Szabo, I., Influence of nicotine on progesterone and estradiol production of cultured human granulosa cells, *Early Preg.*, 3, 34–37, 1997.

13 Experimental Models of Environmental Tobacco Smoke Exposure: Structural and Functional Changes in Epithelial Cells of the Respiratory Tract

Walter K. Schlage and Ashok Teredesai

CONTENTS

13.1 BACKGROUND

The health effects of exposure to environmental tobacco smoke (ETS) have been the subject of numerous literature reviews during the past decade. Most of the investigations, which concluded that exposure to ETS may increase the risk of lung cancer, chronic obstructive pulmonary disease (COPD), cardiovascular disease,

0-8493-0311-7/00/$0.00+$.50
© 2001 by CRC Press LLC

intrauterine growth retardation, and Sudden Infant Death Syndrome, are based on epidemiological studies (cf. References 1 to 6). In a notice of proposed rulemaking (NPR), the Occupational Safety and Health Administration (OSHA) concluded that ETS "may produce mucous membrane irritation, pulmonary, cardiovascular, reproductive and carcinogenic effects" and that ETS "presents a serious health risk to workers."[7] Some authors have pointed out, however, that the studies on ETS and lung cancer and heart disease either have not reported a statistically significant increase in risk or have reported risks that are so low that they cannot be assumed to be reliable by normal scientific standards in judging epidemiological studies (cf. References 8 and 9).

A working group of the International Agency for Research on Cancer (IARC) concluded that "passive smoking gives rise to some risk of cancer,"[10] a conclusion drawn on the background of epidemiological data that, according to the same working group, are "compatible either with an increase in or an absence of risk." The IARC working group based their conclusion on the fact that carcinogenic mainstream smoke (MS) and sidestream smoke (SS) constituents are found in the air inhaled by passive smokers and on the assumption that ETS effects can be extrapolated from the effects of high doses of MS in the same way as quantitative dose–effect relationships usually observed following high dose carcinogen exposures can be extrapolated down to low doses. A similar approach was applied when ETS was classified as a class A (known human) carcinogen by the U.S. Environmental Protection Agency (EPA);[1] however, more experimental data from long-term studies and mechanistic investigations are necessary to support the plausibility of this approach and to contribute to a quantitative risk assessment. So long as these data are not available, this extrapolation approach has two major shortcomings. (1) As outlined in earlier reviews,[8,9,11,12] major qualitative and quantitative differences in the composition of MS and SS and in the uptake of smoke constituents from MS and ETS are incompatible with simple extrapolation. (2) The results of numerous experimental carcinogenesis studies, which indicated no consistent induction of malignant tumors in the respiratory tract of rats and mice following MS inhalation, do not correlate with the human epidemiological data (for review, see Coggins[13] and Witschi[14]).

Among the ETS inhalation studies that employed various surrogates for ETS (see below), carcinogenic effects have been demonstrated in the A/J mouse model employed by Witschi and colleagues at extremely high "ETS" (a mix of 89% SS and 11% MS) concentrations.[15] The few subchronic and chronic ETS inhalation studies that use other animal species (rats, hamsters) have failed to prove carcinogenicity so far, and the histomorphological changes occurring in the respiratory tract, e.g., hyperplasia and metaplasia, reversed completely after postinhalation periods.[16–19] These effects can be regarded as adaptive and reactive responses to repeated irritation.[18,20,21]

The lack of a significant induction of neoplastic lesions as an unambiguous endpoint in almost all of the ETS carcinogenicity studies raises the need to study additional mechanistic endpoints and intermediate biomarkers that can provide evidence that some of the histomorphological changes observed may represent either preneoplastic events or nonneoplastic changes. The inclusion of mechanistic endpoints is also in line with U.S. EPA proposed guidelines for risk assessment.[22]

13.2 AIMS AND SCOPE

Because epithelium of the respiratory tract is the initial site of contact with inhaled ETS, the focus of this chapter is to review the structural and functional changes caused by ETS in this epithelium from a cellular perspective. An attempt is made to discuss the changes in cellular structure and function that are determined by histomorphological, histochemical, and biochemical methods with possible application as intermediate biomarkers associated with carcinogenicity. The heterogeneity of the published findings may reflect species differences as well as the variety of experimental ETS surrogates and exposure conditions. These differences have to be considered in any attempt to interpret experimental results with regard to all kinds of extrapolation or risk assessment; therefore, a comparison of ETS models is provided in the first part of this chapter. The aim of the chapter is to summarize the status of the known cellular effects of ETS exposure in the respiratory tract as a platform for designing new experimental approaches toward a better understanding of the carcinogenic risk of ETS exposure. Cardiovascular and other systemic effects of ETS exposure outside the respiratory tract are discussed in other chapters of this book.

13.3 ETS SURROGATES

Experimental ETS atmospheres for inhalation studies should be as realistic as possible, but they also have to be reproducible in their chemical composition and concentration and have to be constant over the experimental period, i.e., over weeks or months in subchronic or chronic settings. This implies a standardized technique for the generation and processing of smoke to achieve an ETS-like composition. Only in acute experimental situations, i.e., for several hours, is the use of true ETS produced by human smokers[23–26] feasible. ETS has been defined as a mixture — highly diluted by the indoor atmosphere — of SS emanating from the burning end of a cigarette and through the paper, notably between puffs, and exhaled MS, which has passed through the smoker's respiratory tract.[27–29] The properties of ETS cannot be regarded simply as the properties of a mixture of machine-generated MS and SS because there are a number of factors that influence the quality of ETS:

- The room ventilation[30]
- Aging and dilution[19,31,32]
- The differential absorption of surfaces in contact with ETS (e.g., furniture, painted wall surfaces, floor materials, people's clothing)[31,32]
- The fraction of particles that do not originate from tobacco smoke: approximately 10 to 50% of the respirable suspended particulates (RSP) measured in real-life ETS are typically derived from nontobacco smoke sources; similarly, many volatile organic compounds are not of tobacco smoke origin[28,30]
- The type of cigarettes smoked, but only to a lesser degree (except for tobacco-heating prototypes, which have a much stronger influence)[25,26,33]

In published investigations, various surrogates for ETS differing in their complexity have been applied. Even diluted MS was repeatedly used as "ETS" (for review and discussion, see Coggins[12]). Examples are given in Table 13.1.

- Fresh SS, which was more or less diluted and generated either by collection between the puffing of MS by a smoking machine (MS discarded)[19,34] or by the static burning of cigarettes in a small-volume chamber[35]
- Aged and diluted SS (ADSS), which was collected from a smoking machine and conditioned in a large chamber with inert surfaces[17]
- Diluted mix of 89% SS and 11% MS (SS/MS), which approximates the estimated contribution of MS to real ETS[15]
- Room-aged SS (RASS), which was collected from a smoking machine and conditioned in a large chamber with non-inert surfaces[19,32]
- Experimental ETS, which was produced by human smokers in an experimental chamber (short-term exposure or analytical sampling only)[24-26,33]

The atmospheres have generally been characterized by the concentration of particles, i.e., total particulate matter (TPM) for experimental atmospheres without interfering particles from other sources, or RSP for real-life ETS; the concentration of carbon monoxide (CO) as an unreactive gas phase component; and the concentration of nicotine as a rapidly adsorbing gas phase component[29,36] (in contrast to MS, where nicotine is found primarily in the particulate phase). Among these experimental atmospheres, diluted MS and fresh SS are possibly the poorest proxies for ETS because major differences in the relative concentrations of constituents, in phase distribution, and in particle size in comparison to ETS have been described for fresh SS as well as for MS.[28,29,33,37]

13.3.1 QUANTITATIVE ASPECTS

ETS concentrations in field measurements as well as in experimental ETS models have been characterized predominantly by their RSP or TPM concentrations (see Table 13.1, Figure 13.1), although the CO concentrations would provide a more reliable estimate for the number of cigarettes smoked per unit air volume.[21] ETS particle mass concentrations in smoker-occupied residences showed average increases of up to 0.1 mg TPM/m^3 and in restaurants of up to approximately 2 mg RSP/m^3.[1,28,30,36,38,39] The TPM concentrations in ETS inhalation studies ranged up to 380 mg/m^3 or more, which is approximately 1000-fold higher than the maximum of the average concentrations of RSP reported for real ETS in residential settings. Figure 13.1 shows the particle concentration ranges for the various ETS models listed in Table 13.1 compared to real-life ETS and employs the terms "typical" (TPM concentration of 0.1 mg/m^3), "extreme" (1.0 mg/m^3), and "exaggerated" (10 mg/m^3), which have been used by Coggins et al.[18] to describe the experimental conditions. It should be noted that highly diluted experimental ETS atmospheres in the "realistic" concentration range, i.e., below 1 mg TPM/m^3, are very difficult to control (cf. References 17 and 18).

Most of the investigations with the ETS atmospheres listed in Table 13.1 included endpoints related to lung carcinogenesis or to irritative symptoms of the upper and

Text continues on page 237

TABLE 13.1
Examples of Experimental ETS Surrogates Used for Inhalation Studies

Smoke type	Cigarette	Species	Exposure mode	Duration	Particle conc. (mg/m³)	CO conc. (ppm)	Nicotine conc. (µg/m³)	Purpose, interest	Endpoint, result	Ref.
SS	1R4F	Guinea pig	WB	6 h/d, 5 d/week, 5 week	1.0	6.4	144	Asthma, bronchoconstriction, C-fibers	No effect on C-fiber responsiveness to inhaled capsaicin or bradykinin, but increased responsiveness to systemically administered capsaicin or to lung hyperinflation	Mutoh[181]
SS	1R4F	Rat	WB	6 h/d; 5 d/week; 8, 11, or 15 weeks	1.0	6.5	n.i.	Lung function (developing rat)	Reduced airway reactivity to serotonin, but not to metacholine	Joad et al.[203]
SS	n.i.	Guinea pig	WB	6 h/d, 5 d/week, 35 d	1.0	5.7	541	Lung function (developing guinea pig), C-fiber responsiveness	Reduced airway reactivity to systemic capsaicin, but not to substance P	Joad et al.[204]
SS	1R4F	Guinea pig	WB	6 h/d, 5 d/week, 5 weeks	1.0	6.2	345	Asthma, bronchoconstriction, C-fibers	C-fiber responsiveness to inhaled capsaicin decreased, but neurotransmitters, receptors, and number of C-fibers not altered by ETS (unlike MS)	Joad et al.[182]
SS	1R4F	Hamster	WB	6 h/d, 7 d/week, 1–3 weeks, 1 week recovery	1.0	4.0	385	Cell proliferation, nasal cavity, trachea, lung	Nasal respiratory epithelium: initial small increase of BrdU labeling in, decrease below control level during recovery; trachea and lung: no effects	Witschi[78]

TABLE 13.1 (*continued*)
Examples of Experimental ETS Surrogates Used for Inhalation Studies

Smoke type	Cigarette	Species	Exposure mode	Duration	Particle conc. (mg/m³)	CO conc. (ppm)	Nicotine conc. (µg/m³)	Purpose, interest	Endpoint, result	Ref.
SS	1R4F	Mice (A/J and C57BL/6)	WB	6 h/d, 1–5 d	1.0	6.0	549	Cell proliferation, lung, comparison gas phase vs. complete smoke	*C57 mice:* no effects; *A/J mice:* increased BrdU labeling on days 3 and 5 in large airways and terminal bronchioles; *gas phase alone:* no effect	Rajini[155]
SS	1R4F	Human	WB	2 h/d, 4 d	n.i.	15	n.i.	Rhinitis, nasal congestion, tobacco allergy	Increased congestion, no differences between sensitized and nonsensitized persons; methodic variations between acoustic rhinometry and rhinomanometry	Kesavanathan[100]
SS	n.i.	Human	WB	1 h	n.i.	15	n.i.	Rhinitis, nasal mucociliary clearance	Equivocal: ETS induced enhanced clearance (radiotracer) in 6/12 subjects and reduced clearance in 3/12 subjects	Bascom[104]
SS	2R1F	Human	WB	2 × 2h	n.i.	15	n.i.	Irritation, rhinitis, nasal congestion, epithelial permeability, tobacco allergy	*Irritation of eyes, nose, and throat; nasal and pulmonary resistance increased; neutrophils and albumin in nasal lavage:* no significant increase	Willes[103]

TABLE 13.1 (*continued*)
Examples of Experimental ETS Surrogates Used for Inhalation Studies

Smoke type	Cigarette	Species	Exposure mode	Duration	Particle conc. (mg/m³)	CO conc. (ppm)	Nicotine conc. (µg/m³)	Purpose, interest	Endpoint, result	Ref.
SS	1R4F	Human	WB	2h	n.i.	1, 5, 15	73, 305, 958	Irritation, sensitization rhinitis, nasal congestion	*Symptoms* (headache, irritations in eyes, nose) increased, some symptoms stronger in sensitized persons; *nasal resistance* (acoustic rhinometry): significant only at 15 ppm CO, dose related in sensitive persons only	Bascom[101]
SS	1R4F	Cockerel	WB	6 h/d, 5 d/week, 16 weeks	2.5	4.5	92.5	Atherosclerosis, aortic plaque formation	Increased size of plaques in abdominal aorta	Penn et al.[41]
SS	Commercial	Human	WB	1 h	3.1	20.3	397	Irritation, lung function in asthmatics	*Symptoms*: increased irritation of eyes, nose, and throat; *airway responsiveness and lung mechanics*: no changes	Joerres[179]
SS	Commercial	Human	WB	3 h	3.1	20.3	397	Inflammation in asthmatics	Increased symptoms of *irritation*; no *inflammatory mediators* in nasal or bronchioloalveolar lavage, *lung function*: no effects	Nowak[102]

TABLE 13.1 (continued)
Examples of Experimental ETS Surrogates Used for Inhalation Studies

Smoke type	Cigarette	Species	Exposure mode	Duration	Particle conc. (mg/m³)	CO conc. (ppm)	Nicotine conc. (µg/m³)	Purpose, interest	Endpoint, result	Ref.
SS	2R1F	Rat, hamster	WB	10 h/d, 5 d/week, 90 d	4	25–30	1,000	Tumorigenicity, histopathology, adaptive response	*Hamsters*: no effect in respiratory tract, no systemic effects in other organs; *rats*: squamous metaplasia, hyperplasia of respiratory epithelium of nasoturbinates, fully reversible, no other respiratory or systemic effects	Von Meyerinck[16]
SS	n.i.	A/J mouse	WB	6 h/d, 5 d/week, 6 months	4.5	17	1,122	Lung, trachea tumorigenicity, mutagenicity, cell proliferation	No tumorigenicity; *lung*: K-ras mutational spectrum shifted, slight increase in BrdU labeling in the bronchial epithelium in the first week; *nose*: increased BrdU labeling in the respiratory epithelium of the naso- and maxilloturbinates during the first 3 weeks	Witschi[66]
SS	Commercial	Human	WB	1 h	n.i.	23.5	n.i.	Acute lung injury, inflammation	*Lung function*: no effect; *alveolar permeability*: decrease (MS: increase!)	Yates[180]

TABLE 13.1 (*continued*)
Examples of Experimental ETS Surrogates Used for Inhalation Studies

Smoke type	Cigarette	Species	Exposure mode	Duration	Particle conc. (mg/m³)	CO conc. (ppm)	Nicotine conc. (µg/m³)	Purpose, interest	Endpoint, result	Ref.
SS	2R1	Rat, hamster	NHO	7 h/d, 7 d/week, 90 d, 21 d recovery	2.1, 6.0	9, 22	600, 1,300	Histopathology, nose, larynx, trachea, lung	*Rats*: dose-related histomorphological changes: slight reserve cell hyperplasia of respiratory epithelium of nose, fully reversible; hyperplasia, metaplasia in the larynx, not completely reversed within 2 weeks; *hamsters*: no detectable effects	Teredesai[165]
SS	1R4F	Cockerel	WB	6 h/d, 5 d/week, 16 weeks	ca. 8	ca. 35	ca. 270	Atherosclerosis, aortic plaque formation	Increase in size, but not in the number and distribution of plaques	Penn and Snyder[40]
SS	2R1, 1R4F	Rat	WB	6 h/d, 7 d/week, 4 weeks	9.7	36.4	1,034	Genotoxicity, DNA adducts in lung, trachea, heart, bladder; prevention by *N*-acetylcysteine	Specific adducts increased in lung, trachea, heart, and bladder; NAC did not affect levels of most adducts (some were slightly modulated)	Arif et al.[125]
SS (?)	Commercial	Rat	WB	90 min/d, 5 d/week, 3 months	n.i.	35	n.i.	Lung emphysema, BAL cytology	*Morphometry*: enlarged alveolar spaces, decreased thickness of alveolar walls, loss of elastic components; *BAL cells*: increased number of hemosiderin-positive cells, decreased number of PMNs	Escolar et al.[152]

TABLE 13.1 (continued)
Examples of Experimental ETS Surrogates Used for Inhalation Studies

Smoke type	Cigarette	Species	Exposure mode	Duration	Particle conc. (mg/m³)	CO conc. (ppm)	Nicotine conc. (µg/m³)	Purpose, interest	Endpoint, result	Ref.
SS	Commercial	Ferret	NHO	2 h/d, 5 d/week, 15 weeks	1.6, 11.2	n.i., n.i.	n.i., n.i.	Induction of heat shock protein in the lungs	No increase of constitutive level of HSP-70 expression (also negative with MS exposure in earlier investigations)	Wong[171]
SS	n.i.	Human	WB	15 min	>10?	45	n.i.	Rhinitis, nasal congestion, lung function, tobacco allergy	Symptoms increased (headache, irritations in eyes, nose, throat), small lung function changes; increased congestion in sensitive persons only; no inflammatory mediators, no histamine detected, no signs of IgE-mediated allergic mechanism	Bascom[99]
SS	2R1	Rat	NHO	10 min/d, 7 d/week, 4–20 weeks	>>10 (?)	n.i.	n.i.	COPD, polyamine biosynthesis	ODC and AdoMetDC activated in trachea, but not in lungs (MS: also in lungs)	Olson[129]
SS	n.i.	Rabbit	WB	6 h/d, 5 d/week, 10 weeks	32.8	60	1,040	Atherosclerosis, lipid plaque formation under high cholesterol diet	Increase in lipid plaques and platelet aggregation, no changes in serum lipids (HDL, cholesterol, triglycerides)	Zhu[195]

TABLE 13.1 (continued)
Examples of Experimental ETS Surrogates Used for Inhalation Studies

Smoke type	Cigarette	Species	Exposure mode	Duration	Particle conc. (mg/m³)	CO conc. (ppm)	Nicotine conc. (µg/m³)	Purpose, interest	Endpoint, result	Ref.
SS	n.i.	Nude mouse	WB	0.5 h/d, 5 d/week, 2 weeks	>30?; 2 mg/expos.	n.i.	n.i.	Lung function, injury, and oxidative stress; prevention by antioxidants	*Epithelial permeability and conjugated dienes: no effect; pulmonary resistance and lipid fluorescence increased; vitamin E protects against effects*	Zhang,[184]
SS	Commercial	Rabbits	WB	15 min/d, 20 d	n.i.	>60[b]	n.i.	Lung injury, epithelial permeability, inflammation	Increased permeability, increased no. of BAL cells, increased leukotriene E4, eosinophil infiltrates in airway epithelium, neutrophil accumulation in capillary and perivascular spaces	Witten[185]
SS	n.i.	Rat	WB	6 h/d, 5 d/week, 3–6 weeks	60	92	1,100	Myocardial infarct	Increased size of experimentally induced infarcts	Zhu et al.[205]
SS	Commercial	Rabbits	WB.	15 min/d, 20 d	ca. 60[d]	n.i.	n.i.	Tracheal epithelial production of bronchoreactive eicosanoids	*Ex vivo* cultured tracheal epithelial cells from smoke-exposed rabbits exhibit increased synthesis of PGE_2, 6-keto $PGF_{1\alpha}$, and TxB_2 following stimulation with acrolein or endotoxin	Joseph[132]

TABLE 13.1 (continued)
Examples of Experimental ETS Surrogates Used for Inhalation Studies

Smoke type	Cigarette	Species	Exposure mode	Duration	Particle conc. (mg/m³)	CO conc. (ppm)	Nicotine conc. (µg/m³)	Purpose, interest	Endpoint, result	Ref.
SS	2R1	Mouse	WB	1 to 3 × 30 min	ca. 130[a]	n.i.	n.i.	Genotoxicity, oxidative stress in lung, liver, heart	DNA damage and repair: increased level of 8-OHdG in lung, liver, heart	Howard[170]
SS	Commercial	Rat	WB	1–4 cig/d, up to 6 h/d, 1 week	ca. 50–200?[a]	>>100 [b]	n.i.	Embryotoxicity	*Lung:* enhanced apoptosis, mesenchymal changes, hyperplasia of bronchial muscles; *liver, GI tract, and kidney:* moderate histomorphological changes	Nelson[151]
SS	Commercial	Rat	WB	1–4 cig/d, up to 6 h/d, 1 week	ca. 50–200?[a]	>>100 [b]	n.i.	Embryotoxicity	Retardation of ossification, not dose-dependent, no gross teratogenic effects	Nelson[206]
SS, also MS	2R1	Rat, mouse, guinea pig	WB	7 d/week, 16 weeks	Total inhaled doses: 22.5 mg/rat, 5.1 mg/mouse, 499 mg/guinea pig	>200[b]	n.i.	Induction of B(a)P metabolism in the lung	AHH activity in microsomal supernatants of lung homogenates: increased in rats and mice, but not in guinea pigs; SS and MS compared: both induce equally well, although SS has sixfold lower TPM	Gairola[175]
SS, MS	2R1	Mice (C57BL, DBA)	NHO	7 d/week, 65–70 weeks	n.i.	>300[b]	n.i.	Lung DNA adducts, PAH metabolism	12 to 25-fold increase in preexisting lung ³²P adducts; 2 to 3-fold AHH induction (only in C57, not in DBA mice)	Gairola[169]

TABLE 13.1 (*continued*)
Examples of Experimental ETS Surrogates Used for Inhalation Studies

Smoke type	Cigarette	Species	Exposure mode	Duration	Particle conc. (mg/m³)	CO conc. (ppm)	Nicotine conc. (µg/m³)	Purpose, interest	Endpoint, result	Ref.
SS	Commercial	Guinea pig	WB	3 × 10 min/d, 1–4 d	n.i.	2,600	n.i.	Bronchial hyperresponsiveness	*Lung:* enhanced bronchoconstriction response to histamine, but not to acetylcholine; *trachea:* no effect	Omini[187]
SS	n.i.	Rat	WB	45 min/d, 6 d/week, 16 weeks	High: 3 × 1 cig/cage	n.i.	n.i.	Asthma, mediators of bronchoconstriction	Increased synthesis and decreased metabolism of adenosine in BAL fluid	Anand[35]
SS	Commercial	Rat	WB	3 × 10 min/d, 45–90 d	ca. 1,000[?a]	n.i.	n.i.	Lung emphysema: physiology, inflammation, histopathology	Increased residual volume, increased alveolar spaces, decreased lung elastance; inflammation and goblet cell hyperplasia in bronchial epithelium; hypertrophy of bronchial smooth muscle	Cendon[153]
SS	n.i.	Rat	WB	2 × 1 h/d, 8–20 weeks	n.i.	n.i.	n.i.	Lung tumor initiator and promoter mechanism	*Initiation:* K-ras activation (G:C transversion at codon 12) in lung homogenate from 1 of 26 exposed rats; *promotion:* PKC activity transiently increased, subsequently downregulated relative to control level	Maehira[157]
SS, MS	n.i.	Hamster	WB	24 months	n.i.	n.i.	n.i.	Exposure conditions	Max. tolerated dose level leads to ca. 20% COHb. study results not published	Haley et al.[150]

TABLE 13.1 (continued)
Examples of Experimental ETS Surrogates Used for Inhalation Studies

Smoke type	Cigarette	Species	Exposure mode	Duration	Particle conc. (mg/m³)	CO conc. (ppm)	Nicotine conc. (µg/m³)	Purpose, interest	Endpoint, result	Ref.
ADSS	1R4F	Mouse (C57, DBA)	WB	6 h/d, 4 d	1.0	3.4	116	Mechanism of pulmonary CYP450 induction	Induction of CYP1A1 in lung homogenates of C57, but not of DBA mice; mechanism via Ah receptor binding	Gebremichael[176]
ADSS	1R4F	Rat	WB	6 h/d, 7 d/week, 9–16 d	1.0	4.9	344	Embryonic development of bronchiolar epithelial cells	Maternal exposure leads to increased expression of CC10 and CC10 mRNA at birth	Ji et al.[177]
ADSS	1R4F	Rat	WB	6 h/d, 5 d/week, 7–100 d	1.0	6.0	350	Postnatal bronchiolar epithelial cell development	Increased level of CYP450 1A1 and NADPH reductase in bronchiolar Clara cells and in alveolar type II cells, decreased cell proliferation in distal bronchioles; normal expression of CC10 and CYP450 2B	Ji et al.[167]
ADSS	1R4F	Rat	WB	4–6 h/d, 21 d prenatal + 21 d postnatal	1.0	6.9	285	Postnatal increase in bronchial hyper-responsiveness	*Lung function:* perinatal exposure associated with significantly increased bronchial hyperresponsiveness; *immunohistochemistry:* increased number of neuroendocrine cells (statistically not significant), number of mast cells not increased	Joad et al.[183]

TABLE 13.1 (*continued*)
Examples of Experimental ETS Surrogates Used for Inhalation Studies

Smoke type	Cigarette	Species	Exposure mode	Duration	Particle conc. (mg/m³)	CO conc. (ppm)	Nicotine conc. (µg/m³)	Purpose, interest	Endpoint, result	Ref.
ADSS	1R4F	A/J mouse	WB	6 h/d, 5 d/week, 6 months	4.1	17	1,011	Lung histopathology, morphometry, induction of CYP450 isozymes	No histomorphological changes, no increased number of type II cells and alveolar macrophages; no increased expression of P450 isozymes; only effect: increased CYP450 1A1 in capillary endothelial cells	Pinkerton[67]
ADSS	1R4F	Rat	WB, NHO	6 h/d, 14 d + 14 d recovery	0.1, 1.1, 9.8	3.6, 11.3, 57	0.9, 252, 1,708	Adaptive changes, histopathology	*Nose:* mild hyperplasia with inflammatory infiltrate in respiratory epithelium of nasoturbinate and lateral wall; fully reversible; *larynx, lung:* no effects	Coggins[17]
ADSS	1R4F	Rat	WB, NHO	6 h/d, 14 d + 14 d recovery	0.1, 1.1, 9.8	3.6, 11.3, 57	0.9, 252, 1,708	Lung genotoxicity	*DNA adducts* in lung and heart, only in high dose group; *chromosomal aberrations* in lung macrophages: no effect	Lee et al.[168]
ADSS	1R4F	Rat	WB, NHO	6 h/d, x d/week, 90 d + 90 recovery	1, 10	2.9, 9.3, 55	0.39, 272, 2,377	Nose and lung histopathology, adaptive changes	*Nose:* only at 10 mg/m³ slight to mild hyperplasia of respiratory epithelium of nasoturbinates and lateral wall, fully reversible; *lungs:* no effect	Coggins[18]

TABLE 13.1 (continued)
Examples of Experimental ETS Surrogates Used for Inhalation Studies

Smoke type	Cigarette	Species	Exposure mode	Duration	Particle conc. (mg/m³)	CO conc. (ppm)	Nicotine conc. (µg/m³)	Purpose, interest	Endpoint, result	Ref.
ADSS	1R4F	Rat	WB, NHO	6 h/d, x d/week, 90 d + 90 d recovery	0.1, 1, 10	2.9, 9.3, 55	0.39, 272, 2,377	Cell proliferation in nose, larynx, trachea, lung	*Nose*: elevated rates of BrdU incorporation in the respiratory and cuboidal epithelium at 10 mg/m³ throughout exposure period, reverted to control level during recovery; *trachea, and larynx*: no effects; *lung, bronchiolar epithelium*: transient, slight increase on day 5 at 10 mg/m³; *alveolar region*: no effects	Ayres et al.[79]
SS	2R1	Rat	NHO	6 h/d, 7 d/week, 90 d + 42 d recovery	8.7	27.8	2,210	Adaptive response, histopathology, B(a)P metabolism; comparison SS vs. RASS	*Reversible dose-related histomorphological changes*: slight reserve cell hyperplasia of respiratory epithelium of nose; hyperplasia, metaplasia in the larynx; *reversible dose-related enhancement of B(a)P metabolism*: nose respiratory epithelium (not olfactory), lung homogenate NOEL 0.6 mg TPM/m³; effects follow TPM, not gas phase concentration!	Haussmann[19]
RASS	2R1	Rat	NHO		2.6	28.7	520			

TABLE 13.1 (continued)
Examples of Experimental ETS Surrogates Used for Inhalation Studies

Smoke type	Cigarette	Species	Exposure mode	Duration	Particle conc. (mg/m³)	CO conc. (ppm)	Nicotine conc. (μg/m³)	Purpose, interest	Endpoint, result	Ref.
RASS	1R4F	Rat	WB	12 h/d, 5 d/week, 12 months	5.9	27	940	Tumorigenicity, histopathology, adaptive changes, biomarkers, mechanism	*General:* noninhalative nicotine uptake in whole body exp.; no tumors; *nasal cavity:* hyperplasia, metaplasia of respiratory epithelium, slight hyperplasia, atrophy, eosinophilic granules in olfactory epithelium *larynx:* pronounced squamous metaplasia and hyperplasia (base of epiglottis); *bronchus:* slight reserve cell hyperplasia; *BAL cells:* no signs of inflammation; *oxidative DNA damage:* slight increase in 8-OHdG in nasal, not lung epithelium; no progression from 3 to 12 months!	Haussmann[21]
			WB	12 h/d, 5 d/week, 12 months	11.9	51	1,950			
			NHO	7 h/d, 5 d/week, 12 months	12.1	42	2,350			

TABLE 13.1 (*continued*)
Examples of Experimental ETS Surrogates Used for Inhalation Studies

Smoke type	Cigarette	Species	Exposure mode	Duration	Particle conc. (mg/m³)	CO conc. (ppm)	Nicotine conc. (µg/m³)	Purpose, interest	Endpoint, result	Ref.
RASS	1R4F	Rat	WB	12 h/d, 5 d/week, 8 d	11.1	52.7	1,900	Adaptive response, histopathology, cell differentiation biomarkers	*Histopathology:* slight hyperplasia and squamous metaplasia of nasal respiratory epithelium; *cytokeratin expression:* expression pattern changes at sites of histomorphologic changes, additional changes at other sites of nasal epithelium, minor changes in trachea and lungs, various cell types involved, sensitive marker	Schlage et al.[64]
RASS	1R4F	Rat	WB	12 h/d, 5 d/week, 12 months	11.9	51	1,950	Nose: adaptive response, histopathology, cell differentiation biomarkers	*Histopathology:* slight hyperplasia and squamous metaplasia of nasal respiratory epithelium; *cytokeratin expression:* expression pattern changes at sites of histomorphologic changes, additional changes at other sites of nasal epithelium	Schlage et al.[90]
RASS	1R4F	Rat	NHO	6 h/d, 7 d/week, 24 months	10, 3.1	38, 12.5	1,833, 561	Carcinogenicity, comparison to diesel engine exhaust	Only analytical characterization of test atmospheres	Stinn et al.[98]

TABLE 13.1 (*continued*)
Examples of Experimental ETS Surrogates Used for Inhalation Studies

Smoke type	Cigarette	Species	Exposure mode	Duration	Particle conc. (mg/m³)	CO conc. (ppm)	Nicotine conc. (μg/m³)	Purpose, interest	Endpoint, result	Ref.
MS/SS mix	1R4F	Rat	WB	2 h, 4 h	3.3	18	615	Cardiovascular effects, atherosclerosis	Plasma from ETS-exposed rats causes increased LDL accumulation, interaction with LDL, not with endothelial cells	Roberts et al.[207]
MS/SS mix	2R1	Ferret	NHO	2 h/d, 5 d/week, 15 weeks	38	n.i.	n.i.	Induction of PAH metabolism (lung)	B(a)P metabolism increased chronically	Sindhu et al.[208]
MS/SS mix or gas phase alone	1R4F	A/J mouse	WB	6 h/d, 5 d/week, 5 months, + 4 months recovery	78.5, 0.1	211, 113	13,400, 3,100	Lung tumorigenicity, histopathology, CYP450 isozyme induction	Increased tumor multiplicity in smoke-exposed mice; gas phase had same effect as unfiltered smoke; no shift in tumor histotype spectrum vs. spontaneous pattern; no induction of CYP1A1, 2B1, 2E1 by gas phase	Witschi[154]

TABLE 13.1 (*continued*)
Examples of Experimental ETS Surrogates Used for Inhalation Studies

Smoke type	Cigarette	Species	Exposure mode	Duration	Particle conc. (mg/m³)	CO conc. (ppm)	Nicotine conc. (µg/m³)	Purpose, interest	Endpoint, result	Ref.
MS/SS mix	1R4F	A/J mouse	WB	6 h/d, 5 d/week, 5 months, + 4 months recovery	87.3, 83.5, 52.6	244, 233, 161	16,100, 18,900, 17,000	Lung tumorigenicity, histopathology, cell proliferation, enzyme induction	*Tumorigenicity:* signif. increase; no additivity to carcinogen-induced tumorigenicity; *histopathology, morphometry:* no effects; *CYP1A1:* reversible increase in airway epithelia and lung parenchyma; *DNA synthesis* transiently increased in the large airways (first 12 weeks), terminal bronchioles (first 6 weeks) and alveolar zone (first 2 weeks)	Witschi et al.[15]
MS/SS mix	1R4F	A/J mouse	WB	6 h/d, 5 d/wk, 5 months	83.5	233	18,900	Nasal cell proliferation (other effects reviewed)	DNA synthesis in respiratory epithelium of maxilloturbinates: strong transient increase during the first 10 weeks	Witschi[14]

TABLE 13.1 (continued)
Examples of Experimental ETS Surrogates Used for Inhalation Studies

Smoke type	Cigarette	Species	Exposure mode	Duration	Particle conc. (mg/m³)	CO conc. (ppm)	Nicotine conc. (µg/m³)	Purpose, interest	Endpoint, result	Ref.
MS/SS mix	2R1	rat	WB	6 h/d, 5 d/week, 1–5 weeks, 1 week recovery	73–93	350	n.i.	Genotoxicity: DNA adduct formation in trachea, lung, liver, heart, testis; oxidative damage, induction of PAH metabolism, GSH protection	High levels of *adducts* in trachea, alveolar macrophages, lung, heart, and bladder, regression during recovery (not in heart and macrophages); lung 8-OHdG increased; lung *AHM and GST* increased, *GSH* was depleted	Izzotti et al.[126]
MS/SS mix		A/J mouse	WB	6 h/d, 5 d/week, 5 months, 4 month recovery	132	n.i.	26,000	Carcinogenicity, chemoprevention	Myoinositol and dexamethasone in diet decrease lung tumor multiplicity	Witschi et al.[68]
MS/SS mix	2R1	Ferret	NHO	2 h	381	n.i.	n.i.	Induction of PAH metabolism (lung, liver)	Acute increase only in the lung; different pathways in lung and liver	Sindhu et al.[173]
MS/SS mix	2R1	Ferret	NHO	2 h/d, 5 d/week, 8 weeks	38, 381	n.i.	n.i.	Induction of PAH metabolism (liver)	*Downregulation* of B(a)P metabolism in liver homogenates (after *initial increase*)	Sindhu et al.[146]
MS/SS mix	2R1	Ferret	NHO	2 h/d, 5 d/week, 15 weeks	38, 381	n.i.	n.i.	Mucociliary clearance in nasal cavity and lung	*Nasal airways:* increased clearance of radiolabeled particles at both concentrations; *thoracic region:* decreased particle clearance at high conc. only	Phalen et al.[147]

TABLE 13.1 (*continued*)
Examples of Experimental ETS Surrogates Used for Inhalation Studies

Smoke type	Cigarette	Species	Exposure mode	Duration	Particle conc. (mg/m³)	CO conc. (ppm)	Nicotine conc. (µg/m³)	Purpose, interest	Endpoint, result	Ref.
ETS exp.	Commercial	Human	WB	8 h	n.i.	10, 25	n.i.	Exposure monitoring, urinary mutagenicity	No significant increase in urinary mutagenicity	Scherer et al.[23]
ETS exp.	Var.	Human	WB	8 h	n.i.	10, 25	60, 120	Exposure monitoring, urinary mutagenicity, detoxification	Gas phase as well as full ETS increase the excretion of urinary thioethers (marker of exposure to electrophilic toxicants); *conclusion:* TPM-bound mutagens/carcinogens are less important	Scherer et al.[24,43]
ETS exp.	n.i.	Human	WB	6 h	n.i.	ca. 25[c]	n.i.	Exposure monitoring, urinary mutagenicity	Slight increase in urinary mutagenicity	Bos et al.[201]

Note: Abbreviations: SS, sidestream smoke; MS, mainstream smoke; ADSS, aged and diluted sidestream smoke, RASS, room-aged sidestream smoke, SS/MS mix, mixture of approximately 85% SS and 15% MS; ETS exp., experimental SS produced by human smokers under controlled conditions; NHO, nose-/head-only exposure; n.i., not indicated.

[a] Calculated from static burning rate and chamber volume.
[b] Estimated from COHb value.
[c] Estimated by Scherer et al.[23]
[d] 2 mg TPM in 33.7 l chamber.

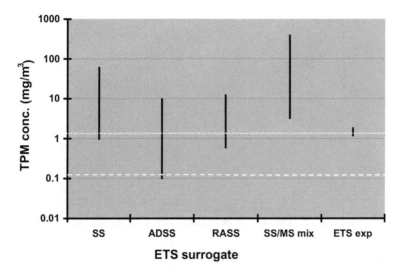

FIGURE 13.1 Published TPM concentration ranges in experimental surrogates of ETS. TPM concentrations are taken from the publications listed in Table 13.1. The dashed line denotes the upper limit of typical particulate concentrations of real-life ETS, and the dotted line denotes the maximum value of *extreme* concentrations of ETS.[1,28,36] Abbreviations: SS, side-stream smoke; MS, mainstream smoke; ADSS, aged and diluted sidestream smoke, RASS, room-aged sidestream smoke, SS/MS mix, mixture of approximately 85% SS and 15% MS; ETS exp., experimental SS produced by human smokers under controlled conditions.

lower respiratory tract (the latter frequently in acute human inhalation experiments), while others were related to cardiovascular or other systemic effects, e.g., chronic SS exposure of cockerels as a model for atherosclerosis,[40,41] or were performed to generate ETS for analytical purposes.

13.3.2 QUALITATIVE ASPECTS

In Figure 13.2, the relative amounts of CO, TPM, and nicotine in the various ETS surrogates are shown (as far as they were indicated in the publications listed in Table 13.1); others have used the nicotine to RSP ratio to characterize ETS atmospheres.[31] The TPM to CO and nicotine to CO ratios vary considerably between the ETS surrogates as well as within a given ETS surrogate (Figures 13.2a and 13.2b). The TPM to CO ratios vary by a factor of 4 within the SS atmospheres, by a factor of 2.5 within ADSS, by a factor of 2 within the SS/MS mix, by a factor of 3 within RASS, and by a factor of 1.3 for experimental ETS produced by human smokers (Figure 13.2a). For the nicotine to CO ratios, the variations are even higher: 6.5-fold for SS, 3-fold for ADSS, 2-fold for the SS/MS mix, 3.3-fold for RASS, and 5-fold for "human" ETS (but the ratios tend to be lower than for machine-generated ETS models; see Figure 13.2b). This distribution represents the entire broad spectrum of setups and concentrations. Within a particular setup and concentration range, the reproducibility can be very good, and it has been shown that the brand-to-brand

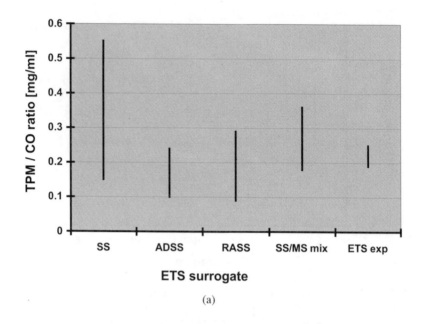

(a)

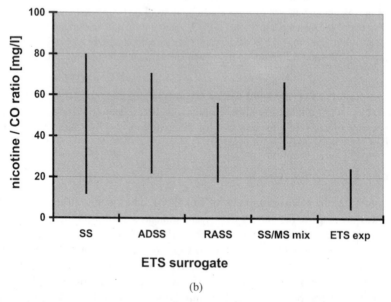

(b)

FIGURE 13.2 Differences in the relative amounts of CO, TPM, and nicotine in experimental surrogates of ETS: (a) TPM to CO ratios, values taken from the publications listed in Table 13.1; (b) nicotine to CO ratios, values taken from the publications listed in Table 13.1. Abbreviations: SS, sidestream smoke; MS, mainstream smoke; ADSS, aged and diluted sidestream smoke, RASS, room-aged sidestream smoke; SS/MS mix, mixture of approximately 85% SS and 15% MS; ETS exp., experimental SS produced by human smokers under controlled conditions.

variations and the variations between different conventional cigarette types have relatively little influence on these ETS components.[25,26]

In addition to these variations in the composition of ETS concerning CO, TPM, and nicotine, similar or even higher variations should be considered for other important components of ETS, such as formaldehyde and ammonia, which are not regularly included in the characterization of the test atmospheres. This was demonstrated by ageing diluted SS in the same experimental chamber in the presence of either inert surfaces (equivalent to ADSS) or typical floor and wall surface materials, fabrics, and furniture (equivalent to RASS) and comparing the concentrations of TPM and additional gas phase components in these ETS surrogates with fresh SS:[32] roughly, acetaldehyde dropped by only 20%, but TPM dropped by 70%, formaldehyde by 80%, nicotine by 90%, and ammonia by 95%. Other analytes that occur at low concentrations, but are thought by many to be representative of the classes of compounds (nitrosamines and polyaromatic hydrocarbons [PAHs]) that contribute to the carcinogenic potential of ETS, e.g., NNK, BaP,[1,42] have only been sporadically determined in ETS inhalation studies (cf. References 19, 21, and 43), but can be expected to vary in a similar manner.[11]

13.4 EXPOSURE MODES

The influence of exposure modes (whole body vs. head/nose only, see Table 13.1) on the amount of smoke deposition in the lung has been investigated in a comparative rat inhalation study with MS.[44] These authors observed the most efficient particulate dose delivery to the lungs of whole-body exposed rats and suggested that this may be due to a reduction in stress and irritation (these investigations used MS at 100 and 200 mg TPM/m^3). This advantage of whole-body exposure may not hold true for ETS inhalation at much lower smoke concentrations. In a chronic inhalation study in which RASS was administered at a concentration of 12 mg TPM/m^3 by whole-body and head/nose-only routes,[21] the authors observed that there was considerable uptake of smoke components by noninhalative routes, as well as losses of TPM (approximately 20%), nicotine (30%), and formaldehyde (70%) in the whole-body chambers; and they concluded that "head-only exposure … is generally considered preferable … over whole body exposure to avoid artificial changes in smoke composition and the noninhalative uptake of smoke constituents."

13.5 EFFECTS OF EXPERIMENTAL ETS EXPOSURE

13.5.1 Nasal Cavity

In both humans and obligatory nose-breathing experimental animals, the nasal cavity is the initial site of exposure to ETS and all other kind of pollutants in the air.[45,46] Animal models are thus more closely related to the human situation for ETS exposure than is the case for MS exposure, i.e., mouth-breathing smokers vs. nose-breathing animals. Anatomical differences between humans/primates and rodents, for example, with regard to the shape and inner structure of the nasal passages are obvious. The

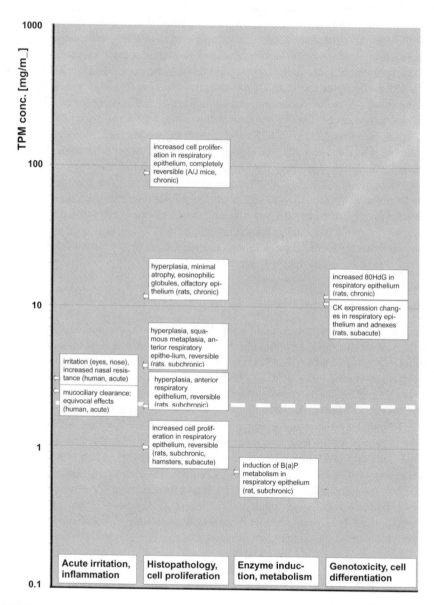

FIGURE 13.3 Effects of exposure to various ETS atmospheres in the nasal cavity. The lowest observed effect levels (LOELs) for various irritational, histopathological, biochemical, and genotoxic endpoints are summarized, regardless of the ETS surrogate (for details, see text). The dotted line indicates the upper limit of RSP concentrations in real ETS. For duration of daily exposures, see Table 13.1.

influence of the higher complexity and enlarged surface of the rodent nasal mucosa on particle deposition and gas phase absorption has been investigated thoroughly by Morgan et al.[47,48] Despite the geometric differences, there are similarities in the epithelial composition of the nasal mucosa:[49] large surface areas are covered by respiratory epithelium, which provides the mucociliary functions by means of its mucus-producing goblet cells and its mucus-transporting ciliated cells, and by olfactory epithelium consisting of the sensory cells for smell perception and their supporting (sustentacular) cells. The respiratory epithelium is located primarily in the anterior region of the nasal cavity, and the olfactory epithelium is in the posterior/dorsal region; the proportions vary considerably in different species.[50,51] In addition, transitional epithelium (between squamous and respiratory epithelium), squamous epithelium (in the nasal vestibulum), glandular epithelia (in the submucosa and in adnexes of the nasal cavity), and nasal-associated lymphoid tissue (NALT)[52] are present in varying amounts (for reviews, see References 53 and 54). Using morphological criteria,[55,56] ultrastructural criteria,[57] immunohistochemical criteria,[58,59] and enzyme histochemical criteria,[60,61] more than a dozen different cell types have been identified in the nasal mucosa, e.g., of the rat. This heterogeneity may reflect the functional importance of the nasal cavity, which is located at the site of entry of the highest volume of an environmental medium (the air for breathing) into the body, for recognizing dangerous compounds by the trigeminal and olfactory sense and for prefiltering the air before it passes into the lungs.[62,63]

In this chapter, we focus on the structural and functional epithelial changes that appear to be preneoplastic lesions and/or may indicate a carcinogenic risk (Table 13.1, Figure 13.3). For the upper respiratory tract, such a risk has not been recognized in human nonsmokers exposed to ETS. In addition, some irritative effects, as far as they concern the function of mucosal epithelial cells, e.g., mucociliary function or local production of inflammatory mediators, have also been considered (Table 13.1, Figure 13.3).

Short-term exposure (e.g., for several days) to *exaggerated* concentrations of "ETS" can cause a histopathologically detectable stress response seen as structural changes in the nasal respiratory epithelium, which constitutes the mucociliary apparatus (Table 13.1), (for acute effects on mucociliary functions, see Section 13.5.1 on acute irritation in the nasal cavity). Exposure to RASS (11 mg TPM/m^3) for 8 d caused slight basal cell hyperplasia and squamous metaplasia of the anterior respiratory epithelium at the tips of the nasoturbinates and the lateral wall.[64] Similarly, Coggins et al.[17] observed slight to mild hyperplasia and inflammation at these specific sites following inhalation of ADSS (10 mg TPM/m^3) for 14 d. These findings reversed completely in a 14-d recovery period, and no effects were seen at 1 or 0.1 mg TPM/m^3.

In four published subchronic (90-d) ETS inhalation studies, similar histomorphological changes were observed in the nasal cavity of rats: Von Meyerinck et al.[16] found fully reversible hyperplasia and metaplasia of the respiratory epithelium of the nasoturbinates at 4 mg TPM/m^3 of SS, but no effects at the same concentration in hamsters. After exposing rats and hamsters to SS at 2 and 6 mg TPM/m^3, Teredesai and Prühs found no effects in hamsters and reserve cell hyperplasia of the nasal respiratory epithelium in rats, which reversed completely within 2 weeks.[65] Haussmann et al.[19]

compared fresh SS and RASS and observed slight reserve cell hyperplasia in 8 of 20 rats exposed to 8.7 mg TPM/m^3 SS and in 1 of 19 rats exposed to 2.6 mg TPM/m^3 RASS as the only nasal histopathological findings, with a no observed effect level (NOEL) of 3.6 mg/m^3 for SS and 1.2 mg/m^3 for RASS. This is also in concordance with the results reported for ADSS,[18] namely slight to mild hyperplasia of the respiratory epithelium of nasoturbinates and lateral wall, exclusively at 10 mg TPM/m^3, with the NOEL at 1 mg/m^3. These changes also reversed completely within a 90-d recovery period. In a subsequent chronic RASS inhalation study (12 months) on rats, Haussmann et al.[21] used 6 and 12 mg TPM/m^3 and observed moderate reserve cell hyperplasia and slight squamous metaplasia of the respiratory epithelium, as well as a slight increase in eosinophilic globules in the sustentacular cells, and minimal reserve cell hyperplasia and atrophy of the olfactory epithelium. No progression in the severity of lesions was seen after12 months compared to 90 d.[19,21] This was also the case in the ADSS studies after 90 d compared to 2 weeks.[17,18]

In the chronic "ETS" inhalation studies on A/J mice,[14,15,66-68] histopathological examination of the nasal cavity was not included. Reversible hyperplasia and metaplasia of the respiratory epithelium seen following MS and "ETS" exposure are common protective and adaptive changes in response to repeated irritation.[50,66-65] Hyperplasia in the nasal mucosa (i.e., an increase in cell number) can affect the "reserve cells," e.g., basal cells of the respiratory and olfactory epithelium, or the mucus-producing goblet cells. Goblet cell hyperplasia is a common, reversible effect following exposure to various irritants and MS.[50,69-71] Increases in eosinophilic globules may indicate an increased intracellular pool of metabolic enzymes.[72] Smoke-induced atrophy of the olfactory epithelium following exposure to high MS concentrations (249 mg TPM/m^3) was shown to be reversible within 2 weeks.[73] More severe effects would be degeneration or even ulceration, e.g., as seen after inhalation of chlorine or acrolein, but still reversible after discontinuation of exposure.[50] Extensive, preneoplastic, keratinized squamous metaplasia of the respiratory epithelium, together with inflammatory reactions and/or dysplasia, and progression to malignant tumors were observed, e.g., after lifetime exposure of rats to high concentrations of acetaldehyde or formaldehyde; and similar effects were seen in humans occupationally exposed to formaldehyde, but the tumor data were not unequivocal (for a detailed review, see Woutersen et al.[54]). As a carcinogenic effect in the nasal cavity following chronic exposure to MS, one nasal adenocarcinoma and one nasal squamous carcinoma in rats have been described in the literature;[74] but as discussed by Coggins,[13] this result could not be reproduced in two subsequent studies by the same working group.[75,76] In a total of 14 chronic MS inhalation studies on mice and rats, no significant increases in the rate of neoplastic tumors in the nose or other parts of the respiratory tract were produced by smoke inhalation,[13] nor have nasal tumors been observed following exposure to ETS surrogates, as reviewed by Witschi et al.[4] and shown in Table 13.1. In addition to the histomorphological changes, a few inhalation studies included immunohistochemical and biochemical endpoints to monitor changes in cellular function and differentiation status in the nasal mucosa. They found enhanced cell proliferation, the induction of enzyme activity involved in PAH metabolism, DNA damage and repair, and changes in the cytokeratin (CK) expression patterns following "ETS" exposure.

More frequently, such mechanistic endpoints have been investigated in the lower respiratory tract (Table 13.1, Figures 13.4 and 13.5).

Epithelial cell proliferation is necessary to replace damaged cells, to generate additional cells in order to shield against irritative agents (reserve cell hyperplasia and squamous metaplasia), or to increase the number of mucus-producing cells (goblet cell hyperplasia). In addition to its protective implications, cell proliferation in these normally quiescent cells may also enhance the propensity of existing or misrepaired DNA lesions to become fixed as mutations or may act in a tumor promoter-like fashion on a population of mutated cells, i.e., initiated cells (for discussion, see Brown et al.[77]). Enhanced BrdU labeling indicative of replicative DNA synthesis was seen in the nasal respiratory epithelium of hamsters[78] and rats[79] as a weak transient increase following inhalation of 1 mg TPM/m^3 of SS (hamsters) or ADSS (rats). A sustained elevation of BrdU labeling was seen in rats exposed to 10 mg TPM/m^3 of ADSS for 90 d, but this decreased to control level after 90 d of recovery. In A/J mice chronically exposed to SS (4.5 mg TPM/m^3), a transient, initial increase (first 3 weeks) in BrdU-labeled cells was observed in the respiratory epithelium of the naso- and maxilloturbinates;[66] following chronic exposure to very high concentrations of ADSS (83.5 mg TPM/m^3), the initial labeling in the maxilloturbinates was stronger and remained significantly enhanced over the control for 10 weeks.[15] The fact that even at this concentration the replicative DNA synthesis reverts during the postexposure period makes it likely that it correlates to the reversible adaptive changes in the nasal respiratory epithelium.

Cytochrome P450 (CYP450) isozymes are the key enzymes for the detoxification of a variety of toxic compounds as well as their metabolic activation from procarcinogens to the ultimate carcinogens, e.g., PAHs and nitrosamines. The CYP450 expression can be induced or enhanced by exposure to their substrate molecules. The presence of CYP450 in the mammalian nasal mucosa has been demonstrated, e.g., in rats, mice, and dogs. In the rat nasal cavity, CYP450 PB-B and NADPH-CYP450 reductase have been localized to the basal and apical zone of the olfactory epithelium, all cells of the respiratory epithelium, and the submucosal glands of the respiratory and olfactory region using immunohistochemistry.[58] In the subchronic inhalation study by Haussmann et al.,[19] a reversible, dose-related increase in B(a)P metabolism was seen in homogenates of the respiratory epithelium, but not the olfactory epithelium. This effect, as well as the histopathological changes, was obviously dependent on the concentration of TPM and not on the concentration of the gas phase. In humans, metabolic phenotypes characterized by higher inducibility of aryl hydrocarbon hydroxylase and the presence of CYP450 isozymes with extensive PAH metabolism have in the past been associated with a higher lung cancer risk (genetic susceptibility) in smokers;[80–82] however, this association is currently being revisited.[83]

The formation of DNA adducts is a critical step in mutagenesis: the adducted base can either be repaired correctly or give rise to the fixation of a definitive mutation in the following DNA replication. The attack of oxygen free radicals on DNA, which can be generated from endogenous processes as well as from external oxidative stress, can elicit base oxidation leading to the formation of adducts, e.g., 8-hydroxydeoxyguanosine (8-OHdG). The increase in spontaneous 8-OHdG levels observed with age may result from an increased rate of oxidant generation as well as from an increase in the susceptibility of tissues to oxidative damage, possibly due to a decreased level of protective endogenous antioxidants.[84] Oxygen free radicals play an important role in carcinogenesis, and a direct correlation between the formation of 8-OHdG in DNA and carcinogenesis has been shown.[85] The role of oxidative damage in lung carcinogenesis was also demonstrated following exposure of rats to the tobacco-specific nitrosamine, NNK, which primarily generates methyl and bulky DNA adducts.[42,86] DNA damage was investigated in the nasal cavity in only one "ETS" inhalation study. Oxidative DNA damage was measured as the amount of 8-OHdG in enzymatically digested DNA samples from rat nasal respiratory and olfactory epithelium. Following 6 and 12 months of exposure to RASS (12 mg TPM/m^3),[21] a slight increase in 8-OHdG was seen in the nasal respiratory epithelium, but in the olfactory epithelium only after 6 months. Other adducts, e.g., from PAH and nitrosamine metabolites (^{32}P-postlabeling method), were only investigated in the trachea and lung following ETS exposure (see later). The tobacco-smoke-specific, nitrosamine-related pyridyloxobutylate adducts have been investigated in the rat nasal mucosa following exposure to the nitrosamines only (not to cigarette smoke) and are considered by Hecht to be more relevant biomarkers of cigarette smoke exposure.[42] The NNK doses in these investigations were usually two to five orders of magnitude higher than those observed in heavy smokers.[87–89]

Epithelial cell differentiation changes in the rat respiratory tract following RASS exposure have been investigated by the immunohistochemical analysis of the expression patterns of CK polypeptides in the rat respiratory tract.[64,90] The intermediate filament cytoskeleton of epithelial cells is composed of several CK polypeptides from a total of approximately 20 different CKs. The combination patterns are organ specific and specific for the type of epithelium, the cell type within an epithelium, and the differentiation status of a given cell type. Changes in the CK expression pattern have been associated with a variety of neoplastic and non-neoplastic diseases;[91–93] however, the main application for CK expression is the differential diagnosis of epithelial neoplasias. According to the biomarker concept,[94] CK expression changes have also been correlated with preneoplastic and adaptive responses, e.g., squamous metaplasia in the trachea of vitamin A-deficient rats,[95] or with tissue repair following acute injury, e.g., in the rat lung following bleomycin instillation[96] or ionizing irradiation.[97]

In the rat nasal cavity, we investigated CK expression changes at the anterior nasal cavity levels 1 and 2 (according to Young[55]) following an 8-d inhalation of RASS[64] and at level 1 also following a 1-year inhalation of 12 mg TPM/m^3 RASS.[90] In the areas of histomorphological changes — hyperplasia and slight metaplasia of the respiratory epithelium of the naso- and maxilloturbinates and lateral wall at nasal cavity level 1 — we saw the typical CK changes expected in association with the thickening and stratification of the pseudostratified epithelium, for example, the suprabasal expression of CK14 in hyperplastic areas, which in normal pseudostratified epithelium is restricted to basal cells. The CK expression changes seen at nasal cavity level 1 were similar after 1 year of exposure and did not seem to progress as compared to the 8-d exposure, as was seen for the histomorphological changes. In addition, several small changes in CK expression were observed at sites where no histomorphological changes were observed, e.g., in the submucosal glands; in the olfactory epithelium at nasal cavity level 2 (so far only evaluated in the short-term study); and in the adnexes of the nasal cavity, nasolacrimal duct, and vomeronasal organ. The significance of these changes, perhaps reflecting physiological reactions to irritation, is not yet understood. In these initial applications of CK expression as an indicator of epithelial cell differentiation changes, the sensitivity of this novel endpoint in inhalation studies has been demonstrated. For its use as an intermediate biomarker of late-occurring adaptive or preneoplastic changes, the method is currently being validated with histological specimens from the 1-year RASS study[21] and from a recently completed 30-month inhalation study.[98]

Irritative ETS effects in the nasopharyngeal region have frequently been investigated experimentally in human volunteers. This is facilitated by the fact that exposure durations can be acute and the nasal cavity is accessible for minimum invasive or noninvasive investigations, e.g., nasal lavage,[99] and acoustic rhinometry.[100,101] The subjects were exposed for 1 to 3 h to SS at *extreme* concentrations of 15 to 20 ppm CO, which may be equivalent to approximately 3 mg TPM/m³, or for 15 min to concentrations as high as 45 ppm CO; in all cases, symptoms of irritation in the eyes, nose, and throat (burning, congestion) were reported (Table 13.1). Increased nasal resistance was measured by rhinomanometry and acoustic rhinometry, and no gross differences were seen between ETS-sensitive and -insensitive individuals after 2 h of exposure.[100,101] After 15 min of exposure to the highest concentration (45 ppm CO), congestion was noted only in ETS-sensitive individuals.[99] The analysis of nasal lavage fluid revealed no signs of increased epithelial permeability (albumin), neutrophils, or inflammatory mediators, nor signs of histamine- or IgE-mediated allergic mechanisms.[99,102,103] The cellular basis of this local, acute irritational response is not evident, and a neuronal signal-mediated acute increase in vascular permeability was seen following MS exposure in rodent models only (for detailed discussion, see Bascom et al.[104]). With regard to the mucociliary clearance, an equivocal finding was reported with a radiotracer method: clearance was enhanced in 6 of 12 subjects, unchanged in 3 of 12 subjects, and reduced in 3 of 12 subjects.[104] Previous studies with exhaled MS in nonsmokers[105] and in donkeys[106] also reported that the nasal mucociliary clearance was not impaired, whereas others described effects of cigarette smoke on the viscoelasticity of the mucous blanket[107] and disturbances of the ciliary beating[108,109] in airway mucociliary epithelium.

The details of exposure conditions are listed in Table 13.1. For an overview of the various ETS effects on the epithelial cells of the nasal cavity and the lowest observed effect levels (LOEL), see Figure 13.3.

13.5.2 LARYNX

The larynx, especially of the rat, has proven to be a sensitive site for the detection of epithelial reactions to inhaled toxicants including cigarette smoke.[19–21,70,110] In hamsters, chronic MS inhalation caused the formation of neoplastic tumors of the larynx in specific inbred strains.[111–113] Smoke exposure-related laryngeal tumors in other experimental animal species have not been reported; however, in humans, smoking, alcohol abuse, and occupational toxicants have been associated with an increased risk of laryngeal cancer.[114,115] The anatomical and histological structure of the rodent larynx is very complex: eight different epithelial cell types have been identified by scanning electron microscopy;[110] therefore, exactly defined sectioning levels[116–118] are required for the quantitative comparison of the corresponding epithelial sites from treated and untreated animals.

The larynx was included in the histopathological evaluation in only six ETS inhalation studies, and the results were conflicting: no effects were seen in hamsters exposed for 90 d to SS at concentrations of 4 mg TPM/m^3,[16] or 2.1 and 6 mg/m^3,[65] nor were any effects seen in rats exposed to ADSS for 2 weeks and 90 d at concentrations of up to 10 mg/m^3.[17,79] In rats exposed to SS for 90 d at 2.1 and 6 mg/m^3, dose-related hyperplasia and squamous metaplasia were seen.[65] This was also the case for rats exposed for 90 d to SS at 8.7 mg/m^3 or to RASS at 2.6 mg/m^3, the severity being related to the TPM concentration rather than to the gas phase concentration.[19] These results were confirmed in a 12-month inhalation study on rats exposed to RASS at concentrations of 6 and 12 mg/m^3. The authors explained their positive results by the selection of a particularly sensitive evaluation site at the base of the epiglottis, whereas little or no effect was seen at other epithelial locations in the larynx.[21] Because of their reversibility within 42 d (though not complete within 21 d) these effects, i.e., hyperplastic thickening and metaplastic replacement of columnar epithelial cells by multilayered squamous epithelial cells, have been regarded as an adaptive response,[19,20] which has been ascribed to the particulate phase.[119]

Biochemical or other intermediate biomarkers for the effects of "ETS" exposure in the larynx have not been published to date. The details of exposure conditions are listed in Table 13.1, and the findings and their LOELs are summarized in Figure 13.4.

13.5.3 Trachea

The tracheal epithelium, unlike the epithelia of the nasal cavity and the larynx, is a more or less uniform respiratory epithelium consisting primarily of three cell types: ciliated cells, goblet cells, and basal cells. In epidemiological studies, no association between tracheal tumors and ETS exposure has been reported. In inhalation studies with MS exposure, histomorphological changes such as goblet cell hyperplasia, reserve cell hyperplasia, and squamous metaplasia have been observed.[110,120] No such effects were reported following exposure to various ETS surrogates in the following studies: two 90-d SS inhalation studies on hamsters at 4 and at 6 mg TPM/m^3;[16,65] three 90-d studies on rats at 4, 6, and 8.7 mg TPM/m^3;[19,65] and two RASS inhalation studies, 90 d at 2.6 mg/m^3 [19] and 1 year at 6 and 12 mg TPM/m^3.[21] In mice, histopat! ological examination of the tracheal epithelium following ETS exposure has not been reported. Following exposure to SS or ADSS at concentrations that were effective in inducing enhanced replicative DNA synthesis in the nasal respiratory and/or the pulmonary epithelium, no effects were observed in the tracheal epithelium in three inhalation studies: SS at 1 mg TPM/m^3 for 3 weeks on hamsters,[78] SS at 4.5 mg TPM/m^3 for 6 months on mice (A/J and B6),[66] and ADSS at 0.1 to 10 mg TPM/m^3.[79]

The formation of DNA adducts is generally recognized as a critical step in the generation of mutations, which represent the first step (initiation) in multistage carcinogenesis models.[121,122] "Bulky" DNA adducts, i.e., the binding of PAH and

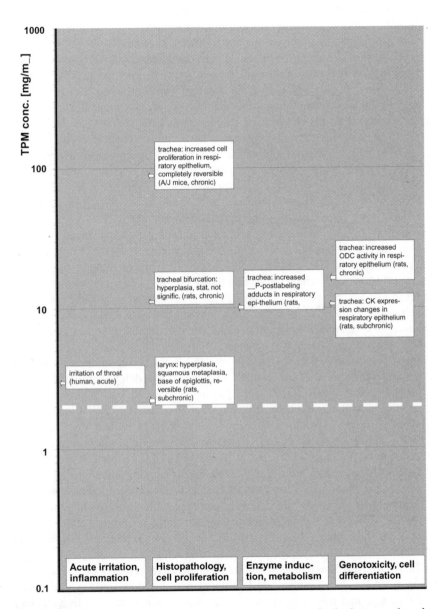

FIGURE 13.4 Effects of exposure to various ETS atmospheres in the larynx and trachea. The LOELs for various irritational, histopathological, biochemical, and genotoxic endpoints are summarized, regardless of the ETS surrogate (for details, see text). The dotted line indicates the upper limit of RSP concentrations in real ETS. For duration of daily exposures, see Table 13.1.

nitrosamine metabolites to the DNA, were found in increased levels in human smokers.[123,124] The ^{32}P-postlabeling method was sensitive enough to detect increased levels of DNA adducts in the rat trachea following exposure to SS for 4 weeks at 10 mg TPM/m^3 (a concentration ineffective in inducing histomorphological changes), and dietary N-acetylcysteine was effective in partially preventing adduct formation.[125] A time-related increase in DNA adduct formation during 1 to 5 weeks of exposure to high concentrations of an SS/MS mix (73 to 93 mg TPM/m^3) with a partial regression during 1 week of recovery was also observed in rats.[126] The relevance of such adducts as exposure-related, pro-mutagenic events remains unclear.[42] In mice exposed to high levels of MS or SS (leading to COHb values of 17 and 34%, respectively), only the enhancement of preexisting DNA adducts, which are not related to typical PAHs, e.g., B(a)P,[127] was observed. When a different methodology was used, tobacco-smoke-specific, nitrosamine-derived pyrydiloxobutyl adducts, which are thought to be unique to smoke exposure, were detected in experimental animals exposed to these nitrosamines and in the lungs of human smokers; however, no data from animals experimentally exposed to MS or SS have been published so far (for review, see Reference 42). While no tumorigenic response was observed in the tracheas of ETS-exposed rats following exposure to MS, an increased rate of "enhanced growth variants" was obtained in isolated rat tracheal epithelial cells cultured in vitro.[44,128] This finding was interpreted by the authors as a preneoplastic change in this ex vivo model, possibly representing an initiated state in the concept of multistage carcinogenesis.

Ornithine decarboxylase (ODC) is involved in the regulation of cell differentiation and proliferation,[129] and its mRNA is induced as a delayed early response.[130] In tracheal epithelial cells, ODC activity could be induced, e.g., by asbestos.[131] In the tracheal epithelium of rats, significantly increased activity of ODC was induced following exposure to presumably high concentrations of SS and MS:[129] based on a comparison of the reported COHb values, the exposure concentrations can be estimated to be in the range of at least 10 mg TPM/m^3 for SS and 30 mg TPM/m^3 for MS. ODC increases were seen at the higher concentration level following 8 weeks of exposure, and after 20 weeks at the lower concentration as well, thus this endpoint is more sensitive in the trachea than in the lung. With regard to CK expression, CK15 was increased in ciliated cells of the respiratory epithelium and decreased in the submucosal gland cells following 8 d of exposure to RASS (11 mg TPM/m^3).[64]

Endpoints related to irritation and inflammation have rarely been investigated in the trachea following experimental ETS inhalation. In an ex vivo study, explanted tracheal epithelium from SS-exposed rabbits was shown to respond to a challenge with acrolein or endotoxin with an unexpectedly enhanced production of bronchorelaxant eicosanoids (PGE$_2$, 6-keto PGF$_{1\alpha}$, and TxB$_2$).[132] Tracheal hyperresponsiveness with regard to constriction of the trachea following acute exposure of guinea pigs to SS was not observed under conditions that elicited bronchial hyperresponsiveness to histamine.[133]

The details of exposure conditions are listed in Table 13.1, and the effects of ETS exposure in the trachea are summarized in Figure 13.4.

13.5.4 LUNG

The lung is the most relevant site for the health effects of cigarette smoke because the human diseases most frequently associated with smoking are COPD and lung cancer (for a review, see Murin et al.[3]). ETS effects on asthmatic diseases and immune functions as well as cardiovascular effects are addressed in the accompanying chapters of this book; therefore, here we concentrate on those reactions and lesions that have been observed in the lung epithelia and are associated with possible neoplastic development, and focus on some COPD-related papers which address irritative and inflammatory effects. In contrast to the large number of MS studies, relatively little has been published about ETS effects on pulmonary macrophages, inflammation, epithelial permeability, and submucosal reactions, e.g., in the airway smooth muscles or the vasculature.

With regard to comparisons and extrapolations, it should be noted that there are some important anatomical and histological differences between the lungs of different species, e.g., rodents and primates/humans. Some examples are the smaller number of generations of the cartilage-enforced subsegmental bronchial branches in the rat,[134] the different morphology in the terminal bronchiolar regions of laboratory animals,[135] and the different proportions of the cell types in a given structural entity, e.g., the percentage of Clara cells in the bronchiolar epithelium of rats (25%), rabbits (61%), cats (100%), and bonnet monkeys (92%).[136] Finally, there are also interspecies differences with regard to the ultrastructural organelle composition within the same cell type, e.g., in Clara cells of the rabbit approximately 45% of the cell volume is occupied by smooth endoplasmic reticulum (sER), in the cat approximately 10%, and in the bonnet monkey less than 4%. This may also imply functional differences because the sER, as well as the plasma membrane, is the primary site of the CYP450 monooxygenase system.[136] A total of more than 40 different cell types, 17 of them epithelial related, have been identified in the mammalian lung,[137] and detailed information on various major lung cell types has been reviewed, e.g., airway epithelial cells,[138] airway basal cells,[139] Clara cells,[136] alveolar type I cells,[140] type II cells,[141] mesothelial cells,[142] neuroendocrine cells (APUD cells),[143,144] and alveolar macrophages.[145] Some authors have mentioned that the ferret lung resembles the airway situation in humans more closely than the rodent models.[146,147]

The interpretation of the prospective value of ETS-induced epithelial lesions in the lung, and whether they are related to neoplastic development or just represent adaptive changes, is still open. Several investigators have argued that the onset of stratification/cornification in squamous metaplastic areas of the pulmonary respiratory epithelium may constitute a restriction point between nonneoplastic and neoplastic development. These changes may be determined either by morphological criteria alone[148] or supported by a "biochemical grading," e.g., by using CK expression as a differentiation marker.[149] How these concepts can be related to the biological effects of experimental ETS exposure is discussed later.

In hamsters, no observable histopathological changes in the lung as well as in other respiratory tract tissues and nonrespiratory organs following a 90-d inhalation exposure of SS at 4 and 6 mg TPM/m^3 have been reported.[16,65] One chronic SS inhalation study with hamsters has been published,[150] but no histological data have been reported.

In rats, no observable effects have been reported for subchronic inhalation of SS at concentrations of 4 to 8.7 mg/m³ [16,19,65] and of ADSS at concentrations of 0.1 to 10 mg/m³.[18] The only effect in the rat lung was observed following chronic inhalation (12 months) of RASS at 12 mg TPM/m³, namely minimal reserve cell hyperplasia of the bronchial respiratory epithelium. No effects were seen at 6 mg/m³.[21] In newborn rats, systemic exposure *in utero* to high SS concentrations (not given by the authors, approximately 50 to 200 mg TPM/m³ as calculated from the static burning rate of the cigarettes) for 1 week during pregnancy led to histo-morphological changes in the lung — mesenchymal changes, hyperplasia of the bronchial muscles, and enhanced apoptosis. Histomorphological changes also occurred in the gastrointestinal tract, liver, and kidney.[151] In one morphometric study, signs of emphysema (increased alveolar spaces, decreased alveolar wall thickness, loss of elastic compounds) were observed in rats exposed to cigarette smoke at a concentration of 35 ppm CO (TPM concentration not indicated); however, smoke generation was not typical — a constant air flow was apparently applied through the cigarette from the filter to the burning tip.[152] Similar findings were reported in rats exposed to probably much higher concentrations (not indicated by the authors, estimated at ≥1000 mg TPM/l from the static burning rate) of SS for up to 90 d, with additional findings of inflammation and goblet cell hyperplasia of the bronchial epithelium and bronchial smooth muscle hypertrophy.[153]

In A/J mice, no histopathological or morphometric effects were detected in the alveolar epithelium following 6 months of exposure to ADSS at 4 mg/m³ or to an MS/SS mix at very high concentrations of 78.5 or 87 mg/m³.[15,67] But at these and even higher concentrations of the SS/MS mix (130 mg TPM/m³), a reproducible enhancement of the multiplicity of the lung tumors also occurring spontaneously in control mice was observed, provided that the 5-month exposure period was followed by a 10- to 20-week postexposure period.[15,68,154] Tumor formation in the A/J mouse model was induced at equal rates by both the gas phase alone (residual TPM as low as 0.1 mg/m³) and the unfiltered SS/MS mix,[14,154] while cell proliferation was not induced by the gas phase.[155] The histotypes of the A/J mouse lung tumors were bronchioloalveolar adenoma (80%) and adenocarcinoma (20%), and the proportion of the tumor types was not altered by smoke exposure (MS or SS/MS mix).[14] The only difference observed between spontaneous and smoke-related tumors was a shift in the mutational spectrum of K-ras mutations toward codon 61,[66] which may parallel the finding of codon 61 mutations in the mouse lung following treatment with urethane.[156] In the lung homogenate of one SS-exposed rat (out of 26), K-ras mutation was also observed.[157] In human lung adenoma and squamous cell carcinoma, 27 to 60% of the tumors harbored K-ras mutations, most frequently in codons 12, 13, and 61 (for review, see References 158 and 159). Some investigators observed associations of lung tumor K-ras mutations, particularly in codons 12 and 13, with heavy smoking (e.g., References 160 to 162), while others did not (e.g., Reference 163). With regard to the cell type of origin of bronchioloalveolar adenoma and adenocarcinoma in the mouse and rat lung, conflicting results that suggest either bronchiolar Clara cells or alveolar type II cells have been published.[164–166] Histological investigations on the bronchial epithelium of A/J mice have not been reported.

BrdU labeling of cells undergoing replicative DNA synthesis has proven to be a very sensitive endpoint for cellular effects not only in the upper respiratory tract, but in the lung as well. Subacute exposure to SS at 1 mg TPM/m^3 for 1 to 5 d elicited increased labeling in the large airways and terminal bronchioles in strain A/J mice, but not in C57 Black mice,[155] while no effects were seen in the lungs of hamsters exposed to SS for 1 to 3 weeks or in rats exposed to ADSS for 90 d at the same concentration level.[78,79] An unexpected phenomenon was observed in the lungs of young rats exposed to ADSS (1 mg TPM/m^3) for 100 d from birth: the proliferative DNA synthesis measured by ^{3}H-thymidine incorporation in the distal bronchiolar epithelium was lower than in controls during the first 2 weeks of life (for a discussion of possible explanations, see References 167). At a somewhat higher concentration of SS (4.5 mg/m^3) in a 6-month exposure study on A/J mice, Witschi et al.[66] observed a transient increase in replicative DNA synthesis during the first week in the bronchial respiratory epithelium, but not in the bronchiolar and alveolar epithelium. In rats, a slight, transient increase was observed in the bronchiolar respiratory epithelium in the 90-d ADSS study at 10 mg TPM/m^3; yet no effects were seen in the alveolar epithelium.[79] Only in A/J mice exposed to an MS/SS mix at 83.5 mg TPM/m^3 was a transient increase in BrdU labeling also seen in the alveolar epithelium during the first 2 weeks of exposure. At this concentration, the transient increase in replicative DNA synthesis in the respiratory epithelium of the large airways was significantly enhanced over control values during the first 3 months of exposure (maximum during weeks 1 and 2), which is much longer than the 1-week increase seen after 4.5 mg TPM/m^3 (see earlier), while in the respiratory epithelium of the bronchioles, the transient increase in labeling was significantly enhanced during the first 6 weeks only.[15] To summarize (cf. Figure 13.5), it has been shown that with increased smoke concentration, the enhancement of cell proliferation proceeded deeper into the lung and persisted longer. In any case, it remained transient even at high exposure concentrations and did not increase again during the 10-week recovery period in which the growth of the spontaneous and exposure-related tumors occurred.[15] This finding, together with the histomorphological findings of slight reversible reserve cell hyperplasia in the bronchial epithelium only, is in accordance with the view that enhanced cell proliferation and hyperplasia in the lung following chronic ETS inhalation are signs of an adaptive response to repeated irritation. The development of tumors requires cell proliferation, yet no signs of enhanced cell proliferation were observed in the alveolar epithelium of A/J mice both at the location and at the time point before and during growth of the adenomas and adenocarcinomas (it should be noted that the evaluation procedures in the tumorigenicity studies were macroscopic, and no routine histopathological evaluation on serial sections was done).

Investigations on DNA adducts and genotoxic damage following ETS exposure are still few. With the ^{32}P-postlabeling method, exposure-related adducts in DNA from lung and heart homogenates were detected following exposure of rats to ADSS at 9.7 mg TPM/m^3 for 2 weeks or to SS at 10 mg TPM/m^3 for 4 weeks[168] and also in the trachea and bladder;[125] the NOEL was 1 mg TPM/m^3.[168] Following 5 weeks of exposure to almost 10-fold higher levels of an MS/SS mix (73 to 93 mg TPM/m³), high levels of adducts were seen in the lung as well as in lung macrophages isolated from bronchioloalveolar lavage (BAL) fluid. Within 1 week of recovery, these levels reversed in the lung homogenate, but not in the macrophages.[126] In a chronic SS and MS inhalation study on mice (65 to 70 weeks) with very high SS concentrations (the COHb value of 30 to 35% indicates approximately 500 ppm CO), a 12- to 25-fold increase in the amount of preexisting lung DNA adducts was observed.[169] The quality and quantity of adducts was comparable for MS and SS and for both mouse strains tested (C57BL and DBA). All adducts were also detectable in the lungs of control mice, and no additional, smoke-specific adducts were observed. The relevance of such ³²P-postlabeling adducts is unclear (for discussion, see Section 13.5.3). Increased levels of 8-OHdG were not seen in the rat lung, but were seen in the nasal epithelium (see earlier) following 1 year of exposure to 12 mg TPM/m³ of RASS,[21] though after only 5 weeks of exposure to MS/SS mix at 73 to 93 mg TPM/m³.[126] In mice, acute exposure to SS at a very high concentration (not given by the authors, approximately 130 mg TPM/m³) for 30 min one to three times a week also elicited increased levels of 8-OHdG in the lungs, liver, and heart.[170] Another possible marker of oxidative (and other environmental) stress, the induction of the protective heat shock protein HSP-70, was not elevated in lung homogenates from ferrets exposed to SS (11.2 mg TPM/m³) for 15 weeks.[171] Chromosome aberrations were investigated in BAL macrophages from rats exposed for 14 d to ADSS at concentrations of 0.1 to 10 mg TPM/m³ and no effects were detectable,[168] while in rats exposed to MS at higher concentrations effects were seen.[172]

The most sensitive endpoint in lung epithelium, as well as in nasal respiratory epithelium, was the induction of enhanced B(a)P metabolism. Dose-related increases in the B(a)P metabolism were already seen in rat lung homogenates at relatively low exposure concentrations of SS and RASS following 90 d of exposure to concentrations of 0.6 to 8.7 mg TPM/m³. The effects were reversible within a 42-d recovery period. As expected, the magnitude of the increase depended on the TPM concentration and not on the gas phase concentration, as was observed for the nasal respiratory epithelium and for the histomorphological changes in the nasal and laryngeal epithelia.[19] In the ferret, exposure to an SS/MS mix at 381 mg TPM/m³ for 2 h or at 38 mg TPM/m³ for 15 weeks also induced an increased B(a)P metabolism.[173,174] The aryl hydrocarbon monooxygenase (AHM) — synonym: aryl hydrocarbon hydroxylase (AHH) — activity, which is a measure for the ability of one or several CYP450 isozymes to hydroxylate PAH substrates, was investigated comparatively in lung homogenates from rats, mice, and guinea pigs exposed for 16 weeks to SS at very high concentrations (not given by authors, at least 200 ppm CO estimated from COHb values) or to MS. The AHM activity was increased for both SS and MS in rats and mice, but not in guinea pigs.[175] Interestingly, the effects followed the concentration of the gas phase, and not of the TPM concentration (contrary to what was reported for B(a)P metabolism by Haussmann et al.[19]). In a study with MS and extremely high SS concentrations (not specified, but probably 500 ppm CO) in which two strains of mice (DBA and C57BL) were exposed for 65 to 70 weeks, MS enhanced the pulmonary AHH activity (assayed as B(a)P metabolism) about 2.5-fold in both strains, whereas SS induced a statistically significant AHH increase of the same magnitude in C57BL mice only.[169] Using ethoxyresorufin o-deethylase (EROD) as a selective substrate for CYP1A, enzyme induction via an Ah-receptor binding mechanism was observed at much lower concentrations of ADSS (1 mg TPM/m³) after a 4-d exposure; again, the induction occurred in C57BL, but not in DBA mice.[176] In rats exposed to ADSS at 73 to 93 mg TPM/m³ for 1 to 5 weeks, the lung AHM activity was also increased concomitant with an increase in GST activity and a depletion of the GSH level.[126]

In contrast to the aforementioned studies, which used tissue homogenates that do not allow correlation of increased metabolic enzyme activity with individual cell types, there were also three studies in which the cellular localization and intensity changes of the metabolic enzymes in the lung were determined immunohistochemically. In a study on young rats exposed from day of birth for 100 d to ADSS at 1 mg TPM/m³, the PAH-metabolizing isozyme CYP450 1A1 was distinctly increased in bronchiolar Clara cells and to a lesser degree in alveolar type II cells, and there was also a slight transient increase in NADPH reductase, but no effects were seen for the Clara-cell-specific marker CC10 and for CYP450 2B.[167] The authors took the results to indicate a delay in some, but not all, aspects of Clara cell maturation during postnatal lung development, yet their results are also consistent with the hypothesis that the ADSS-related, delayed decrease of CYP450 1A1 from high values in the newborn to lower values in the adult rat lung could reflect an adaptation to the PAH challenge because the other parameters of maturation/differentiation were unaltered. In A/J mice exposed to ADSS at 4 mg TPM/m³ for 6 months, the authors observed increased staining for CYP450 1A1 in the capillary endothelial cells of the lung parenchyma, but not in Clara or type II cells.[67] In the third study, A/J mice were exposed to an SS/MS mix for 5 months at 87.3 mg TPM/m³, and under these drastic conditions, CYP450 1A1 staining was increased in the airway epithelia as well as in the alveolar epithelium.[15] The effects reversed within 4 months of recovery.[15] In the A/J model, the induction of CYP450 isozymes was shown to depend on the concentration of the particulate phase by using the gas phase (0.1 mg TPM/m³) obtained by filtering an SS/MS mix of 78.5 mg TPM/m³, which did not induce CYP1A1, 2B1, and 2E1.[154]

In rats exposed to RASS at 11 mg TPM/m³ for 8 d, only two minor changes in CK expression were observed: a slight decrease in CK8 expression in nonciliated cells of the bronchus and an increase in CK7 expression in ciliated cells in the bronchiolar epithelium.[64]

ODC activity was investigated in the lungs of rats exposed to SS at concentrations estimated to be greater than 10 mg TPM/m³ for up to 20 weeks, and no effects were seen, which is in contrast to the ODC activation in tracheal epithelium. Following MS exposure, ODC activation was also seen in the lung.[129]

The expression of the Clara-cell-specific surfactant protein CC10 was increased in the bronchiolar epithelium of newborn rats following exposure to ADSS at 1 mg TPM/m³ *in utero*.[177] As already mentioned, the authors had also exposed neonate rats for 100 d to the same concentration of ADSS with no changes observed in CC10, but with an increased induction of P450 1A1.[167]

Oncogene activation and p53 mutations as intermediate biomarkers in ETS inhalation studies have not been published; however, K-ras mutations were investigated in order to characterize lung tumors in A/J mice,[66] and p53 overexpression was investigated in lung tumors of nonsmokers.[178] Other possible intermediate biomarkers used in human investigations[149] have either not yet been established for animal inhalation models, e.g., the oncofetal protein CEA, or have been applied in animal models in which ETS-induced tumor formation as the final result of preneoplastic development was not observed, e.g., CK expression changes in the rat respiratory tract.[64,90] Obviously, there is a need for more mechanistic and molecular investigations for a better understanding of why there has been no ETS-induced tumorigenicity observed in most of the animal models, even though they are sensitive to TPM concentrations of 0.6 mg/m^3 and up, and why the only animal model (A/J mice) with a tumorigenic response in the lung lacks all the typical signs of preneoplastic development as described for human lung cancer.

With regard to asthma and COPD, acute (1 to 3 h) experimental ETS exposure studies on healthy and asthmatic human volunteers, in addition to animal models for long-term exposure studies (for review, see Witschi et al.[4]), have been reported (see also nasal cavity section). From the numerous "challenging" studies, several more recent ones have been included in this chapter (for a more detailed review, see Reference 3).

At concentrations of machine-generated SS between 15 and 23.5 ppm CO (around 3 mg TPM/m^3), which induced symptoms such as headache, eye irritation, and irritation of the throat, there were no observable effects on lung function in three studies,[102,179,180] while an increased pulmonary and nasal resistance was measured in only one study.[103] There were also no detectable inflammatory mediators in the BAL fluid,[99,102] no histamine release, and no signs of an IgE-mediated allergic mechanism, even in historically "smoke-sensitive" individuals exposed to the very high SS concentration of 45 ppm CO.[99] An unexpected result was observed for the alveolar epithelial permeability — it decreased after 1 h of exposure to SS (23.5 ppm CO) — although it is known to increase following exposure to higher TPM doses of MS.[180] In animal models, inhomogenous results with regard to hyperresponsiveness and C-fiber-mediated bronchoconstriction have been observed: in guinea pigs exposed for 5 weeks to SS at 1 mg TPM/m^3, Mutoh et al.[181] found no effect on C-fiber responsiveness to inhaled capsaicin or bradykinin, but an increased responsiveness to systemically administered capsaicin and to lung hyperinflation; Joad et al.[182] observed a decreased C-fiber responsiveness to inhaled capsaicin, but no change in several neurotransmitters, receptors, or the number of C-fibers in the nociceptive nerves (unlike the situation after MS exposure). In the latter study, almost identical concentrations of TPM (1.0 mg/m^3) and CO (6.2 vs. 6.4 ppm), but a more than twofold higher nicotine concentration (345 vs. 144 µg/m^3) were measured. In a subsequent study with perinatal exposure (3 weeks prenatal, 3 weeks postnatal) of rats to ADSS with similar concentrations, a significant increase in bronchial hyperresponsiveness, but no significant increase in the number of neuroendocrine cells and mast cells present in the epithelium, was found.[183]

At much higher SS concentrations (not given by the authors, but presumably around or above 10 mg TPM/m^3) for up to 20 weeks, the polyamine biosynthesis, which may be involved in the development of COPD, was not enhanced in the rat lung, but was enhanced in the trachea.[129] Even higher SS concentrations (not indicated by the authors, but estimated to be in the range of 30 mg TPM/m^3 or above) for 2 weeks still did not increase epithelial permeability or the level of conjugated dienes in nude mice; but there was an increased pulmonary resistance and lipid fluorescence, both of which were partially protected against by dietary vitamin E.[184] After 16 weeks of exposure of rats to seemingly very high SS concentrations (SS from three cigarettes in one cage), an increased synthesis and decreased metabolism of bronchiolar adenosine, which is able to mediate bronchoconstriction, was seen.[35]

Similarly, high SS concentrations in short daily exposure periods (15 min/d) were applied to rabbits for 20 d. Under these conditions, signs of epithelial injury and inflammation, e.g., increased alveolar permeability and neutrophil infiltrates in the perivascular and capillary spaces, increased levels of leukotriene E4 in the BAL fluid, and eosinophil infiltrates in the airway mucosa were observed; and the macrophages from the BAL fluid exhibited increased superoxide production upon lipopolysaccharide stimulation *in vitro*.[185] For comparison, minimal alveolar hyperplasia and mild inflammation, but no increase in lung tumor rate, were observed in A/J mice after 6 months of inhalation exposure to MS at 248 mg TPM/m^3/230 ppm CO,[186] but no tumors were seen in B6C3F1 mice exposed at approximately the same concentration of MS.[234] Also, under conditions of short, bolus-like (3×10 min/d) exposures to SS at unrealistically high concentrations (2600 ppm CO, more than 100-fold higher than the values of extremely high real ETS concentrations), bronchial hyperresponsiveness to histamine stimulation, but not to acetylcholine, was observed in guinea pigs.[187] These authors also observed an increased number of alveolar macrophages, but no mucus hypersecretion. Although the total SS doses delivered (as calculated by time × concentration) in such bolus-like exposure regimens may not be so far away from the doses of the more frequently applied 5 to 7 h/d exposure regimens, which use lower concentrations, the latter may reflect the situation of involuntary ETS exposure more realistically. The highly concentrated exposure pulses, particularly of the gas phase irritants like aldehydes, obviously induce alveolar injury and inflammation. This was not observed at lower though still *exaggerated* concentrations, even after chronic exposure to experimental ETS for longer daily exposure periods, e.g., 1 year exposure of rats to RASS at 12 mg TPM/m^3/50 ppm CO for 12 h/d.[21] In A/J mice exposed to ADSS at 4.1 mg/m^3 for 6 months, a morphometric analysis of the alveolar tissue also revealed no increase in the number of alveolar macrophages.[67] It has also been demonstrated for the hyperplasia- and metaplasia-inducing effects of formaldehyde exposure in the rat nasal epithelium that splitting the daily dose into short pulses with higher concentrations was more effective than the same dose given continuously at half the concentration.[188]

The details of exposure conditions are listed in Table 13.1. For a compilation of ETS effects in the lung and their LOELs, see Figure 13.5.

13.6   *IN VITRO* INVESTIGATIONS

With regard to *in vitro* testing of the effects of ETS exposure on airway or lung epithelial cells, two major restrictions are obvious: (1) only a limited number of respiratory tract cell lines and primary cell culture models that are able to express organotypical functionality in combination with a reproducible sensitivity and quality of toxicological response are available (Table 13.2); and (2) exposure of *in vitro* cultures to aerosols requires specialized devices designed to avoid or at least minimize particle precipitation in the conducting tubing, contamination, and dehydration of the cultures (Table 13.3).

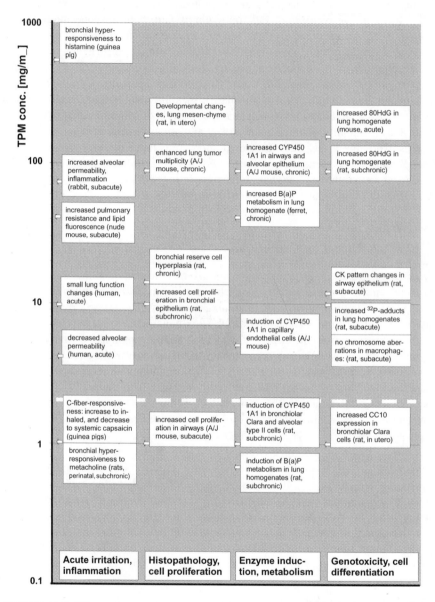

FIGURE 13.5 Effects of exposure to various ETS atmospheres in the lungs. The LOELs for various irritational, histopathological, biochemical, and genotoxic endpoints are summarized, regardless of the ETS (for details, see text). The dotted line indicates the upper limit of RSP concentrations in real ETS. For duration of daily exposures, see Table 13.1.

TABLE 13.2
Examples of Organotypically Differentiated *In Vitro* Cultures from Respiratory Tract Epithelia

Organ, tissue	Cell differentiation	Species	Cell line	Differentiation induced by	Ref.
Trachea, respiratory epithelium	Mucociliary	Rat	— (Primary)	Cultivation at air–liquid interface	Gray et al.[209]
Trachea, respiratory epithelium	Mucociliary	Rat	— (Primary)	Cultivation on collagen substrate at air–liquid interface	Davenport and Nettesheim[210]
Trachea, respiratory epithelium	Mucociliary	Rabbit	— (Primary)	Cultivation on collagen gel matrix under a thin layer of culture medium	Baeza-Squiban et al.[211,212]
Lung (embryonic)	(1) Clara, (2) alveolar type II, (3) neuroendocrine	Hamster	M3E3C3	(1) On collagen gel in hormone- and vitamin A-supplemented medium, (2) under an agar overlay, (3) under conditions of hypoxia and increased CO_2 levels	Emura et al.[213–215]
Adult lung, bronchial epithelium	(1) Ciliated, (2) mucous (goblet cell)	Human	Primary	(1) Cell aggregates in collagen gel matrix, (2) mixed cell aggregates of epithelial cells and MRC-5 fibroblasts in collagen gel matrix	Emura et al.[216]
Lung, bronchial epithelium	Not characterized	Human	BEAS-2B	Cocultivation with endothelial or fibroblastic cells in two-compartment chamber	Moegel,[192] Lang[193]
Fetal lung, bronchial epithelium	Secretory	Human	HFBE (finite lifespan)	Cultivation on collagen gel in vitamin A-supplemented medium	Ochiai et al.[217]
Lung, bronchial epithelium	Competent for PAH metabolism	Rat	R3/1	Cultivation on filter membranes	Knebel et al.[218,219]

TABLE 13.3
Examples of Smoke Preparations and Exposure Systems Used for *In Vitro* Testing of Cigarette Smoke

Smoke type	Smoke fraction/preparation	Exposure system	Assay, endpoint	Cell line/tissue	Ref.
MS	Condensate	Submersed monolayers, condensate suspended in culture medium	Morphological transformation (1 stage)	C3H 10 T1/2 mouse fibroblast line	Benedict et al.[220]
MS	Condensate	Submersed monolayers, condensate suspended in culture medium	Morphological transformation (2 stage)	C3H 10 T1/2 mouse fibroblast line	Schlage et al.[221]
MS	Condensate	Submersed monolayers, condensate suspended in culture medium	Gap junctional intercellular communication	Primary hamster tracheal epithelial cells	Rutten et al.[222]
MS	Condensate	Tissue submersed intermittently, condensate suspended in culture medium	Cell differentiation: cytokeratin expression	Organ culture of hamster tracheal epithelial rings	Rutten et al.[223]
MS	Diluted smoke	Cells cultured on filter membranes, direct aerosol exposure of cell surface	Cell proliferation, micronucleus induction	V79 hamster lung fibroblast line	Ramm,[224] Massey[225]
MS	Diluted smoke	Puffing smoke into culture flasks on a rocking plate with cell monolayers covered intermittently by culture medium	Cytotoxicity, sister chromatid exchanges	CHO hamster ovary fibroblast line	Bombick et al.[226]
SS, MS	Condensate	Submersed monolayers, condensate suspended in culture medium	Genotoxicity: (1) DNA repair, (2) HPRT mutation assay, (3) sister chromatid exchanges, (4) chromosome aberrations	(1) Rat hepatocytes, (2) to (4) CHO hamster ovary fibroblast line	Doolittle et al.[227]
SS, MS	Diluted smoke	Puffing smoke into culture flasks with cell monolayers covered by culture medium	Inhibition of interferon production	L929 mouse fibroblast cell line	Sonnenfeld and Hudgens[228]
ETS	Condensate	Submersed monolayers, condensate suspended in culture medium	Sister chromatid exchanges	CHO hamster ovary fibroblast line	Chen and Lee[229]

TABLE 13.3 (*continued*)
Examples of Smoke Preparations and Exposure Systems Used for *In Vitro* Testing of Cigarette Smoke

Smoke type	Smoke fraction/ preparation	Exposure system	Assay, endpoint	Cell line/tissue	Ref.
ETS	Solvent extract from particulate matter collected on filter	Submersed monolayers, condensate suspended in culture medium	Gap junctional intercellular communication	Primary rat alveolar type II cells	Heussen and Alink[230]
MS	Aqueous trapping	Submersed monolayers, "smoke bubbled PBS" diluted in culture medium	Oncogene activation (c-fos)	3T3, mouse fibroblast line	Müller[231]
MS	Aqueous trapping	Submersed monolayers, "smoke bubbled PBS" diluted in culture medium	DNA damage (COMET-assay)	Raji, human lymphoblastoma line	Yang et al.[232]
SS, MS	Aqueous trapping	"Smoke bubbled PBS" pipetted onto epithelium *in situ*	Irritation (HET-CAM test)	Chick embryo chorioallantoic membrane *in situ*	Schlage et al.[233]

Note: Abbreviations: SS, sidestream smoke; MS, mainstream smoke; ETS, real environmental tobacco smoke trapped on a filter or in an impactor; n.i., not indicated.

In vitro assays are an invaluable tool for the comparative screening of specific cytotoxic, genotoxic, and physiological effects of substances, as well as for the elucidation of the structural basis and molecular mechanisms of cellular function. The main advantages of cell-culture-based investigations — notably when established cell lines are used — are the relative independence of complex tissue-related and organism-derived systemic effects, the direct access of test substances to the cells at defined concentrations, and the long-term stability and reproducibility of culture conditions and cell properties. One drawback, however, is the partial or complete loss of differentiated, tissue-specific cellular functions, which is derived from the artificial culture conditions and the constant pressure on rapid cell division. The gap between the findings seen with commonly used cell lines and *in situ* mechanisms might be bridged to a certain extent by the use of organ cultures or "organotypic" cell culture systems with freshly isolated primary cells or cells from a permanent, established cell line, which were induced under "complex culture conditions" to keep or to re-express their differentiated functions. For the respiratory tract, only a limited number of suitable organotypic *in vitro* models are available (for examples, see Table 13.2).

Examples of smoke preparations and *in vitro* exposure methods are given in Table 13.3. Direct aerosol exposure is difficult, especially if the exposed cells are not protected by a differentiated surface structure, e.g., the mucus blanket, as they are *in vivo*. It has been demonstrated that the presence of a thin layer of culture medium on the cell surface or the presence of surfactant *in vivo* drastically modulates the availability of soluble, reactive gaseous substances, e.g., ozone, to the cells.[189,190] Obviously, the difficulties in maintaining appropriate epithelial *in vitro* cultures and the problems associated with aerosol exposure are the main reasons why tobacco smoke (particularly ETS) was tested largely in cell lines not derived from the respiratory tract epithelia, and why it was not applied as an aerosol. The number of such "simplified" investigations is large, and their discussion could be the topic of a separate review article (for examples, see Table 13.3). The most frequently applied smoke preparations are condensates, solvent extracts of the TPM trapped on a filter, and aqueous suspensions obtained by "bubbling" the smoke through a wash bottle with buffer or culture medium. While the first two methods will partially or completely miss the gas phase components, the aqueous trapping method primarily recovers the gas-vapor phase. In the case of SS, the aqueous trapping appears to lead to an exaggeration of nicotine toxicity, as seen in the chick embryo chorioallantoic membrane irritation test. A high cytotoxicity of fresh SS, decreasing rapidly with ageing, was also seen in L929 mouse fibroblasts when they were exposed to a stream of SS in their culture flasks.[191]

Taking these considerations together, the ideal *in vitro* model should consist of an immortalized cell line of rapidly proliferating pulmonary epithelial stem cells — as a qualitatively reproducible and quantitatively unlimited stock — from which the target cells in an assay can be brought to differentiate on a filter support at the air/medium interface, form a functional mucociliar epithelium, and then be exposed directly to the smoke, without dehydration, thus mimicking the effects that occur *in vivo*. An experimental design closely resembling this model has recently been developed for use with immortalized human bronchial epithelial cells (BEAS-2B) cocultured with endothelial cells[192] or with lung fibroblasts[193] in a two-compartment chamber in ozone exposure studies. Although the bronchial epithelial cells were not induced to differentiate to a mucociliary epithelium, it was possible to expose them to ozone at the air–liquid interface for 60 to 90 min without loss of viability. This sophisticated cell culture technology in combination with an aerosol exposure system[194] may prove a promising approach for ETS testing *in vitro*.

13.7 SUMMARY AND CONCLUDING REMARKS

The effects of ETS exposure on the structure and function of epithelial cells of the nasal cavity, larynx, trachea, and lung have been summarized together with their LOELs in Figures 13.3 to 13.5. In general, acute irritational responses (burning, rhinorrhea) but no inflammatory responses were observed in human volunteers at concentration levels above the upper limits of real ETS in heavily polluted smoking areas (approximately 2 mg RSP/m^3) (cf. References 30 and 195), and the first detectable lung function changes were seen at about fivefold of this value. Typical inflammatory or hyperresponsive reactions, as seen in active smokers, were only observed in animal models at approximately 60 mg TPM/m^3 or at CO concentrations that correspond to approximately 500 mg TPM/m^3 (3×10 min/d).[187] These cigarette smoke concentrations are typical for MS inhalation, but not for ETS exposure. Differences between tobacco-smoke-sensitive and -insensitive individuals and healthy and asthmatic subjects were minimal or absent under the experimental conditions realistic for human exposure.

With regard to the possible carcinogenicity of ETS, a variety of histomorphological and biochemical effects have been investigated in animal models. To date, the A/J mouse model is the only relevant ETS inhalation model that exhibits a reproducible tumor response at ETS concentrations *exaggerated* up to 100-fold over the upper limits of real-life ETS; however, the endpoint is an enhancement of the multiplicity of a high spontaneous lung tumor rate, without changes in the histological spectrum and malignancy of the tumors. Morphological and biochemical evidence for the mechanism of tumor formation is still lacking, e.g., enhanced cell proliferation or preneoplastic changes — as seen in the hamster model of MS-induced laryngeal tumors[113] — have not been documented in conjunction with lung tumor formation in the A/J mouse.

The rat inhalation model was used for several subchronic SS/ADSS and one chronic RASS inhalation study, but no indications of tumorigenicity were observed. Although frequently used in MS inhalation studies, rats and mice (except strain A/J) apparently only respond to cigarette smoke exposure (MS or ETS) with hyperplasia and metaplasia and not in a progressive sequence from preneoplastic to neoplastic lesions (hyperplasia → metaplasia → dysplasia → carcinoma *in situ* → malignant carcinoma) in a manner analogous to the neoplastic development assumed for human lung carcinomas,[148] for nonsmokers exposed to ETS,[196] and, in particular, for smokers.[197]

It still remains unclear, however, which of these features distinguish whether a hyperplastic or metaplastic lesion in the human bronchus, which may be derived either from basal cells or from goblet cells, will persist as an adaptive protective structure or proceed to neoplasia.[148,149,198] From a histological viewpoint, a smoke-induced transient increase in cell proliferation and hyperplasia and metaplasia represent adaptive rather than preneoplastic changes: they are reversible, do not progress with time, and occur at sites where smoke-related tumor formation, even after MS inhalation, have not been observed in these animals, e.g., nose, larynx, and bronchus. This is different from the situation of formaldehyde inhalation, for example, where tumor formation is observed in the nasal respiratory epithelium at the sites of hyperplasia and metaplasia; however, severe tissue damage and inflammation is required to elicit the tumor response to these agents.[188,199,200] Whether the apparent inability to develop respiratory tract tumors in response to cigarette smoke is due to anatomical differences, e.g., the more effective filtering in the nasal passages of animals and the almost complete lack of basal cells in the rat bronchus, or to physiological/biochemical differences or simply to the lack of longevity remains to be investigated (cf. Coggins[13]).

Open questions are also evident with regard to the phase distribution of potentially carcinogenic compounds in ETS, e.g., the dose-dependent increase in B(a)P metabolism in the nose and lung of exposed animals correlated with the TPM concentration rather than with the gas phase concentration, as were the histomorphological changes in the nasal and laryngeal epithelia.[19] Investigations on urinary mutagenicity, which is thought to be indicative of the metabolism of TPM-associated compounds, revealed no detectable increase in humans exposed to very high concentrations of ETS (10 and 25 ppm CO),[24,43] whereas a slight increase was seen in a similar study where there was no exact indication of the ETS concentration.[201] The excretion of thioethers, indicating the detoxification of electrophilic toxicants in humans exposed to ETS, however, was caused by the gas phase alone or by complete ETS at the same magnitude.[24] The enhancement of lung tumors in the A/J mouse was also independent of the particulate phase,[14,154] yet the majority of possibly carcinogenic PAHs in ETS are thought to be in the particulate phase.[202] In light of these results, the idea that free radical-mediated oxidative stress might play a more prominent role than the "classical" tobacco smoke-related carcinogens (PAHs, nitrosamines) has attracted new interest.[154,157]

Probably the most intriguing question — how ETS exposure may cause lung cancer — then seems to be a phenomenon that cannot be attributed to a single or a few substances or rate-limiting mechanistic events (unlike in experimental carcinogenesis with defined model carcinogens). In consequence, as long as we cannot attribute the potential carcinogenicity of the complex mixture ETS to individual compounds, it appears advisable (but hardly feasible) to keep as many components in the test atmospheres in the same proportions as is found in real-life ETS. This postulate is most likely in conflict with low-concentration extrapolations because of the possibility that the unrealistically high concentrations of experimental ETS, which are required to elicit measurable biological effects, may cause altered interactions between the smoke constituents that may not occur in the highly diluted aerosol of real-life ETS. This points to another experimental shortcoming of the current state-of-the-art approach: the need for more sensitive biological models capable of expressing neoplastic changes or for intermediate biomarkers unambiguously correlated with carcinogenicity at relevant ETS concentration levels (of relevant constituents).

Probably the best means to clarify the observed discrepancies in the existing data from the various experimental models can be expected from mechanistic studies, where the experimental endpoint can be correlated with an individual component or a family of related compounds, and where, ideally, the concentration of the specific compound(s) in the test atmosphere can be adjusted to match (or at least be measured and correlated with) the relevant concentrations in real-life ETS. The use of cytochemical techniques can provide more precise, in terms of cell type specificity, information than an assay of tissue homogenates, for example, the induction of CYP1A1 was observed immunocytochemically in capillary endothelial, but not in epithelial lung cells;[67] however, this could not be differentiated when lung homogenates were used. In a similar manner, more of the histochemical and molecular biomarkers that have been used, e.g., in human investigations, should be applied in the experimental models to enhance the mechanistic understanding of the findings with respect to extrapolations to the human situation.

ACKNOWLEDGMENTS

We wish to thank our colleagues at INBIFO for helpful discussions and hints, particularly Ms. H. Jachimsky and Mr. H. Lang for help with literature retrieval; Mr. W. Stinn for help with the calculation of estimated exposure concentrations; Drs. H. Haussmann, E. Römer, W. Reininghaus (INBIFO), and T. Sanders (FTR) for their careful reading of the manuscript and their helpful suggestions; and Ms. L. Conroy for her skillful help in preparing and editing the manuscript.

REFERENCES

1. Bayard, S.P. et al., Environmental Protection Agency (EPA), Respiratory Health Effects of Passive Smoking: Lung Cancer and Other Disorders, EPA/600/6-90/00, U.S. EPA, Office of Research and Development, Washington, D.C., 1992.
2. Pirkle, J.L., Flegal, K.M., Bernert, J.T., Brody, D.J., Etzel, R.A., and Maurer, K.R., Exposure of the U.S. population to environmental tobacco smoke: the third national health and nutrition examination survey, 1988 to 1991, *J. Am. Med. Assoc.*, 275, 1233, 1996.
3. Murin, S., Hilbert, J., and Reilly, S.J., Cigarette smoking and the lung, *Clin. Rev. Allergy Immunol.*, 15, 307, 1997.
4. Witschi, H., Joad, J.P., and Pinkerton, K.E., The toxicology of environmental tobacco smoke, *Annu. Rev. Pharmacol. Toxicol.*, 37, 29, 1997.
5. Hackshaw, A.K., Lung cancer and passive smoking, *Stat. Methods Med. Res.*, 7, 119, 1998.
6. Sasco, A.J. and Vainio, H., From in utero and childhood exposure to parental smoking to childhood cancer: a possible link and the need for action, *Hum. Exp. Toxicol.*, 18, 192, 1999.
7. Occupational Safety and Health Administration, Proposed rules on indoor air quality, *Fed. Reg.*, 59, 15968, 1994.
8. Armitage, A.K., Ashford, J.R., Gorrod, I.W., and Sullivan, F.M., Environmental tobacco smoke — is it really a carcinogen?, *Med. Sci. Res.*, 25, 3, 1997.

9. Nilsson, R., Environmental tobacco smoke and lung cancer: a reappraisal, *Ecotoxicol. Environ. Safety,* 34, 2, 1996.

10. International Agency for Research on Cancer (IARC), *IARC Monographs on the Evaluation of the Carcinogenic Risk of Chemicals to Humans. Tobacco Smoking,* IARC, Lyon, 1986.

11. Adlkofer, F.X., Scherer, G., Von Meyerinck, L., Von Maltzan, C., and Jarczyk, L., Exposure to ETS and its biological effects: a review, in *Present and Future of Indoor Air Quality,* Bieva, C.J., Courtois, M., and Govaerts, M., Eds., Elsevier Science Publishers B.V., Amsterdam, 1989, 183.

12. Coggins, C.R.E., The OSHA review of animal inhalation studies with environmental tobacco smoke, *Inhal. Toxicol.,* 8, 819, 1996.

13. Coggins, C.R.E., A review of chronic inhalation studies with mainstream cigarette smoke in rats and mice, *Toxicol. Pathol.,* 26, 307, 1998.

14. Witschi, H., Tobacco smoke as a mouse lung carcinogen, *Exp. Lung Res.,* 24, 385, 1998.

15. Witschi, H., Espiritu, I., Peake, J.L., Wu, K., Maronpot, R.R., and Pinkerton, K.E., The carcinogenicity of environmental tobacco smoke, *Carcinogenesis,* 18, 575, 1997.

16. Von Meyerinck, L., Scherer, G., Adlkofer, F., Wenzel-Hartung, R., Brune, H., and Thomas, C., Exposure of rats and hamsters to sidestream smoke from cigarettes in a subchronic inhalation study, *Exp. Pathol.,* 37, 186, 1989.

17. Coggins, C.R.E., Ayres, P.H., Mosberg, A.T., Ogden, M.W., Sagartz, J.W., and Hayes, A.W., Fourteen-day inhalation study in rats, using aged and diluted sidestream smoke from a reference cigarette. I. Inhalation toxicology and histopathology, *Fundam. Appl. Toxicol.,* 19, 133, 1992.

18. Coggins, C.R.E., Ayres, P.H., Mosberg, A.T., Sagartz, J.W., and Hayes, A.W., Subchronic inhalation study in rats using aged and diluted sidestream smoke from a reference cigarette, *Inhal. Toxicol.,* 5, 77, 1993.

19. Haussmann, H.J., Anskeit, E., Becker, D., Kuhl, P., Stinn, W., Teredesai, A., Voncken, P., and Walk, R.A., Comparison of fresh and room-aged cigarette sidestream smoke in a subchronic inhalation study on rats, *Toxic. Sci.,* 41, 100, 1998.

20. Burger, G.T., Renne, R.A., Sagartz, J.W., Ayres, P.H., Coggins, C.R.E., Mosberg, A.T., and Hayes, A.W., Histologic changes in the respiratory tract induced by inhalation of xenobiotics: physiologic adaptation or toxicity?, *Toxicol. Appl. Pharmacol.,* 101, 521, 1989.

21. Haussmann, H.J., Gerstenberg, B., Goecke, W., Kuhl, P., Schepers, G., Stabbert, R., Stinn, W., Teredesai, A., Tewes, F., Anskeit, E., and Terpstra, P., 12-month inhalation study on room aged cigarette sidestream smoke in rats, *Inhal. Toxicol.,* 10, 663, 1998.

22. U.S. Environmental Protection Agency, Proposed guidelines for carcinogen risk assessment, *Fed. Reg.,* 61, 17960, 1996.

23. Scherer, G., Westphal, K., Biber, A., Hoepfner, I., and Adlkofer, F., Urinary mutagenicity after controlled exposure to environmental tobacco smoke (ETS), *Toxicol. Lett.,* 35, 135, 1987.

24. Scherer, G., Westpahl, K., Adlkofer, F., and Sorsa, M., Biomonitoring of exposure to potentially genotoxic substances from environmental tobacco smoke, *Environ. Int.,* 15, 49, 1989.

25. Nelson, P.R., Conrad, F.W., Kelly, S.P., Maiolo, K.C., Richardson, J.D., and Ogden, M.W., Composition of environmental tobacco smoke (ETS) from international cigarettes part II: nine country follow-up, *Environ. Int.,* 24, 251, 1998.

26. Nelson, P.R., Kelly, S.P., and Conrad, F.W., Studies of environmental tobacco smoke generated by different cigarettes, *J. Air Waste Manage. Assoc.,* 48, 336, 1998.
27. First, M., Constituents of sidestream and mainstream tobacco smoke and markers to quantify exposure to them, in *Indoor Air and Human Health,* Gammage, R.B. and Kaye, S.V., Eds., Lewis Publishers, Chelsea, MI, 1985, 195.
28. Guerin, M.R., Jenkins, R.A., and Tomkins, B.A., *The Chemistry of Environmental Tobacco Smoke: Composition and Measurement,* Lewis Publishers, Chelsea, MI, 1992.
29. Baker, R.R. and Proctor, C.J., The origins and properties of environmental tobacco smoke, *Environ. Int.,* 16, 231, 1990.
30. Repace, J.L., Indoor concentrations of environmental tobacco smoke: models dealing with effects of ventilation and room size, *IARC Sci. Publ.,* 81, 25, 1987.
31. Eatough, D.J., Hansen, L.D., and Lewis, E.A., The chemical characterization of environmental tobacco smoke, *Environ. Technol.,* 11, 1071, 1990.
32. Voncken, P., Stinn, W., Haussmann, H.-J., and Anskeit, E., Influence of aging and surface contact on the composition of cigarette sidestream smoke. Models for environmental tobacco smoke, in *Toxic and Carcinogenic Effects of Solid Particles in the Respiratory Tract,* Mohr, U., Dungworth, D.L., Mauderly, J.L., and Oberdörster, G., Eds., ILSI Press, Washington, D.C., 1994, 637.
33. Martin, P., Heavner, D.L., Nelson, P.R., Maiolo, K.C., Risner, C.H., Simmons, P.S., Morgan, W.T., and Ogden, M.W., Environmental tobacco smoke (ETS): a market cigarette study, *Environ. Int.,* 23, 75, 1997.
34. Griffith, R.B. and Hancock, R., Simultaneous mainstream-sidestream smoke exposure systems 1. Equipment and procedures, *Toxicology,* 34, 123, 1985.
35. Anand, U., Anand, C.V., Agarwal, R., and Kanaka, R., Effect of exposure of rats to cigarette smoke on 5′ nucleotidase and adenosine deaminase in bronchoalveolar lavage, *Med. Sci. Res.,* 27, 189, 1999.
36. Loefroth, G., Environmental tobacco smoke: multicomponent analysis and room-to-room distribution in homes, *Tob. Control,* 2, 222, 1993.
37. Rodgman, A., Environmental tobacco smoke, *Regul. Toxicol. Pharmacol.,* 16, 223, 1992.
38. Jenkins, R.A., Palausky, M.A., Counts, W.R., Guerin, M.R., Dindal, A.B., and Bayne, C.K., Determination of personal exposure of non-smokers to environmental tobacco smoke in the United States, *Lung Cancer,* 14, S195, 1996.
39. Jenkins, R.A., Palausky, A., Counts, R.W., Bayne, C.K., Dindal, A.B., and Guerin, M.R., Exposure to environmental tobacco-smoke in 16 cities in the United States as determined by personal breathing zone air sampling, *J. Expos. Anal. Environ. Epidemiol.,* 6, 473, 1996.
40. Penn, A. and Snyder, C.A., Inhalation of sidestream cigarette smoke accelerates development of arteriosclerotic plaques, *Circulation,* 88, 1820, 1993.
41. Penn, A., Chen, L.C., and Snyder, C.A., Inhalation of steady-state sidestream smoke from one cigarette promotes arteriosclerotic plaque development, *Circulation,* 90, 1363, 1994.
42. Hecht, S.S., DNA adduct formation from tobacco-specific N-nitrosamines, *Mutat. Res. Fund. Mol. Mech. Mutagenesis,* 424, 127, 1999.
43. Scherer, G., Von Maltzan, C., Von Meyerinck, L., Westphal, K., and Adlkofer, F., Biomonitoring after controlled exposure to environmental tobacco smoke (ETS), *Exp. Pathol.,* 37, 158, 1989.

44. Mauderly, J.L., Bechtold, W.E., Bond, J.A., Brooks, A.L., Chen, B.T., Harkema, J.R., Henderson, R.F., Johnson, N.F., Rithidech, K., and Thomassen, D.G., Comparison of three methods of exposing rats to cigarette smoke, *Exp. Pathol.*, 37, 194, 1989.

45. Proctor, D.F. and Chang J.C.F., Comparative anatomy and physiology of the nasal cavity, in *Nasal Tumors in Animals and Man*, Vol. 1, Reznik, G. and Stinson, S.F., Eds., CRC Press, Boca Raton, FL, 1984, 1.

46. Calderon-Garciduenas, L., Rodriguez-Alcaraz, A., Villareal-Calderon, A., Lyght, O., Janszen, D., and Morgan, K.T., Nasal epithelium as a sentinel for airborne environmental polution, *Toxicol. Sci.*, 46, 352, 1999.

47. Kimbell, J.S. and Morgan K.T., Upper respiratory tract toxicology, airflow modeling, and supercomputers, *CIIT Act.*, 10, 1, 1990.

48. Morgan, K.T., Kimbell, J.S., Monticello, T.M., Patra, A.L., and Fleishman, A., Studies of inspiratory airflow patterns in the nasal passages of the f344 rat and rhesus monkey using nasal molds — relevance to formaldehyde toxicity, *Toxicol. Appl. Pharmacol.*, 110, 223, 1991.

49. Monticello, T.M. and Morgan, K.T., Cell proliferation in rat nasal respiratory epithelium following three months exposure to formaldehyde gas, *Toxicologist*, 10, 181, 1990.

50. Jiang, X.Z., Morgan, K.T., and Beauchamp, R.O., Histopathology of acute and subacute nasal toxicity, in *Toxicology of the Nasal Passages*, Barrow, C.S., Ed., Hemisphere Publishing, Washington, D.C., 1986, 51.

51. Gross, E.A., Swenberg, J.A., Fields, S., and Popp, J.A., Comparative morphometry of the nasal cavity in rats and mice, *J. Anat.*, 135, 83, 1982.

52. Kuper, C.F., Reuzel, P.G.J., Feron, V.J., and Verschuuren, H., Chronic inhalation toxicity and carcinogenicity study of propylene oxide in Wistar rats, *Food Chem. Toxicol.*, 26, 159, 1988.

53. Harkema, J., Comparative aspects of nasal airway anatomy: relevance to inhalation toxicology, *Toxicol. Pathol.*, 19, 321, 1991.

54. Woutersen, R.A., Van Garderen-Hoetmer, A., Slootweg, P.J., and Feron, V.J., Upper respiratory tract carcinogenesis in experimental animals and in humans, in *Carcinogenesis*, Waalkes, M.P. and Ward, J.M., Eds., Raven Press, New York, 1994, 215.

55. Young, J.T., Light microscopic examination of the rat nasal passages: preparation and morphologic features, in *Toxicology of the Nasal Passages*, Barrow, C.S., Ed., Hemisphere Publishing, Washington, D.C., 1986, 27.

56. Farbman, A.I., Cellular interactions in the development of the vertebrate olfactory system, in *Molecular Neurobiology of the Olfactory System*, Margolis, F.I. and Getchell, T.V., Eds., Plenum Press, New York, 1988, 319.

57. Monteiro-Riviere, N.A. and Popp J.A., Ultrastructural characterization of the nasal respiratory epithelium in the rat, *Am. J. Anat.*, 169, 31, 1984.

58. Voigt, J.M., Guengerich, F.P., and Baron, J., Localization of cytochrome P-450 isozyme (cytochrome P-450 PB-b) and NADPH-cytochrome P-450 reductase in rat nasal mucosa, *Cancer Lett.*, 27, 241, 1985.

59. Schlage, W.K., Bülles, H., Friedrichs, D., Kuhn, M., and Teredesai, A., Cytokeratin expression patterns in the rat respiratory tract as markers of epithelial differentiation in inhalation toxicology. I. Determination of normal cytokeratin expression patterns in nose, larynx, trachea, and lung, *Toxicol. Pathol.*, 26, 324, 1998.

60. Bogdanffy, M.S., Randall, H.W., and Morgan, K.T., Histochemical identification of biotransformation enzymes in the nasal passages, *CIIT Act.*, 6, 3, 1986.

61. Randall, H.W., Bogdanffy, M.S., and Morgan, K.T., Enzyme histochemistry of the rat nasal mucosa embedded in cold glycol methacrylate, *Am. J. Anat.*, 179, 10, 1987.

62. Alarie, Y., Sensory irritation by airborne chemicals, *Crit. Rev. Toxicol.*, 2, 299, 1973.
63. Schreider, J.P., Comparative anatomy and function of nasal passages, in *Toxicology of the Nasal Passages,* Barrow, C.S., Ed., Hemisphere Publishing, Washington, D.C., 1986, 1.
64. Schlage, W.K., Bülles, H., Friedrichs, D., Kuhn, M., Teredesai, A., and Terpstra, P.M., Cytokeratin expression patterns in the rat respiratory tract as markers of epithelial differentiation in inhalation toxicology. II. Changes in Cytokeratin expression patterns following 8-day exposure to room-aged cigarette sidestream smoke, *Toxicol. Pathol.*, 26, 344, 1998.
65. Teredesai, A. and Prühs D., Histopathological findings in the rat and hamster respiratory tract in a 90-day inhalation study using fresh sidestream smoke of the standard reference cigarette 2R1, in *Toxic and Carcinogenic Effects of Solid Particles in the Respiratory Tract,* Mohr, U., Dungworth, D.L., Mauderly, J.L., and Oberdörster, G., Eds., ILSI Press, Washington, D.C., 1994, 629.
66. Witschi, H., Oreffo, V.I.C., and Pinkerton, K.E., Six-month exposure of strain A/J mice to cigarette sidestream smoke: cell kinetics and lung tumor data, *Fundam. Appl. Toxicol.*, 26, 32, 1995.
67. Pinkerton, K.E., Peake, J.L., Espiritu, I., Goldsmith, M., and Witschi, H., Quantitative histology and cytochrome-P-450 immunocytochemistry of the lung parenchyma following 6 months of exposure of strain A/J mice to cigarette sidestream smoke, *Inhal. Toxicol.*, 8, 927, 1996.
68. Witschi, H.P., Espiritu, I., and Uyeminami, D., Chemoprevention of tobacco smoke-induced lung tumors in A/J- strain mice with dietary myo-inositol and dexamethasone, *Carcinogenesis*, 20, 1375, 1999.
69. Gopinath, C., Prentice, D.E., and Lewis, D.J., *Atlas of Experimental Toxicological Pathology,* MTP Press, Lancaster, 1987.
70. Coggins, C.R.E., Morgan, K.T., Lam, R., and Fouillet, X.L.M., Cigarette smoke induced pathology of the rat respiratory tract: a comparison of the effects of the particulate and vapour phases, *Toxicology*, 16, 83, 1980.
71. Hotchkiss, J.A., Evans, W.A., Chen, B.T., Finch, G.L., and Harkema, J.R., Regional differences in the effects of mainstream cigarette smoke on stored mucosubstances and DNA synthesis in F344 rat nasal respiratory eoithelium, *Toxicol. Appl. Pharmacol.*, 131, 316, 1995.
72. Lewis, J.L., Nikula, J., and Sachetti, L.A., Induced xenobiotic-metabolizing enzymes localized to eosinophilic globules in olfactory epithelium of toxicant-exposed F-344 rats, in *Nasal Toxicity and Dosimetry of Inhaled Xenobiotics. Implications for Human Health*, Taylor & Francis, London, 1995, 422.
73. Maples, K.R., Nikula, K.J., Chen, B.T., Finch, G.L., Griffith, W.C., and Harkema, J.R., Effects of cigarette smoke on the glutathione status of the upper and lower respiratory tract of rats, *Inhal. Toxicol.*, 5, 389, 1993.
74. Dalbey, W.E., Nettesheim, P., Griesemer, R., Caton, J.E., and Guerin, M.R., Chronic inhalation of cigarette smoke by F344 rats, *J. Natl. Cancer Inst.*, 64, 383, 1980.
75. Heckman, C.A. and Dalbey, W.E., Pathogenesis of lesions induced in rat lung by chronic tobacco smoke inhalation, *J. Natl. Cancer Inst.*, 69, 117, 1982.
76. Heckman, C.A. and Lehman, G.L., Ultrastructure and distribution of intracellular spicules in rat lung following chronic tobacco smoke exposure, *J. Natl. Cancer Inst.*, 74, 647, 1985.
77. Brown, B.G., Bombick, B.R., McKarns, S.C., Lee, C.K., Ayres, P.H., and Doolittle, D.J., Molecular toxicology endpoints in rodent inhalation studies, *Exp. Toxicol. Pathol.*, 47, 183, 1995.

78. Witschi, H. and Rajini, P., Cell kinetics in the respiratory tract of hamsters exposed to cigarette sidestream smoke, *Inhal. Toxicol.*, 6, 321, 1994.

79. Ayres, P.H., McKarns, S.C., Coggins, C.R.E., Doolittle, D.J., Sagartz, J.E., Payne, V.M., and Mosberg, A.T., Replicative DNA synthesis in tissues of the rat exposed to aged and diluted sidestream smoke, *Inhal. Toxicol.*, 7, 1225, 1995.

80. Kellermann, G., Trell, E., Shaw, C.R., and Luyten-Kellermann, M., Aryl hydrocarbon hydroxylase inducibility and bronchogenic carcinoma, *N. Engl. J. Med.*, 289, 934, 1973.

81. Caporaso, N.E., Tucker, M.A., Hoover, R.N., Hayes, R.B., Pickle, L.W., Issaq, H.J., Muschik, G.M., Green-Gallo, L., Buivys, D., Aisner, S., Resau, J.H., Trump, B.F., Tollerud, D., Weston, A., and Harris, C.C., Lung cancer and the debrisoquine metabolic phenotype, *J. Natl. Cancer Inst.*, 82, 1264, 1990.

82. Bouchardy, C., Benhamou, S., and Dayer, P., The effect of tobacco on lung cancer risk depends on CYP2D6, *Cancer Res.*, 56, 251, 1996.

83. Caporaso, N., Genetics of smoking-related cancer and mutagen sensitivity, *J. Natl. Cancer Inst.*, 91, 1097, 1999.

84. Sohal, R.S., Agarwal, S., and Sohal, B.H., Oxidative stress and aging in the Mongolian gerbil (Meriones unguiculatus), *Mech. Ageing Dev.*, 81, 15, 1995.

85. Floyd, R.A., The role of 8-hydroxyguanine in carcinogenesis, *Carcinogenesis*, 11, 1447, 1990.

86. Chung, F.L. and Xu, Y., Increased 8-oxodeoxyguanosine levels in lung DNA of A/J mice and F344 rats treated with the tobacco-specific nitrosamine 4-(methylnitrosamine)-1-(3-pyridyl)-1-butanone, *Carcinogenesis*, 13, 1269, 1992.

87. Belinsky, S.A., Walker, V.E., Maronpot, R.R., Swenberg, J.A., and Anderson, M.W., Molecular dosimetry of DNA adduct formation and cell toxicity in rat nasal mucosa following exposure to the tobacco specific nitrosamine 4-(N-methyl-N-nitrosamino)-1-(3-pyridyl)-1-butanone and their relationship to induction of neoplasia, *Cancer Res.*, 47, 6058, 1987.

88. Belinsky, S.A. and Anderson, M.W., Tissue and cellular specificity for DNA adduct formation and persistence following in vivo exposure to chemicals, *Prog. Exp. Tumor Res.*, 31, 11, 1987.

89. Belinsky, S.A., White, C.M., Devereux, T.R., Swenberg, A.E., and Anderson, M.W., Cell selective alkylation of DNA in rat lung following low dose exposure to the tobacco specific carcinogen 4-(n-methyl-n-nitrosamino)-1-(3-pyridyl)-1-butanone, *Cancer Res.*, 47, 1143, 1987.

90. Schlage, W.K., Bülles, H., Friedrichs, D., Kuhn, M., Teredesai, A., and Terpstra, P., Tobacco smoke-induced alterations of cytokeratin expression in the rat nasal cavity following chronic inhalation of room-aged sidestream smoke, *Toxicol. Lett.*, 96/97, 309, 1998.

91. Fuchs, E. and Weber, K., Intermediate filaments: structure, dynamics, function, and disease, *Annu. Rev. Biochem.*, 63, 345, 1994.

92. Nagle, R.B., A review of intermediate filament biology and their use in pathologic diagnosis, *Mol. Biol. Rep.*, 19, 3, 1994.

93. Schaafsma, H.E. and Ramaekers, F.C.S., Cytokeratin subtyping in normal and neoplastic epithelium: basic principles and diagnostic applications, *Pathol. Annu.*, 29, 21, 1994.

94. Henderson, R.F., Strategies for use of biological markers of exposure, *Toxicol. Lett.*, 82/83, 379, 1995.

95. Gijbels, M.J.J., Van Der Ham, F., Van Bennekum, A.M., Hendriks, H.F., and Roholl, P.J., Alterations in cytokeratin expression precede histological changes in epithelia of vitamin A-deficient rats, *Cell Tissue Res.*, 269, 197, 1992.

96. Woodcock-Mitchell, J.L., Burkhardt, A.L., Mitchell, J.J., Rannels, S.R., Rannels, D.E., Chiu, J.F., and Low, R.B., Keratin species in type II pneumocytes in culture and during lung injury, *Am. Rev. Respir. Dis.,* 134, 566, 1986.

97. Kasper, M., Rudolf, T., Haase, M., Schuh, D., and Müller, M., Changes in cytokeratin, vimentin and desmoplakin distribution during the repair of irradiation-induced lung injury in adult rats, *Virchows Arch. B. Cell Pathol.,* 64, 271, 1993.

98. Stinn, W., Schnell, P., Anskeit, E., Rustemeier, K., Schepers, G., and Haussmann, H.J., Chronic Inhalation of Room-Aged Cigarette Sidestream Smoke and Diesel Engine Exhaust in Rats: Study Design, Aerosol Characterization, and Biomonitoring, 7th International Symposium On Particle Toxicology, Maastricht, October 1999, 1.

99. Bascom, R., Kulle, T., Kagey-Sobotka, A., and Proud, D., Upper respiratory tract environmental tobacco smoke sensitivity, *Am. Rev. Respir. Dis.,* 143, 1304, 1991.

100. Kesavanathan, J., Swift, D.L., Fitzgerald, T.K., Permutt, T., and Bascom, R., Evaluation of acoustic rhinometry and posterior rhinomanometry as tools for inhalation challenge studies, *J. Toxicol. Environ. Health,* 48, 295, 1996.

101. Bascom, R., Kesavanathan, J., Permutt, T., Fitzgerald, T.K., Sauder, L., and Swift, D.L., Tobacco smoke upper respiratory response relationships in healthy nonsmokers, *Fundam. Appl. Toxicol.,* 29, 86, 1996.

102. Nowak, D., Joerres, R., Martinez-Mueller, L., Grimminger, F., Seeger, W., Koops, F., and Magnussen, H., Effect of 3 hours of passive smoke exposure in the evening on inflammatory markers in bronchoalveolar and nasal lavage fluid in subjects with mild asthma, *Int. Arch. Occup. Environ. Health,* 70, 85, 1997.

103. Willes, S.R., Fitzgerald, T.K., Permutt, T., Proud, D., Haley, N.J., and Bascom, R., Acute respiratory repsonse to prolonged, moderate levels of sidestream tobacco smoke, *J. Toxicol. Environ. Health A,* 53, 193, 1998.

104. Bascom, R., Kesavanathan, J., Fitzgerald, T.K., Cheng, K.-H., and Swift, D.L., Sidestream tobacco smoke exposure acutely alters human nasal mucociliary clearance, *Environ. Health Perspect.,* 103, 1026, 1995.

105. Stanley, P.J., Wilson, R., Greenstone, M.A., Macwilliam, L., and Cole, P.J., Effect of cigarette smoking on nasal mucociliary clearance and ciliary beat frequency, *Thorax,* 41, 519, 1986.

106. Weiss, S., Tager, I., Schenker, M., and Speizer, F., State of the art: the health effects of involuntary smoking, *Am. Rev. Respir. Dis.,* 128, 933, 1983.

107. Morgan, K.T., Patterson, D.L., and Gross, E.A., Responses of the nasal mucociliary apparatus to airborne irritants, in *Toxicology of the Nasal Passages,* Barrow, C.S., Ed., Hemisphere Publishing, Washington, D.C., 1986, 123.

108. Iravani, J., Effects of cigarette smoke on the ciliated respiratory epithelium of rats, *Respiration,* 29, 480, 1972.

109. Iravani, J. and Melville, G.N., Long-term effect of cigarette smoke on mucociliary function in animals, *Respiration,* 31, 358, 1974.

110. Jeffery, P.K., Brain, A.P.R., Shields, P.A., Quinn, B.P., and Betts, T., Response of laryngeal and tracheobronchial surface lining to inhaled cigarette-smoke in normal and vitamin-A-deficient rats — a scanning electron microscopic study, *Scanning Microsc.,* 2, 545, 1988.

111. Bernfeld, P., Homburger, F., and Russfield, A.B., Strain differences in the response of inbred Syrian hamsters to cigarette smoke inhalation, *J. Natl. Cancer Inst.,* 53, 1141, 1974.

112. Bernfeld, P., Russfield, A.B., and Homburger, F., Cigarette smoke-induced cancer of the larynx in hamsters (CINCH): a method to assay the carcinogenicity of cigarette smoke, *Prog. Exp. Tumor Res.,* 24, 315, 1979.

113. Dontenwill, W., Chevalier, H.-J., Harke, H.-P., Lafrenz, U., Reckzeh, G., and Schneider, B., Investigations on the effects of chronic cigarette-smoke inhalation in Syrian golden hamsters, *J. Natl. Cancer Inst.*, 51, 1781, 1973.

114. International Agency for Research on Cancer (IARC), Biological data relevant to the evaluation of carcinogenic risk to humans, in *IARC Monographs on the Evaluation of the Carcinogenic Risk of Chemicals to Humans. Tobacco Smoking*, Vol. 38, IARC, Lyon, 1986, 127.

115. International Agency for Research on Cancer (IARC), *Alcohol Drinking*, International Agency for Research on Cancer, WHO, Lyon, 1988.

116. Lewis, D.J. and Prentice, D.E., The ultrastructure of rat laryngeal epithelia, *J. Anat.*, 130, 617, 1980.

117. Lewis, D.J., Experimental Pathology of the Rat Larynx Following Exposure to Tobacco Smoke, Ph.D. thesis, University of Surrey, 1980.

118. Lewis, D.J., Mitotic indices of rat laryngeal epithelia, *J. Anat.*, 132, 419, 1981.

119. Walker, D., Wilton, L.V., and Binns, R., Inhalation toxicity studies on cigarette smoke (VII) 6-week comparative experiments using modified flue-cured cigarettes: histopathology of the conducting airways, *Toxicology*, 10, 241, 1978.

120. Shields, P.A. and Jeffery, P.K., The combined effects of vitamin A-deficiency and cigarette-smoke on rat tracheal epithelium, *Br. J. Exp. Pathol.*, 68, 705, 1987.

121. Miller, D.G., Principles of early detection of cancer, *Cancer*, 47, 1142, 1981.

122. Singer, B. and Grunberger, D., *Molecular Biology of Mutagens and Carcinogens*, Plenum Press, New York, 1983.

123. Randerath, E., Miller, R.H., Mittal, D., Avitts, T.A., Dunsford, H.A., and Randerath, K., Covalent DNA damage in tissues of cigarette smokers as determined by P32-postlabeling assay, *J. Natl. Cancer Inst.*, 81, 341, 1989.

124. Phillips, D.H., Hewer, A., and Martin, C.N., Correlation of DNA adduct levels in human lung with cigarette smoking, *Nature*, 336, 790, 1988.

125. Arif, J.M., Gairola, C.G., Glauert, H.P., Kelloff, G.J., Lubet, R.A., and Gupta, R.C., Effects of dietary supplementation of N-acetylcysteine on cigarette smoke-related DNA adducts in rat tissues, *Int. J. Oncol.*, 11, 1227, 1997.

126. Izzotti, A., Bagnasco, M., D'Agostini, F., Cartiglia, C., Lubet, R.A., Kelloff, G.J., and De Flora, S., Formation and persistence of nucleotide alterations in rats exposed whole-body to environmental cigarette smoke, *Carcinogenesis*, 20, 1499, 1999.

127. Gupta, R.C., Arif, J.M., and Gairola, C.G., Enhancement of pre-existing DNA adducts in rodents exposed to cigarette smoke, *Mutat. Res. Fund. Mol. Mech. Mutagenesis*, 424, 195, 1999.

128. Thomassen, D.G., Chen, B.T., Mauderly, J.L., Johnson, N.F., and Griffith, W.C., Inhaled cigarette smoke induces preneoplastic changes in rat tracheal epithelial cells, *Carcinogenesis*, 10, 2359, 1989.

129. Olson, J.W., Chronic cigarette sidestream smoke exposure increases rat trachea ornithine decarboxylase activity, *Life Sci.*, 37, 2165, 1985.

130. Donohue, P.J., Hsu, D.K., Guo, Y., Burgess, W.H., and Winkles, J.A., Fibroblast growth factor-1 induction of delayed-early mRNA expression in NIH 3T3 cells is prolonged by heparin addition, *Exp. Cell Res.*, 234, 139, 1997.

131. Marsh, J.P. and Mossman, B.T., Mechanism of induction of ornithine decarboxylase activity in tracheal epithelial cells by asbestiform minerals, *Cancer Res.*, 48, 709, 1988.

132. Joseph, P.M., Witten, M.L., Burke, C.H., and Hales, C.A., The effects of chronic sidestream cigarette smoke exposure on eicosanoid production by tracheal epithelium, *Exp. Lung Res.*, 22, 317, 1996.

133. Omini, C., Hernandez, A., Zuccari, G., Clavenna, G., and Daffonchio, L., Passive cigarette smoke exposure induces airway hyperreactivity to histamine but not to acetylcholine in guinea-pigs, *Pulm. Pharmacol.*, 3, 145, 1990.
134. Reznik-Schüller, H.M., The respiratory tract in rodents, in *Spontaneous Respiratory Tract Carcinogenesis*, Vol. I, Reznik-Schüller, H.M., Ed., CRC Press, Boca Raton, FL, 1983, 79.
135. Bal, H.S. and Ghoshal, N.G., Morphology of the terminal bronchiolar region of common laboratory animals, *Lab. Anim.*, 22, 76, 1988.
136. Plopper, C.G., Hyde, D.M., and Buckpitt, A.R., Clara cells, in *The Lung: Scientific Foundations*, Vols. 1 and 2, Crystal, R.G. and West, J.B., Eds., Raven Press, New York, 1991, 229.
137. Sorokin, S.P., The cells of the lungs, in *Morphology of Experimental Respiratory Carcinogenesis, AEC Symposium Series, 21*, NTIS U.S. Dept. of Commerce, Springfield, VA, 1970, 3.
138. Rennard, S.I., Beckmann, J.D., and Robbins, R.A., Biology of airway epithelial cells, in *The Lung: Scientific Foundations*, Vols. 1 and 2, Crystal, R.G. and West, J.B., Eds., Raven Press, New York, 1991, 157.
139. Evans, M.J. and Moller, P.C., Biology of airway basal cells, *Exp. Lung Res.*, 17, 513, 1991.
140. Schneeberger, E.E., Alveolar type I cells, in *The Lung: Scientific Foundations*, Vols. 1 and 2, Crystal, R.G. and West, J.B., Eds., Raven Press, New York, 1991, 229.
141. Mason, R.J. and Williams, M.C., Alveolar type II cells, in *The Lung: Scientific Foundations*, Vols. 1and 2, Crystal, R.G. and West, J.B., Eds., Raven Press, New York, 1991, 235.
142. Bernaudin, J.F., Jaurand, M.C., and Fleury, J., Mesothelial cells, in *The Lung: Scientific Foundations*, Vols. 1 and 2, Crystal, R.G. and West, J.B., Eds., Raven Press, New York, 1991, 631.
143. Polak, J.M., Becker, K.L., and Cutz, E., Lung endocrine cell markers, peptides, and amines, *Anat. Rec.*, 236, 169, 1993.
144. Seldeslagh, K.A., Structure and Function of the Pulmonary Diffuse Neuroendocrine System, Ph.D. Thesis, Rijksuniversiteit Groningen, the Netherlands, 1996, 1–23, ISBN 90 367 0608 4.
145. Crystal, R.G., Alveolar macrophages, in *The Lung: Scientific Foundations*, Vols. 1 and 2, Crystal, R.G. and West, J.B., Eds., Raven Press, New York, 1991, 527.
146. Sindhu, R.K., Rasmussen, R.E., Yamamoto, R., Fujita, I., and Kikkawa, Y., Depression of hepatic cytochrome p450 monooxygenases after chronic environmental tobacco smoke exposure of young ferrets, *Toxicol. Lett.*, 76, 227, 1995.
147. Phalen, R.F., Rasmussen, R.E., Mannix, R.C., and Oldham, M.J., Effects of concentrated cigarette smoke on respiratory tract clearance in the ferret, *Inhal. Toxicol.*, 6, 125, 1994.
148. Trump, B.F., McDowell, E.M., Glavin, F., Barrett, L.A., Becci, P.J., Schuerch, W., Kaiser, H.E., and Harris, C.C., The respiratory epithelium: 3. Histogenesis of epidermoid metaplasia and carcinoma in situ in the human, *J. Natl. Cancer Inst.*, 61, 563, 1978.
149. Leube, R.E. and Rustad, T.J., Squamous cell metaplasia in human lung: molecular characteristics of epithelial stratification, *Virchows Arch. B. Cell Pathol.*, 61, 227, 1991.
150. Haley, N.J., Adams, J.D., Alzofon, J., and Hoffmann, D., Uptake of sidestream smoke by syrian golden hamsters, *Toxicol. Lett.*, 35, 83, 1987.

151. Nelson, E., Goubet-Wiemers, C., Guo, Y., and Jodscheit, K., Maternal passive smoking during pregnancy and foetal developmental toxicity. Part 2: histological changes, *Hum. Exp. Toxicol.,* 18, 257, 1999.
152. Escolar, J.D., Martinez, M.N., Rodriguez, F.J., Gonzalo, C., Escolar, M.A., and Roche, P.A., Emphysema as a result of involuntary exposure to tobacco smoke: morphometricalstudy of the rat, *Exp. Lung Res.,* 21, 255, 1995.
153. Cendon, S.P., Battlehner, C., Lorenzi-Filho, G., Dohlnikoff, M., Pereira, P.M., Conceicao, G.M.S., Beppu, O.S., and Saldiva, P.H.N., Pulmonary emphysema induced by passive smoking: an experimental study in rats, *Braz. J. Med. Biol. Res.,* 30, 1241, 1997.
154. Witschi, H., Espiritu, I., Maronpot, R.R., Pinkerton, K.E., and Jones, A.D., The carcinogenic potential of gas phase of environmental tobacco smoke, *Carcinogenesis,* 18, 2035, 1997.
155. Rajini, P. and Witschi, H., Short-term effects of sidestream smoke on respiratory epithelium in mice: cell kinetics, *Fundam. Appl. Toxicol.,* 22, 405, 1994.
156. Ichikawa, T., Yano, Y., Uchida, M., Otani, S., Hagiwara, K., and Yano, T., The activation of K-*ras* gene at an early stage of lung tumorigenesis in mice, *Cancer Lett.,* 107, 165, 1996.
157. Maehira, F., Miyagi, I., Asato, T., Eguchi, Y., Takei, H., Nakatsuki, K., Fukuoka, M., and Zaha, F., Alterations of protein kinase C, 8-hydroxydeoxyguanosine, and k-ras oncogene in rat lungs exposed to passive smoking, *Clin. Chim. Acta,* 289, 133, 1999.
158. Kiaris, H. and Spandidos, D.A., Mutations of ras genes in human tumours (Review), *Int. J. Oncol.,* 7, 413, 1995.
159. Vachtenheim, J., Occurrence of ras mutations in human lung cancer, *Neoplasma,* 44, 145, 1997.
160. Rodenhuis, S., Slebos, R.J.C., Boot, A.J.M., Evers, S.G., Mooi, W.J., Wagenaar, S.S., Van Bodegom, P.C., and Bos, J.L., Incidence and possible clinical significance of K-ras oncogene activation in adenocarcinoma of the human lung, *Cancer Res.,* 48, 5738, 1988.
161. Vainio, H., Husgafvel-Pursiainen, K., Anttila, S., Karjalainen, A., Hackman, P., and Partanen, T., Interaction between smoking and asbestos in human lung adenocarcinoma: Role of K-ras mutations, *Environ. Health Perspect. Suppl.,* 101(Suppl. 3), 189, 1993.
162. Rodenhuis, S. and Slebos, R.C., Clinical significance of ras oncogene activation in human lung cancer, *Cancer Res.,* 52, 2665s, 1992.
163. Gao, H.-G., Chen, J.-K., Stewart, J., Song, B., Rayappa, C., Whong, W.-Z., and Ong, T., Distribution of p53 and K-ras mutations in human lung cancer tissues, *Carcinogenesis,* 18, 473, 1997.
164. Gunning, W.T., Castonguay, A., Goldblatt, P.J., and Stoner, G.D., Strain A/J mouse lung adenoma growth patterns vary when induced by different carcinogens, *Toxicol. Pathol.,* 19, 168, 1991.
165. Palmer, K., Clara cell adenomas of the mouse lung. Interaction with alveolar type II cells, *Am. J. Pathol.,* 120, 455, 1985.
166. Belinsky, S.A., Foley, J.F., White, C.M., Anderson, M.W., and Maronpot, R.R., Dose-response relationship between O6-methylguanine formation in Clara cells and induction of pulmonary neoplasia in the rat by 4-(methylnitrosamino)-1-(3-pyridyl)-1-butanone, *Cancer Res.,* 50, 3772, 1990.
167. Ji, C.M., Plopper, C.G., Witschi, H.P., and Pinkerton, K.E., Exposure to sidestream cigarette smoke alters bronchial epithelial cell differentiation in the postnatal rat lung, *Am. J. Respir. Cell Mol. Biol.,* 11, 312, 1994.

168. Lee, C.K., Brown, B.G., Reed, B.A., Rahn, C.A., Coggins, C.R.E., Doolittle, D.J., and Hayes, A.W., Fourteen-day inhalation study in rats, using aged and diluted sidestream smoke from a reference cigarette. 2. DNA adducts and alveolar macrophage cytogenetics, *Fundam. Appl. Toxicol.*, 19, 141, 1992.

169. Gairola, C.G., Wu, H.P., Gupta, R.C., and Diana, J.N., Mainstream and sidestream cigarette-smoke-induced DNA adducts in C7Bl and DBA mice, *Environ. Health Perspect.*, 99, 253, 1993.

170. Howard, D.J., Briggs, L.A., and Pritsos, C.A., Oxidative DNA damage in mouse heart, liver, and lung tissue due to acute side-stream tobacco smoke exposure, *Arch. Biochem. Biophys.*, 352, 293, 1998.

171. Wong, C.G., Bonakdar, M., and Rasmussen, R.E., Effects of repeated sidestream cigarette smoke inhalation on stress-inducible heat shock protein 70 in the ferret lung, *Inhal. Toxicol.*, 9, 133, 1997.

172. Rithidech, K., Chen, B.T., Mauderly, J.L., Worton, E.B., and Brooks, A.L., Cytogenetic effects of cigarette smoke on pulmonary alveolar macrophages of the rat, *Environ. Mol. Mutagenesis*, 14, 27, 1989.

173. Sindhu, R.K., Rasmussen, R.E., and Kikkawa, Y., Effect of environmental tobacco smoke on the metabolism of(-)-trans-benzo[a]pyrene-7,8-dihydrodiol in juvenile ferret lung and liver, *J. Toxicol. Environ. Health*, 45, 453, 1995.

174. Sindhu, R.K., Rasmussen, R.E., and Kikkawa, Y., Exposure to environmental tobacco smoke results in an increased production of (+)-anti-benzo[a]pyrene-7,8-dihydrodiol-9,10-epoxide in juvenile ferret lung homogenates, *J. Toxicol. Environ. Health*, 47, 523, 1996.

175. Gairola, C.G., Pulmonary aryl hydrocarbon hydroxylase activity of mice, rats and guinea pigs following long term exposure to mainstream and sidestream cigarette smoke, *Toxicology*, 45, 177, 1987.

176. Gebremichael, A., Tullis, K., Denison, M.S., Cheek, J.M., and Pinkerton, K.E., Ah-receptor-dependent modulation of gene-expression by aged and diluted sidestream cigarette-smoke, *Toxicol. Appl. Pharmacol.*, 141, 76, 1996.

177. Ji, C.M., Royce, F.H., Truong, U., Plopper, C.G., Singh, G., and Pinkerton, K.E., Maternal exposure to environmental tobacco smoke alters Clara cell secretory protein expression in fetal rat lung, *Am. J. Physiol. (Lung Cell Mol. Physiol. 19)*, 275, L870, 1998.

178. Muscat, J.E., Citron, M., Wang, C.X., White, A., and Lazarus, P., P53 protein overexpression in lung adenocarcinomas in non-smokers: possible association with environmental tobacco smoke, *Oncol. Rep.*, 4, 825, 1997.

179. Joerres, R. and Magnussen, H., Influence of short-term passive smoking on symptoms, lung mechanics and airway responsiveness in asthmatic subjects and healthy controls, *Eur. Respir. J.*, 5, 936, 1992.

180. Yates, D.H., Havill, K., Thompson, M.M., Rittano, A.B., Chu, J., and Glanville, A.R., Sidestream smoke-inhalation decreases respiratory clearance of Tc-99m-DTPA acutely, *Aust. N. Z. J. Med.*, 26, 513, 1996.

181. Mutoh, T., Bonham, A.C., Kott, K.S., and Joad, J.P., Chronic exposure to sidestream tobacco smoke augments lung C-fiber responsiveness in young guinea pigs, *J. Appl. Physiol.*, 87, 757, 1999.

182. Joad, J.P., Avadhanam, K.P., Watt, K.C., Kott, K.S., Bric, J.M., and Pinkerton, K.E., Effects of extended sidestream smoke exposure on components of the C-fiber axon reflex, *Toxicology*, 112, 195, 1996.

183. Joad, J.P., Bric, J.M., Peake, J.L., and Pinkerton, K.E., Perinatal exposure to aged and diluted sidestream cigarette smoke produces airway hyperresponsiveness in older rats, *Toxicol. Appl. Pharmacol.*, 155, 253, 1999.

184. Zhang, Z., Araghiniknam, M., Inserra, P., Jiang, S., Lee, J., Chow, S., Breceda, V., Balagtas, M., Witten, M., and Watson, R.R., Vitamin E supplementation prevents lung dysfunction and lipid peroxidation in nude mice exposed to side-stream cigarette smoke, *Nutr. Res.,* 19, 75, 1999.

185. Witten, M.L., Joseph, P.M., Lantz, R.C., Jung, W.K., and Hales, C.A., Chronic sidestream cigarette smoke exposure causes lung injury in rabbits, *Indoor Environ.,* 1, 341, 1992.

186. Finch, G.L., Nikula, K.J., Belinsky, S.A., Barr, E.B., Stoner, G.D., and Lechner, J.F., Failure of cigarette smoke to induce or promote lung cancer in the A/J mouse, *Cancer Lett.,* 99, 161, 1996.

187. Omini, C., Hernandez, A., Zuccari, G., Clavenna, G., and Daffonchio, L., Passive cigarette smoke exposure induces airway hyperreactivity to histamine but not not to acetylcholine in guinea-pigs, *Pulm. Pharmacol.,* 3, 145, 1990.

188. Wilmer, J.W.G.M., Woutersen, R.A., Appelman, L.M., Leeman, W.R., and Feron, V.J., Subchronic (13-week) inhalation toxicity study of formaldehyde in male rats: 8-hour intermittent versus 8-hour continuous exposures, *Toxicol. Lett.,* 47, 287, 1989.

189. Behnke, W., Levsen, K., Scheer, V., and Zetzsch, C., Interactions of gaseous pollutants with cultured cells and at the liquid interface of the lung: an exploratory study on the uptake of ozone, in *Relationships Between Respiratory Disease and Exposure to Air Pollution,* Mohr, U., Dungworth, D.L., Brain, J.D., Driscoll, K., Grafstroem, R.C., and Harris, C.C., Eds., ILSI Press, Washington, D.C., 1998, 314.

190. Uppu, R.M., What does ozone react with at the air/lung interface? Model studies using human red blood cell membranes, *Arch. Biochem. Biophys.,* 319, 257, 1995.

191. Sonnenfeld, G. and Wilson, D.M., The effect of smoke age and dilution on the cytotoxicity of sidestream (passive) smoke, *Toxicol. Lett.,* 35, 89, 1987.

192. Moegel, M., Krueger, E., Krug, H.F., and Seidel, A., Ozone-induced elevation of IL-8 in a coculture model of airway epithelial and endothelial cells, in *Relationships Between Respiratory Disease and Exposure to Air Pollution,* Mohr, U., Dungworth, D.L., Brain, J.D., Driscoll, K., Grafstroem, R.C., and Harris, C.C., Eds., ILSI Press, Washington, D.C., 1998, 383.

193. Lang, D.S., Joerres, R.A., Muecke, M., Siegfried, W., and Magnussen, H., Interactions between human bronchoepithelial cells and lung fibroblasts after ozone exposure in vitro, *Toxicol. Lett.,* 96/97, 13, 1998.

194. Aufderheide, M. and Mohr, U., Cultex — a new system and technique for the cultivation and exposure of cells at the air/liquid interface, *Exp. Toxicol. Pathol.,* 51, 489, 1999.

195. Zhu, B.-Q., Sun, Y.P., Sievers, R.E., Isenberg, W.M., Glantz, S.A., and Parmley, W.W., Passive smoking increases experimental atherosclerosis in cholesterol-fed rabbits, *J. Am. Coll. Cardiol.,* 21, 225, 1993.

196. Agapitos, E., Mollo, F., Tomatis, L., Katsouyanni, K., Lipworth, L., Delsedime, L., Kalandidi, A., Karakatsani, A., Riboli, E., Saracci, R., and Trichopoulos, D., Epithelial, possibly precancerous, lesions of the lung in relation to smoking, passive smoking, and sociodemographic variables, *Scand. J. Soc. Med.,* 24, 259, 1996.

197. Auerbach, O., Hammond, E.C., and Garfinkel, L., Changes in bronchial epithelium in relation to cigarette smoking, 1955-1960 vs. 1970-1977, *N. Engl. J. Med.,* 300, 381, 1979.

198. McDowell, E.M., McLaughlin, J.S., Merenyi, D.K., Kieffer, R.F., Harris, C.C., and Trump, B.F., The respiratory epithelium: 5. Histogenesis of lung carcinomas in the human, *J. Natl. Cancer Inst.,* 61, 587, 1978.

199. Morgan, K.T., A brief review of formaldehyde carcinogenesis in relation to rat nasal pathology and human health risk assessment, *Toxicol. Pathol.*, 25, 291, 1997.

200. Woutersen, R.A., Van Garderen-Hoetmer, A., Bruijntjes, J.P., Zwart, A., and Feron, V.J., Nasal tumours in rats after severe injury to the nasal mucosa and prolonged exposure to 10ppm formaldehyde, *J. Appl. Toxicol.*, 9, 39, 1989.

201. Bos, R.P., Theuws, J.L.G., and Henderson, P.T., Excretion of mutagens in human urine after passive smoking, *Cancer Lett.*, 19, 85, 1983.

202. Pasquini, R., Sforzolini, G.S., Savino, A., Angeli, G., and Monarca, S., Enzyme induction in rat lung and liver by condensates and fractions from main-stream and side-stream cigarette smoke, *Environ. Res.*, 44, 302, 1987.

203. Joad, J.P., Pinkerton, K.E., and Bric, J.M., Effects of sidestream smoke exposure and age on pulmonary function and airway reactivity in developing rats, *Pediatr. Pulmonol.*, 16, 281, 1993.

204. Joad, J.P. and Bric, J.M., Effects of chronic sidestream smoke (SS) exposure on airway function and reactivity to capsaicin and substance-P (SP) in developing guinea pigs, *Am. Rev. Respir. Dis.*, 147, A214, 1993.

205. Zhu, B., Sun, Y., Sievers, R.E., Glantz, S.A., Parmley, W.W., and Wolfe, C.L., Exposure to environmental tobacco smoke increases myocardial infarct size in rats, *Circulation*, 89, 1282, 1994.

206. Nelson, E., Jodscheit, K., and Guo, Y., Maternal passive smoking during pregnancy and fetal developmental toxicity. Part 1: gross morphological effects, *Hum. Exp. Toxicol.*, 18, 252, 1999.

207. Roberts, K.A., Rezai, A.A., Pinkerton, K.E., and Rutledge, J.C., Effect of environmental tobacco smoke on LDL accumulation in the artery wall, *Circulation*, 94, 2248, 1996.

208. Sindhu, R.K., Rasmussen, R.E., and Kikkawa, Y., Exposure to environmental tobacco smoke results in an increased production of (+)-anti-benzo[a]pyrene-7,8-dihydrodiol-9,10-epoxide in juvenile ferret lung homogenates, *J. Toxicol. Environ. Health*, 47, 523, 1995.

209. Gray, T.E., Guzman, K., Davis, C.W., Abdullah, L.H., and Nettesheim, P., Mucociliary differentiation of serially passaged normal human tracheobronchial epithelial cells, *Am. J. Respir. Cell Mol. Biol.*, 14, 104, 1996.

210. Davenport, E.A. and Nettesheim, P., Regulation of mucociliary differentiation of rat tracheal epithelial cells by type I collagen gel substratum, *Am. J. Respir. Cell Mol. Biol.*, 14, 19, 1996.

211. Baeza-Squiban, A., Romet, S., Moreau, A., and Marano, F., Progress in outgrowth culture from rabbit tracheal explants: balance between proliferation and maintenance of differentiated state in epithelial cells, *In Vitro Cell. Dev. Biol.*, 27A, 453, 1991.

212. Baeza-Squiban, A., Boisvieux-Ulrich, E., Guilianelli, C., Oucine, O., Geraud, G., Guennou, C., and Marano, F., Extracellular matrix-dependent differentiation of rabbit tracheal epithelial cells in primary culture, *In Vitro Cell. Dev. Biol.*, 30A, 56, 1994.

213. Emura, M., Stem cells of the respiratory epithelium and their in vitro cultivation, *In Vitro Cell. Dev. Biol.*, 33, 3, 1997.

214. Emura, M., Mohr, U., Riebe, M., Aufderheide, M., and Dungworth, D.L., Predisposition of cloned fetal hamster lung epithelial cells to transformation by a pre-carcinogen, benzo(a)pyrene, using growth hormone supplementation and collagen gel substratum, *Cancer Res.*, 47, 1155, 1987.

215. Emura, M., Riebe, M., Ochiai, A., Aufderheide, M., Germann, P., and Mohr, U., New functional cell-culture approach to pulmonary carcinogenesis and toxicology, *J. Cancer. Res. Clin. Oncol.*, 116, 557, 1990.

216. Emura, M., Ochiai, A., Singh, G., Katyla, S.L., and Hirohashi, S., In vitro reconstitution of human respiratory epithelium, *In Vitro Cell. Dev. Biol. Anim.*, 33, 602, 1997.
217. Ochiai, A., Emura, M., Mohr, U., Tahara, E., and Dungworth, D.L., Induction and characterization of secretory differentiation in human fetal bronchial epithelial-cell line (HFBE) cultured on collagen gel in growth-hormone and vitamin-A-supplemented medium, *Exp. Pathol.*, 41, 157, 1991.
218. Knebel, J.W., Ritter, D., and Aufderheide, M., Development of an in vitro system for studying effects of native and photochemically defined gaseous compounds using an air/liquid interface culture technique, *Toxicol. Lett.*, 96/97, 1, 1998.
219. Knebel, J.W., Aufderheide, M., and Emura, M., Comparison of biological effects of different polycyclic aromatic hydrocarbons in lung cells of hamster and rat in vitro, *Toxicol. Lett.*, 72, 65, 1994.
220. Benedict, W.F., Rucker, N., Faust, J., and Kouri, R.E., Malignant transformation of mouse cells by cigarette smoke condensate, *Cancer Res.*, 35, 857, 1975.
221. Schlage, W.K., Buelles, H., Friedrichs, D., and Kurkowsky, B., Two-stage transformation assay for cigarette smoke condensates using murine C3H-10T1/2 fibroblasts, *Toxicol. In Vitro*, 13, 823, 1999.
222. Rutten, A.A.J.J.L., Jongen, W.M.F., De Haan, L.H.J., Hendriksen, E.G.J., and Koeman, J.H., Effect of retinol and cigarette-smoke condensate on dye-coupled intercellular communication between hamster tracheal epithelial cells, *Carcinogenesis*, 9, 315, 1988.
223. Rutten, A.A.J.J.L., Bruyntjes, J.P., and Ramaekers, F.C.S., Effect of vitamin A deprivation and cigarette smoke condensate on keratin expression patterns in cultured hamster tracheal epithelium, *Virchows Arch. B Cell Pathol.*, 56, 111, 1988.
224. Ramm, D., Hübsch, U., Koch, W., Lödding, H., Pohlmann, G., Windt, H., Westphal, K., and Massey, E.D., A new system for direct exposure of cells maintained in vitro to cigarette smoke, in Abstracts of the 5th International Inhalation Symposium, Hannover, 20 to 24 February 1995: Correlations Between In Vitro and In Vivo Investigations in Inhalation Toxicology, Hannover Medical School, Hannover, 1995.
225. Massey, E., Micronucleus induction in V79 cells after direct exposure to whole cigarette smoke, *Mutagenesis*, 13, 145, 1998.
226. Bombick, D.W., Ayres, P.H., Putnam, K., Reed Bombick, B., and Doolittle, D.J., Chemical and biological studies of a new cigarette that primarily heats tobacco. Part 3. In vitro toxicity of whole smoke, *Food Chem. Toxicol.*, 36, 191, 1998.
227. Doolittle, D.J., Lee, C.K., Ivett, J.L., Mirsalis, J.C., Riccio, E., Rudd, C.J., Burger, G.T., and Hayes, A.W., Genetic toxicology studies comparing the activity of sidestream smoke from cigarettes which burn or only heat tobacco, *Mutat. Res.*, 240, 59, 1990.
228. Sonnenfeld, G. and Hudgens, R.W., Effect of sidestream and mainstream smoke exposure on in vitro interferon-alpha/beta production by L-929 cells, *Cancer Res.*, 46, 2779, 1986.
229. Chen, C.C. and Lee, H., Genotoxicity and DNA adduct formation of incense smoke condensates: comparison with environmental tobacco smoke condensates, *Mutat. Res.*, 367, 105, 1996.
230. Heussen, G.A.H. and Alink, G.M., Inhibition of gap junctional intercellular communication by TPA and airborne particulate matter in primary cultures of rat alveolar type II cells, *Carcinogenesis*, 13, 719, 1992.
231. Müller, T., Expression of C-fos in quiescent Swiss 3T3 cells exposed to aqueous cigarette smoke fractions, *Cancer Res.*, 55, 1927, 1995.
232. Yang, Q., Hergenhahn, M., Weninger, A., and Bartsch, H., Cigarette smoke induces direct DNA damage in human B-lymphoid cell line Raji, *Carcinogenesis*, 20, 1769, 1999.

233. Schlage, W.K., Buelles, H., and Kurkowsky, B., Use of the HET-CAM test for the determination of the irritant potential of cigarette sidestream smoke, *Toxicol. In Vitro*, 13, 829, 1999.

234. March, T.H., Barr, E.B., Finch, G.L., Hahn, F.F., Hobbs, C.H., Menache, M.G., and Nikula, K.J., Cigarette Smoke exposure produces more evidence of emphysema in B6C3F1 mice than in F344 rats, *Toxicol. Sci.*, 51, 289–299, 1999.

14 Environmental Tobacco Smoke in Restaurants: Exposures and Health Effects

Helen Dimich-Ward and Michael Brauer

CONTENTS

14.1 INTRODUCTION

Restaurant and bar smoking restrictions and smoking bans are being implemented with increasing frequency in North American municipalities.[6] Despite extensive analysis indicating that smoking restrictions do not have any negative impacts on restaurants sales,[19,37,38] there is often resistance within the hospitality industry and within segments of the population to increased restriction or elimination of smoking in public eating establishments. While compliance with a total smoking ban will certainly eliminate exposure to environmental tobacco smoke (ETS) in restaurants, claims are often made that less restrictive policies are also effective. As the introduction of smoking restrictions in restaurants is currently an issue of important public health policy, it is surprising how little data are available to indicate the effectiveness of various levels of smoking restrictions on ETS exposures and health impacts.

14.2 CONCENTRATIONS AND EXPOSURES

Several surveys have indicated that restaurants and bars present high levels of ETS, although there is considerable variability between establishments. Due to the complex chemical nature of ETS, most measurements have focused on particles, often combined with one of several tracers for ETS. Carbon monoxide levels have also

0-8493-0311-7/00/$0.00+$.50
© 2001 by CRC Press LLC

been measured in several surveys, although levels are generally not significantly higher than ambient concentrations and appear to be less important as potential health hazards than particles. Other contaminants that have been measured include acrolein, benzo(a)pyrene and other polycyclic aromatic hydrocarbons, sulfur dioxide, and nitrogen oxides.

A criticism of using particulate matter as a tracer is its lack of specificity, due to its many non-ETS sources. This may be an important consideration in restaurants, where cooking also produces substantial particulate concentrations.[4] Therefore, background levels are desirable to avoid overestimation.[20] Nicotine, on the other hand, has the advantage of specificity, but due to its absorptivity on surfaces in indoor environments, it has significant limitations as a marker for ETS in indoor air. A recent large survey of personal exposures to ETS constituents suggests that there are reasonably high correlations between ETS marker compounds such as nicotine and respirable particulate matter for locations where relatively high ETS levels are present.[25]

Siegel summarized measurements of respirable suspended particles (RSP) in bars, restaurants, offices, and residences with at least one smoker.[39] Levels of ETS in restaurants were 1.6 to 2.0 times higher than in offices and 1.5 times higher than in residences with at least one smoker. Levels in bars were three times higher than in restaurants. Siegel concluded that ETS was a significant occupational health hazard for food-service workers. In a review of indoor nicotine measurements collected in the U.S., Hammond reports mean nicotine concentrations in the range of 3 to 8 $\mu g/m^3$ in restaurants and 10 to 40 $\mu g/m^3$ in bars. In comparison, mean nicotine levels in offices were generally between 2 to 6 $\mu g/m^3$ and in the range of 1 to 3 $\mu g/m^3$ in the homes of smokers.[21]

Miesner and colleagues measured fine particulate matter smaller than 2.5 μm ($PM_{2.5}$) in restaurants and bars, as well as in various other public places.[30] In most of the public facilities and office buildings where there was no smoking, particulate concentrations were less than 30 $\mu g/m^3$, while restaurants and bars had levels of 30 to 140 $\mu g/m^3$. An earlier review of indoor tobacco combustion byproducts in various public places[40] showed that mean levels of the following contaminants were generally higher in bars, taverns, and nightclubs than in restaurants: acrolein (10 vs. 7 to 8 ppb), carbon monoxide (5 to 17 ppm vs. 3 to 10 ppm), nicotine (10 vs. 5 $\mu g/m^3$), nitrogen oxides (195 vs. 80 to 120 ppb), particles (230 to 990 vs. 80 to 110 $\mu g/m^3$), and SO_2 (30 vs. 13 to 20 ppb). Comparison control locations such as restaurants where smoking was prohibited or even outdoor levels were not consistently available.

Collett and colleagues measured vapor phase nicotine and RSP in 13 nightclubs, 8 taverns, and 10 neighborhood pubs.[7] Nightclubs were characterized by higher occupant density, while taverns offered less extensive food service than neighborhood pubs. Mean RSP levels were 151, 93, and 95 $\mu g/m^3$ for the nightclubs, taverns, and neighborhood pubs, respectively. Nicotine levels in neighborhood pubs were lower than in taverns, which were lower than in nightclubs. The higher RSP and nicotine levels in the nightclubs were associated with higher occupant densities, higher cigarette counts, and higher CO_2 concentrations. Measurement of CO_2 as an indicator of ventilation adequacy suggested that ventilation levels did not meet the American

Society of Heating, Refrigeration, and Air Conditioning Engineers ASHRAE 62-1989 levels designed to provide "acceptable indoor air quality."

A recent paper reported on two studies of personal exposures to ETS (respirable particles and nicotine) measured in 1993 to 1994.[25] Bartenders who worked in single-room facilities were found to have the highest personal exposures of all occupations studied (including professional, service industry, clerical, and other white-collar workers). Exposures of bartenders who worked in multi-room and restaurant facilities (approximately 50% of the study group) were approximately ten times lower. As a group, bartenders had exposures to nicotine and respirable particles which were three to four times higher than exposures of restaurant/tavern servers. Exposures of restaurant/tavern servers were similar to those of other service workers. When broad occupational groups were compared, service workers, including restaurant and bar workers, were the group with the highest exposures to ETS components.[25]

Lambert and colleagues compared RSP and nicotine levels in the nonsmoking and smoking sections of seven restaurants.[29] The mean concentrations of RSP and nicotine were 40 and 65% lower, respectively, in the nonsmoking than in the smoking sections. The authors concluded that the simple separation of smokers and nonsmokers may reduce, but does not eliminate, the exposure of nonsmokers to ETS.

A cross-sectional survey of the impact of smoking restrictions conducted by Brauer and 't Mannetje[4] also found that smoking restrictions resulted in significantly lower ETS levels. In this survey, fine particles and particulate cadmium were measured in 20 restaurants and bars in Vancouver, British Columbia. The restaurants were divided into three categories based on their smoking policy: 5 nonsmoking restaurants, 11 restaurants with both nonsmoking and smoking sections (restricted smoking), and 4 bars (with food service) where smoking was unrestricted. Regulations at the time of this study required at least 40% of seating to be nonsmoking in restaurants, whereas there were no restrictions applicable to bars. High particle concentrations were measured in all restaurant types (5 min average peak $PM_{2.5}$ concentrations above 400 $\mu g/m^3$). Fine particle concentrations were significantly higher in establishments with no smoking restrictions (mean $PM_{2.5}$ concentration = 190 $\mu g/m^3$, range: 47 to 253 ppm) than in restaurants with partial smoking restrictions (mean $PM_{2.5}$ concentration = 57 $\mu g/m^3$, range: 11 to 163 ppm). Concentrations in nonsmoking restaurants were reduced by an additional 20 to 30%. Measurements of cadmium, a more specific tracer of ETS than $PM_{2.5}$, implicated ETS as the major source of indoor particulate in restaurants where smoking was allowed, with cooking an additional source.

The contribution of additional sources other than ETS to particulate levels in restaurants was also suggested by Crouse et al. who (in collaboration with the tobacco industry) conducted a survey of 42 restaurants without smoking restrictions.[8] Vapor phase nicotine, ultraviolet particulate matter, and RSP were measured. The survey measured 1-h average RSP levels of 16 to 221 $\mu g/m^3$ (mean concentration of 81 $\mu g/m^3$). Since nicotine levels were not significantly correlated with RSP levels, the authors suggested that other sources in addition to ETS contributed to the high RSP levels.

In one of the most important analyses of air quality in relation to restaurant smoking restriction policies conducted to date, Ott and colleagues measured RSP inside a sports tavern before and after the prohibition of smoking.[34] During the smoking period, the average RSP concentration was 57 μg/m^3 above the outdoor concentration, and it decreased by 77% to 13 μg/m^3 above the outdoor concentration after the prohibition of smoking. There was no change in the number of customers following the smoking restriction. It is noteworthy that this tavern had a relatively high air exchange rate of 7.5/h, indicating that a volume of air equal to that of the tavern was replaced every 8 min. This work estimated that each cigarette contributed, on average, 37.5 μg/m^3 to the RSP (PM$_{3.5}$) concentration of this 521-m^3 tavern. This contribution was similar to the 25 μg/m^3 PM$_{2.5}$ increase attributable to each cigarette reported by Brauer and 't Mannetje[4] for a series of restaurants and bars of different volumes. Recently, Ott developed and validated a simple mathematical model to predict respirable particulate levels if the number of active smokers the indoor volume and the ventilation rate are known.[33]

14.3   BIOLOGICAL MARKERS OF EXPOSURE

In addition to air concentration or personal exposure measurements, biological markers have often been used as measures of exposure to ETS for risk assessment or epidemiological studies. ETS biomarkers have been reviewed recently by Benowitz.[1] The presence of nicotine and its metabolites (cotinine) in body fluids, e.g., urine, serum, and saliva, is highly specific to ETS exposure.[9] Nicotine is distributed rapidly to tissues and is rapidly metabolized, thus representing exposure within the past few hours to ETS. Cotinine has a longer elimination half-life than nicotine, of about 12 to 20 h, and is an indicator of the integrated exposure to ETS over the previous 1 to 2 d.[2] Very high levels of cotinine have been found in urine samples of nonsmoking restaurant workers, where the mean level of 56 ng/ml contrasted to 8.3 ng/ml found for staff not exposed to ETS at work.[23] In comparison to their previous studies showing an average of 2.5 ng/ml in nonsmokers exposed to ETS at home or work and 0.7 ng/ml in controls, Jarvis et al.[24] found median saliva samples of 7.95 ng/ml for 42 nonsmoking bar staff. An increase in saliva nicotine was shown after only a few hours of exposure in nonsmoking nightclub musicians;[3] however, neither saliva nicotine or cotinine were correlated with air monitoring of particles and nicotine or perceived smokiness. According to a model developed by Repace and colleagues,[35] a low average salivary cotinine level of 0.4 ng/ml corresponds to an increased lifetime mortality risk of 1/1000 for lung cancer and 1/100 for heart disease. A case-control study of women nonsmokers in the Netherlands did demonstrate that their lung cancer risk was correlated with historical urinary cotinine excretion values measured from 1977 to 1991.[10]

Measurement of constituents of tobacco smoke in hair has the advantage of representing longer term exposure, dependent upon the length of hair analyzed (hair grows at approximately ½ in./month). Based on a GC/MS analysis of 3 in. (20 mg) hair samples from 26 subjects, a gradient of exposure was shown such that, among nonsmokers, bar workers had the highest levels of nicotine in hair (averaging 1.74 ng/mg), followed by hotel workers (0.71 ng/mg) and then controls who were unexposed to ETS (0.19 ng/mg).[12] A positive correlation of 0.74 was found between hair nicotine and questionnaire estimates of hours of exposure to ETS per week.

This expands upon earlier findings of a gradient of exposure in hair nicotine related to household sources of ETS.[32]

14.4 HEALTH EFFECTS FROM EXPOSURE TO ETS IN RESTAURANTS AND BARS

Although much of the public health concern regarding ETS exposure in bars and restaurants is directed toward the public who visit these establishments, employees are potentially at greatest risk of adverse health effects due to their prolonged and repeated exposures.

The issue of exposure to ETS in the workplace is of high priority, as ETS has been classified as a human (Group A) lung carcinogen by the U.S. Environmental Protection Agency.[26] The association of ETS and lung cancer has biological plausibility, in that sidestream smoke contains carcinogens and there appears to be no documented threshold dose for respiratory carcinogens for active smokers.[36] A tobacco-specific lung carcinogen, 4-(methylnitrosamino)-1-(3-pyridyl)-1-butanol, has been detected in urine samples of never smokers exposed to cigarette smoke.[22] Bartenders and servers (waitresses, waiters, cocktail waitresses, and barmaids) have consistently been found to have elevated risks for respiratory cancers, based on epidemiological evidence from proportional mortality, case-control, and retrospective cohort studies.

An elevated risk of proportionate mortality for lung cancer was observed in bartenders, waiters, and waitresses in British Columbia, Canada.[14] The PMR for males was 133 (95% CI: 117, 164) and the PCMR for females was 213 (95% CI: 116, 351). A California study of occupational mortality in women[15] found female waitresses to have high risks for lung cancer (SMR = 368) and respiratory tract disease (SMR = 430). A SIR of 2.3 for lung cancer was observed in a study of a large cohort of waitresses in Norway.[27] Furthermore, a ratio of 2.2 in lung cancer risk was found when comparing waitresses according to whether they had worked in restaurants where alcohol was served or not. Because these occupational groups are known to have a high percentage of cigarette smokers,[41] it is important to discern the effects of active smoking from involuntary smoking.

A review of five case-control studies and one historical cohort study, in which personal smoking habits and other confounders were controlled for, have showed, on average, a 50% excess of lung cancer (ranging from 10 to 90%) among restaurant workers.[39] Cooking fumes were suggested to be of much less importance than exposure to ETS.[42] For instance, a cohort study of Norwegian cooks showed an odds ratio (OR) of 0.7 (95% CI: 0.2 to 1.7) for lung cancer, whereas for waiters it was 2.0 (95% CI: 1.3 to 2.9).[27] The lifetime excess risk of death from lung cancer for U.S. females exposed to ETS from working in bars was calculated to be 9×10^{-4}, a hundred times higher that an acceptable risk level of 10^{-6}.[31]

Many of the studies on noncarcinogenic respiratory health effects related to exposure to ETS have focused on children, and only limited information exists for adults, particularly for occupational exposures to ETS such as in restaurants or bars. The most common acute effects due to exposure to ETS are tissue irritation of the eyes, nose, throat, and airways.[45] A cross-sectional survey of hospitality workers in Greater Vancouver, B.C., Canada showed that the prevalence of respiratory symptoms and sensory irritation among never smokers was highest for the group working in restaurants or bars with at least 70% of the premises set aside for smokers.[13] Significant differences were found for symptoms when comparing the highest ETS exposure group vs. hospitality workers with low occupational exposure (smoking permitted in less than 10% of the workplace), including wheeze (19.4 vs. 7.6%), woken short of breath (14.9 vs. 3.3%), and itchy, burning eyes (66.2 vs. 26.3%).

A small study of bar workers in Vancouver, Canada showed work shift decreases in lung function that were not observed in unexposed servers.[12] However, differences in the timing of the work shift could have influenced the magnitude of changes in lung function. A California study of 53 bartenders found decreases in the respiratory symptoms of wheeze, cough, and phlegm and in the sensory irritation symptoms of the eye, nose, and throat 1 month after implementation of a state-wide legislative ban on smoking in bars and taverns.[16] As well, an improvement was shown in mean (FEV_1, 1.2%) and forced vital capacity (FVC, 4.2%) compared to baseline levels taken in the month preceding the smoking ban. Because there were no comparative controls to evaluate learning effects and other influences, and with 45% of the subjects being smokers, it is arguable whether the health effects could be attributable to the cessation of exposure. However, after stratifying for personal smoking, similar estimates of improvement occurred in nonsmokers for FVC and FEV_1, although the latter change was no longer statistically significant due to the small numbers.

14.5 SUMMARY

Although there is no perfect measure of ETS, a number of airborne contaminants and biological markers have been shown to be highly related to ETS exposure. These include airborne nicotine and particles involving cadmium and measurements of cotinine in bodily fluids and nicotine in hair. Personal exposure measurements and are samples have been collected in a wide variety of bars and restaurants. The indisputable findings are that restaurants and bars with no or few restrictions on smoking have indoor concentrations of tobacco byproducts that are much higher than concentrations found in similar indoor environments where smoking is prohibited. Similarly, higher levels of ETS markers such as urinary and salivary cotinine and hair nicotine have been found in workers exposed to ETS in restaurants and bars than in controls.

The generally high levels of ETS which may be encountered in bars and restaurants are a concern for many different groups of individuals, including children, pregnant women, asthmatics, atopic individuals, and for the nonsmokers who work in these facilities and are subjected to smoke from patrons. Yet scientific evaluation of health effects associated with exposure of workers to ETS during the entire workday has generally been sparse. The risk of lung cancer attributed to occupational exposure to ETS has been more thoroughly studied. Elevated risks for lung and other respiratory cancers have been commonly shown in case-control, mortality, and cohort studies of servers and bar workers. There are a limited number of studies indicating a relationship between occupational exposure to ETS and increased respiratory symptoms or decreased lung function, but very few studies have focused on respiratory health risks for waiters, waitresses and bar workers. Despite this, prohibition of smoking at work has been introduced in many jurisdictions throughout Canada and the U.S. due, in large measure, to the issue of legal liability and recognition of the potential for adverse health consequences.

REFERENCES

1. Benowitz, N.L., Biomarkers of environmental tobacco smoke exposure, *Environ. Health Perspect.*, 107(Suppl. 2), 349–355, 1999.
2. Benowitz, N.L., Cotinine as a biomarker for environmental tobacco smoke exposure, *Epidemiol. Rev.*, 18, 188–204, 1996.
3. Bergman, T.A., Johnson, D.L., Boatright, D.T., Smallwood, K.G., and Rando, R.J., Occupational exposure of nonsmoking nightclub musicians to environmental tobacco smoke, *Am. Ind. Hyg. J.*, 57, 746–752, 1996.
4. Brauer, M. and 't Mannetje, A., Restaurant smoking restrictions and environmental tobacco smoke exposure, *Am. J. Public Health.*, 88(12), 1834–1836, 1998.
5. Broder, I., Pilger, C., and Corey, P., Environment and well-being before and following smoking ban in office buildings, *Can. J. Public Health*, 84, 254–258, 1993.
6. Brownson, R.C., Eriksen, M.P., et al., Environmental tobacco smoke: health effects and policies to reduce exposure, *Annu. Rev. Public Health*, 18, 163–185, 1997.
7. Collett, C., Ross, J., et al., Nicotine, RSP and CO2 levels in bars and nightclubs, *Environ. Int.*, 18, 347–352, 1992.
8. Crouse, W., Ireland, M., et al., Results from a Survey of Environmental Tobacco Smoke in Restaurants. Combustion Processes and the Quality of the Indoor Environment, Air and Waste Management Association, Niagara Falls, NY, 1988.
9. Curvall, M. and Vala, E.K., Nicotine and metabolites: analysis and levels in body fluids, in *Nicotine and Related Alkaloids*, Gorrod, J.W. and Wahren, J., Eds., Chapman & Hall, London, 1993.
10. De Waard, F., Kemmeren, J.M., Van Ginkel, A., and Stolker, A.A.M., Urinary cotinine and lung cancer risk in a female cohort, *Br. J. Cancer*, 72, 784–787, 1995.
11. Dimich-Ward, H., Gee, H., Leung, V., and Brauer, M., Analysis of nicotine and cotinine in the hair of hospitality workers exposed to ETS, *J. Occup. Environ. Med.*, 39, 946–948, 1997.
12. Dimich-Ward, H., Lawson, J., and Chan-Yeung, M., Workshift changes in lung function in bar workers exposed to environmental tobacco smoke, *Am. J. Respir. Crit. Care Med.*, 157, A505, 1998.
13. Dimich-Ward, H., Lawson, J., Rousseau, R., and Chan-Yeung, M., Respiratory symptoms and perception of exposure among never smokers working in the hospitality industry, *Am. J. Respir. Crit. Care Med.*, 159(3), A236, 1999.
14. Dimich-Ward, H., Gallagher, R.P., Spinelli, J.J., Threlfall, W.J., and Band, P.R., Occupational mortality among bartenders and waiters, *Can. J. Public Health*, 79, 194–197, 1988.
15. Doebbert, G., Riedmiller, K.R., and Kizer, K.W., Occupational mortality of California women, 1979-1981, *West. J. Med.*, 149, 734–740, 1988.
16. Eisner, M.D., Smith, A.K., and Blanc, P.D., Bartenders respiratory health after establishment of smoker-free bars and taverns, *J. Am. Med. Assoc.*, 20, 1904–1914, 1998.

17. Eliopoulos, C., Klein, J., and Koren, G., Validation of self-reported smoking by analysis of hair for nicotine and cotinine, *Ther. Drug Monit.*, 18(5), 532–536, 1996.

18. Etzel, R.A., A review of the use of saliva cotinine as a marker of tobacco smoke exposure, *Prev. Med.*, 19, 190–197, 1990.

19. Glantz, S. and Smith, L., The effect of ordinances requiring smoke-free restaurants on restaurant sales, *Am. J. Public Health*, 84, 1081–1085, 1994.

20. Goodfellow, H.D., Eyre, S., and Wyattt, J.A.S., Assessing exposures to environmental tobacco smoke, in *Environmental Tobacco Smoke*, Ecobichon, D.J. and Wu, J.M., Eds., Lexington Books, Lexington, MA, 1990.

21. Hammond, S., Exposure of U.S. workers to environmental tobacco smoke, *Environ. Health Perspect.*, 107(Suppl. 2), 329–340, 1999.

22. Hecht, S.S., Carmella, S.G., Murphy, S.E., Akerkar, S., Brunnemann, K.D., and Hoffman, D., A tobacco-specific lung carcinogen in the urine of men exposed to cigarette smoke, *N. Engl. J. Med.*, 329, 1543–1546, 1993.

23. Husgafvel-Pursianen, K., Sorsa, M., and Engstrom, K., Passive smoking at work: biochemical and biological measurement of exposure to environmental tobacco smoke, *Int. Arch. Occup. Environ. Health*, 59, 337–345, 1987.

24. Jarvis, M.J., Foulds, J., and Feyerabend, C., Exposure to passive smoking among bar staff, *J. Addict.*, 87, 111–113, 1992.

25. Jenkins, R. and Counts, R., Occupational exposure to environmental tobacco smoke: results of two personal exposure studies, *Environ. Health Perspect.*, 107(Suppl. 2), 341–348, 1999.

26. Jinot, J. and Bayard, S., Respiratory health effects of passive smoking: EPA's weight of evidence analysis, *J. Clin. Epidemiol.*, 47, 339–349, 1994.

27. Kjaerheim, K. and Andersen, A., Incidence of cancer among male waiters and cooks; two Norwegian cohorts, *Cancer Causes Control*, 4, 419–426, 1993.

28. Kjaeheim, K. and Andersen, A., Cancer incidence among waitresses in Norway, *Cancer Causes Control*, 5, 31–37, 1994.

29. Lambert, W., Samet, J., et al., Environmental tobacco smoke concentrations in no-smoking and smoking sections of restaurants, *Am. J. Public Health*, 83, 1339–1341, 1993.

30. Miesner, E., Rudnick, S., et al., Particulate and nicotine sampling in public facilities and offices, *J. Air Pollut. Control Assoc.*, 39, 1577–1582, 1989.

31. Morris, P.D., Lifetime excess risk of death from lung cancer for a US female never-smoker exposed to ETS, *Environ. Res.*, 68, 3–9, 1995.

32. Nafstad, P., Botten, G., Zahlsen, K., Nilsen, O.G., Silsand, T., and Kongerud, J., Comparison of three methods for estimating environmental tobacco smoke exposure among children aged between 12 and 36 months, *Int. J. Epidemiol.*, 24, 88–94, 1995.

33. Ott, W., Mathematical models for predicting indoor air quality from smoking activity, *Environ. Health Perspect.*, 107(Suppl. 2), 375–381, 1999.

34. Ott, W., Switzer, P., et al., Particle concentrations inside a tavern before and after prohibition of smoking: Evaluating the performance of an indoor air quality model, *J. Air Waste Manage. Assoc.*, 46, 1120–1134, 1996.

35. Repace, J.L., Jinot, J., Bayard, S., Emmons, K., and Hammond, S.K., Air nicotine and saliva cotinine as indicators of workplace passive smoking exposure and risk, *Risk Anal.*, 18(1), 71–83, 1998.

36. Samet, J.M., The epidemiology of lung cancer, *Chest*, 103, 20S–29S, 1993.

37. Sciacca, J., A mandatory smoking ban in restaurants: concerns versus experiences, *J. Commun. Health*, 21, 133–150, 1996.

38. Sciacca, J. and Eckrem, M., Effects of a city ordinance regulating smoking in restaurants and retail stores, *J. Commun. Health*, 18, 175–182, 1993.
39. Siegel, M., Involuntary smoking in the restaurant workplace. A review of employee exposure and health effects, *J. Am. Med. Assoc.*, 270(4), 490–493, 1993.
40. Sterling, T.D., Dimich, H., and Kobayashi, D., Indoor byproduct levels of tobacco smoke: a critical review of the literature, *J. Air Pollut. Control Assoc.*, 32, 250–259, 1982.
41. Sterling, T.D. and Weinkam, J.J., Smoking characteristics by type of employment, *J. Occup. Med.*, 18, 743–754, 1976.
42. Teschke, K., Hertzman, C., Van Netten, C., Lee, E., Morrison, B., Cornista, A., Lau, G., and Hundal, A., Potential exposure of cooks to airborne mutagens and carcinogens, *Environ. Res.*, 50, 296–308, 1989.
43. Trout, D., Decker, J., Mueller, C., Bernert, J.T., and Pirkle, J., Exposure of casino employees to environmental tobacco smoke, *J. Occup. Environ. Med.*, 40(3), 270–276, 1998.
44. U.S. Environmental Protection Agency, Respiratory Health Effects of Passive Smoking: Lung Cancer and Other Disorders, U.S. EPA, Office of Health and Environmental Assessment, Office of Research and Development, Washington, D.C., 1992.
45. Weber, A. and Fischer, T., Passive smoking at work, *Int. Arch. Occup. Environ. Health*, 47, 209–221, 1980.

15 Passive Smoking and Immune Functions

Anna Maria Castellazzi

CONTENTS

ABSTRACT

Passive smoking is a major threat to public health in Europe and the U.S. Children are exposed to environmental smoke when they live with smoking parents. Environmental smoke causes irritation and/or inflammation of the respiratory mucosa and also seems to modulate humoral and cellular immune activity. The immune system is affected by passive smoke in many functions. The author considers the effects of passive smoking on humoral and cellular immune functions, phagocyte's functions, and natural killer-cell activity. The observed reduction of natural killer-cell activity may be related to the increased incidence of lower respiratory tract illness and childhood cancer in infants from smoking mothers. Pregnant women and their relatives are to be encouraged to stop smoking as soon as pregnancy is diagnosed to avoid the consequences of smoke to newborns.

15.1 INTRODUCTION

Cigarette or tobacco smoking is still widespread in many countries: one in three adults in the U.S. smokes[1] and at least one parent, in about 52% of the families, is a current smoker in Italy.[2] Smokers inhale 25% of the smoke from each cigarette, but the remaining 75% is released into the environment. Non-smokers, mostly children, are

0-8493-0311-7/00/$0.00+$.50
© 2001 by CRC Press LLC

exposed to this environmental smoke. According to a study funded by the European Commission within the Anticancer Program, passive smoking causes the same effects on the health as active smoking and is responsible for the death of over 22,000 people each year in the European Union. This research was performed by a group of independent experts under the French "Comité National Contre le Tabagisme" and estimated that about 80% of Europeans under 15 years of age are exposed to environmental smoke and inhale the equivalent of one or more actively smoked cigarettes a day. Therefore, passive smoking is a major threat to public health in Europe and is a risk factor for many illnesses, especially in childhood.[3] The list of diseases that are associated with passive smoking is increasing and varies with the ages at which exposure occurs. The earliest harmful effects of passive smoking arise during intrauterine life. The consequences of maternal smoking during pregnancy are diminished birthweight[4,5] and increased perinatal mortality linked to the Sudden Infant Death Syndrome.[6–8] During infancy and childhood, exposure to passive smoking is associated with an increased risk of acute and chronic respiratory diseases, middle ear infections, and decreased lung function in infants, children, and adolescents.[9–14] The recent finding of altered serum thyroglobulin concentrations in infants whose parents were both smokers suggests that passive smoking also has an effect on thyroid function.[15] Furthermore, the gastrointestinal tract of children may be affected by the smoking habits of their parents. A case-control study determined that parental cigarette smoking is a risk factor for the development of esophagitis in children,[16] and an association has been ascribed between passive smoking and the development, in childhood, of inflammatory bowel disease, particularly Crohn's disease,[17] and also of acute appendicitis.[18] Bacterial meningitis in children should also be added to the growing list of illnesses associated with passive exposure to cigarette smoke.[19]

In adults, passive smoking seems to be one of the main risk factors for cardiovascular diseases[20, 21] and lung cancer.[22,23]

15.2   WHAT IS PASSIVE SMOKING?

Passive smoking, also called involuntary, secondhand, or environmental smoking, is the exposure of non-smokers to the tobacco smoke released by smokers or by smoldering cigarettes when the smoker is not puffing. The physicochemical composition of tobacco smoke, and notably its contents in toxic and carcinogenic substances, is the same in the secondary stream between puffs as in the primary stream released by smokers.[24] Tobacco contains irritant and carcinogenic substances such as polycyclic aromatic hydrocarbons, heterocyclic nitrates, phenols, nitrosamines, and metals like nickel. Environmental smoke is also an important source of toxic products: it releases nitric oxide, nitrosamines, and twice the amount of carbon monoxide and four times that of benzopyrene found in polluted air in industrialized environments.[25] Environmental tobacco smoke causes irritation and/or inflammation of the respiratory mucosa and also seems to modulate humoral and cellular immune activity.

15.3   IMMUNE FUNCTION AND SMOKE

Many studies in active smokers have demonstrated the immunosuppressive effects of tobacco smoke on both humoral and cellular immunity.[26,27] However, little information is available on the effects of passive smoking on immune function. Many reports are available of experimental studies performed in animals,[28,29] but human studies are relatively few, perhaps because of the difficulty in quantifying the exposure to environmental smoke. A significant number of pregnant women smoke. Moreover, since smoking is frequent in the home where young children spend most of their time, many of them have significant exposure to environmental tobacco smoke. The children of smoking parents are a natural model for studying the effects of exposure to environmental smoke. The fetus is exposed directly to the components of tobacco smoke absorbed from the bloodstream of the mother who smokes during pregnancy. After birth, infants are exposed via inhalation when their mother or other household members smoke in the home.[30] Children from families where both parents smoke more than 20 cigarettes per day are exposed to an average of half a cigarette per day based on nicotine equivalents, but 4 to 5 cigarettes per day based on nitrosodimethylamine equivalents.[31]

15.4 HUMORAL IMMUNE FUNCTION AND PASSIVE SMOKING

Environmental smoke effects on the immune system are similar in animals and humans. In animals exposed to tobacco smoke for only a few weeks, evidence of slight immunostimulation has generally been found; however, studies of subchronic and chronic exposure have revealed that immunosuppressive changes develop. Antibody production can be suppressed, and, as a consequence, circulating immunoglobulin levels decrease, except for IgE which increases.[32,33] In a study of effects of parental smoking on IgE and IgD levels in cord blood and the development of infant allergy, Magnusson showed that maternal smoking was associated with significantly higher levels of both IgE and IgD in cord blood and the increases were greater in newborns with a negative biparental history for allergy.[34] Monafo et al.[35] and Ronchetti et al.[36] also described significantly higher total IgE levels in children whose parents were smokers, while Ownby et al.[37] demonstrated that maternal and paternal smoking is associated with a higher level of IgD, but not of IgE in cord serum. Paganelli et al. found no differences in IgG, IgM, and IgG_1 levels in the cord blood from smoking and non-smoking mothers, but suggested that some regulatory mechanism of T-cell proliferation may be suppressed by maternal smoking.[38] A study performed by a group from the University of Dublin evaluated the association between levels of umbilical cord and maternal IgG subclasses with maternal smoking habits in pregnancy and the infection rate in the neonate's first year of life:[39] they observed a higher serum level of IgG_1 in the smoking mothers than the non-smoking mothers, and a weak but significant inverse correlation between umbilical cord IgG_2 levels and the infection rate in the first year of life in the infants of smoking mothers. Wagner et al.[40] determined serum IgG, IgA, IgM, and IgE and salivary (secretory) IgA (sIgA) concentrations of a group of school children in autumn and spring. They found significantly higher IgE levels in the children of families with smokers in the autumn and lower IgA levels in serum, but higher levels of sIgA in the saliva of the children who were passive smokers in the spring.[40] The observation of Kilian et al.[41] that children with a history of atopic disease show significantly higher levels of cleaved secretory IgA in nasopharyngeal secretions is interesting. In fact, these children harbor significantly higher proportions of IgA1 protease-producing bacteria (mainly *Streptococcus mitis* biovar 1) than healthy children. The percentage of IgA1 protease-producing bacteria is significantly related to passive smoking, which may stimulate the precocious and heavier pharyngeal colonization in atopic infants with smoking parents.

15.5   CELLULAR IMMUNE FUNCTION AND PASSIVE SMOKING

Most studies on the effect of environmental smoke or its components on the function of T- and B-lymphocytes have been performed in mice and rats. Exposure of animals to the substances in tobacco smoke can have severe immunosuppressive effects mainly on T-lymphocyte function. T-lymphocytes lose their capacity to cooperate with normal B-lymphocytes and macrophages in response to antigens such as sheep red blood cells. B-lymphocytes are also directly affected by smoke products, but less so than T-lymphocytes.[30] Chronic exposure to cigarette smoke or nicotine induces T-cell unresponsiveness. This apparent T-cell anergy may account for or contribute to the immunosuppressive and anti-inflammatory properties of cigarette smoke/nicotine. Nicotine-induced immunosuppression may result from nicotine's direct effects on lymphocytes and/or to indirect effects on the neuroendocrine system.[42]

In humans, the most interesting studies concern the effects of cigarette smoke on the children of smoking mothers. Paganelli et al. found significantly higher thymidine uptake in phytohemoagglutinin-stimulated cultures of cord blood lymphocytes from smoking mothers and postulated a suppressor effect of maternal smoking on some regulatory mechanism of T-cell proliferation to explain this finding.[38] The effects of cigarette smoking on T-cell subsets have been studied only in active smokers with different findings. While some early reports suggest a decrease in the ratio of T-helper inducer cells to T-suppressor cytotoxic cells,[43] others recently have shown an increase in this ratio[44,45] and also in that of CD4+ memory to naive cells.[46] HIV-1 patients prior to seroconversion demonstrated an increase in the proportion of CD4+ cells. In contrast, HIV-1 seroconversion is associated with a dramatic decrease in CD4+ lymphocyte percentage.[47] The frequency of sister chromatid exchange induction in lymphocytes has been used to assess health in people. This frequency was examined in lymphocytes from blood samples of people working with smokers[48] and was found to be similar to that observed in lymphocytes from non-smokers. However, if the lymphocytes were pretreated with mitomycin C, cells from people exposed to smoke in the environment had a higher frequency of sister chromatid exchange than those not exposed; these results suggest that passive smoke does not induce directly sister chromatid exchange, but may, with other factors, increase the frequency of sister chromatid exchange. Recently, a Swedish group described increased interleukin-16 (IL-16) levels in the airways of tobacco smokers. IL-16, a cytokine released from inflammatory cells and bronchial epithelial cells, recruits and activates CD4+ T-lymphocytes. In this study, the level of IL-16 in the bronchoalveolar lavage of smokers correlated negatively with the percentage of

CD4+ T-cells and led the authors to speculate that IL-16, which is a selective chemoattractant for CD4+ T-cells *in vitro*, may recruit CD4+ T-cells to the airway wall, thereby depleting peripheral blood peripheral blood of CD4+ T-cells.[49]

15.6 PHAGOCYTES AND PASSIVE SMOKING

Inhalation of tobacco smoke produces change in the immune system at the level of the pulmonary system. The lavage fluid from cigarette smoke-exposed animals shows an increase of two to five times in the number of macrophages.[50] Possible explanations for the recruitment of macrophages in the lungs of smoke-exposed animals include migration and differentiation of blood monocytes into the alveoli, a local proliferation and differentiation of monocytes due to macrophage cytokines, a decreased local adherence of the alveolar macrophages, or a reduction in the transport of macrophages out of the lungs.[51] Moreover, chronic exposure to cigarette smoke causes a neutrophilia associated with a sequester of polymorphonuclear (PMN) leukocytes in pulmonary microvessels which can contribute to the alveolar .wall damage associated with smoke-induced lung emphysema in the rabbit model.[52] Phagocytic function was evaluated in mice chronically exposed to cigarette smoke, and the effects of *in vitro* exposure to cigarette smoke on macrophage activity were also assessed. Macrophages exposed to cigarette smoke *in vitro* initially had a depressed phagocytic rate, but if phagocytosis over a prolonged period was measured, it was enhanced over the rate of the control macrophage.[53] Edwards et al.[54] recently studied the effects of short-term exposure to cigarette smoke condensate prepared from sidestream and mainstream cigarette smoke on macrophage basal metabolism and responsiveness to two different stimuli, bacterial lipopolysaccharide and interferon-gamma (IFN-γ). Mainstream and sidestream cigarette smoke had similar effects on macrophages, enhancing macrophage basal metabolism and responsiveness to lipopolysaccharide and suppressing their responsiveness to IFN-γ.

15.7 NATURAL KILLER ACTIVITY AND PASSIVE SMOKING

Natural killer activity represents an important innate immune defense mechanism against viral infections and in tumor surveillance. It is correlated with healthy lifestyle habits in humans, such as avoidance of tobacco smoking.[55,56] Cigarette smoke has been reported to reduce natural killer activity in adults[57] and is associated with a decrease in the number and proportion of circulating natural killer cells,[58] which appears to persist for many years after smoking cessation. This persistence of an apparent toxic effect of cigarette smoking on a putative anti-cancer component of the immune system many years after smoking cessation is of particular concern for a specific effect of smoking on natural killer cells or their progenitor. Little is known about the effect of passive smoking on such an important immune function as natural killer activity. The aim of the study, performed in our laboratory,[59] was to investigate whether maternal smoking may be correlated with an alteration of natural killer-cell activity in cord blood cells obtained from three groups of term newborns with mothers who had different levels of exposure to cigarette smoke: mothers who were never smokers and not exposed to involuntary smoking (controls); mothers who smoked at least 10 cigarettes per day during pregnancy (active smokers); and mothers exposed to tobacco smoke from their partners, who smoked more than 10 cigarettes per day (passive smokers). No significant difference was found in the number of circulating natural killer cells as a percentage of CD16+ or CD56+ cells (Table 15.1).

Our results showed a significant reduction of natural killer-cell activity, assayed in a 4-h ^{51}Cr-release cytotoxicity assay according to the standard procedure, with K562 used as target cells at three different target to effectors cells ratios (1:100, 1:30, 1:10), in cord blood from infants with both active and passive smoking mothers compared with controls (Table 15.2). The difference from controls and active smokers, at the target/effector cells ratio of 1:100, had a $p = .021$, and between controls and passive smokers, $p = .037$; at the target/effector cells ratio of 1:30, $p = .027$ between controls and active smokers, and between controls and passive smokers, $p = .046$; at the target/effector cells ratio of 1:10, $p = .037$ between controls and active smokers.

We evaluated also cytokine production (IL-2, IL-4, IFN-γ, and sIL-2R) by cord blood lymphocytes obtained from the same three groups of newborns and found no difference in *in vitro* production of IL-2 and IL-4 in the three groups, whereas the production of IFN-γ and sIL-2R was significantly higher in infants from active-smoking mothers (Table 15.3) (unpublished data).

The observed reduction of natural killer-cell activity may be correlated to the increased incidence of lower respiratory tract illness and childhood cancer in infants from smoking mothers.

A decreased production of IFN-γ by human neonatal cells has been previously described.[60-62] We found the same results in controls and infants from passive-smoking mothers, whereas in active smokers the amount of IFN-γ was significantly increased. Moreover, the production of sIL-2R in newborns from smoking mothers was significantly increased. We can hypothesize that these data represent the responsiveness of the immune system to the toxic effects of tobacco products.

15.8 CONCLUSIONS

Many studies show direct effects of tobacco smoke on the immune system, and many of the chemical components of tobacco smoke have been demonstrated to have immunomodulatory effects. On the other hand, little is known about the immunological effects of passive smoking. Since it has been determined that many of the chemical components exist in environmental smoke in higher concentrations than in mainstream smoke, it is important to study their immunological effects in experimental models. Many species of animals have been used in determining the effects of tobacco smoke on the immune system.

The effects of tobacco smoke on the immune system are similar in humans and animals. Animals exposed to tobacco smoke for small periods of time generally exhibit a slight immunostimulation. However, subchronic and chronic exposure studies indicate that immunosuppressive changes may develop.

TABLE 15.1

Percentage of CD16+ and CD56+ Cells in Cord Blood of Newborns from Mothers Who Never Smoked or Who Were Active or Passive Smokers

	Controls	Active smokers	Passive smokers
CD16+ cells	13.6 ± 4.3%	12.0 ± 4.9%	12.6 ± 4.3%
CD56+ cells	14.7 ± 4.1%	13.5 ± 5.3%	13.8 ± 4.4%

TABLE 15.2

Natural Killer Cells Activity in Cord Blood

	Controls	Active smokers	Passive smokers	F value (ANOVA)	p value (ANOVA)
NK 100	38.6 ± 16.8	23.5 ± 14.3	26.9 ± 14.4	4.1	.023
NK 30	22.5 ± 12.7	12.0 ± 9.3	14.5 ± 10.8	3.8	.03
NK 10	13.0 ± 10.6	5.0 ± 6.7	8.1 ± 8.0	3.3	.044

Note: Results are expressed as percentage of specific lysis and reported as means ± SD. NK 100 (100 effectors/1 target), NK 30 (30 effectors/1 target), and NK 10 (10 effectors/1 target) are the three effector/target ratios used in the cytotoxity test.

TABLE 15.3

In Vitro **Cytokine Production by Cord Blood Lymphocytes**

	IL-2	IL-4	IFN-γ	sIL-2R
Controls	74.3 ± 73.0	9.4 ± 12.2	99.9 ± 75.9	9,230.0 ± 8,261.9
Active smokers	65.1 ± 40.0	10.9 ± 11.5	387.9 ± 410.9	14,927 ± 8,143.0
Passive smokers	41.1 ± 35.5	8.5 ± 7.7	109.1 ± 90.3	8,687.9 ± 5,764.1

Note: The values are expressed in UI/ml. The production of IFN-γ in infants from active-smoking mothers is higher compared with controls ($p < .05$). The sIL-2R production is also higher ($p < .05$).

The cells from cord blood of newborns from smoking mothers are a good model for studying the effects of passive smoking on immune functions. Few reports suggested the possible effects of cigarette smoking on the fetal immunhematological system, showing a suppressor effect of maternal smoking on the regulatory mechanism of T-lymphocytes proliferation, an increase of cord serum immunoglobulin levels, and a significant lower level of neutrophils that can be the reason for the enhanced incidence of postnatal infections usually observed in these infants. Natural killer activity seems to be the most affected function in newborns from smoking mothers, in agreement with previous observations of a reduced natural killer activity in cigarette-smoking adults. Interestingly, we found the same effect in infants whose mothers were exposed to tobacco smoke from their partners, although the reduction of natural killer activity in infants from passive smokers was less impressive than in newborns of active-smoking mothers. The reduction of natural killer activity observed in our study may be involved in the increased incidence of lower respiratory tract viral infections and childhood cancer reported in infants exposed to passive smoking.

There is now sufficient evidence that health problems in children may be related to maternal and, to a lesser degree, paternal smoking during pregnancy and, after birth, to exposure to environmental tobacco smoke in the home and day care centers. Doctors must encourage women and their friends and relatives to stop smoking as soon as pregnancy is diagnosed or, if possible, before pregnancy.

ACKNOWLEDGMENT

I wish to thank Virginia Monafo and Patrizia Comoli for the critical reading of the manuscript and M. Antonietta Avanzini for her suggestions.

REFERENCES

1. Lejeune, H.B. and Cote, D.N., Passive smoking, *J. La State Med. Soc.*, 147(10), 444–447, 1995.
2. S.I.D.R.I.A. (Studi Italiani sui Disturbi Respiratori nell'Infanzia e l'Ambiente), Parental smoking, asthma and wheeze among children and adlescent, *Epid. Prev.*, 22, 146–154.
3. Watson, R., Passive smoking is major threat, *Br. Med. J.*, 316, 9, 1998.
4. Gueguen, C., Lagrue, G., and Janse-Marec, J., Retentissement sur le foetus et l'énfant du tabagisme pendant la grossesse (Effect of smoking on the fetus and the child during pregnancy), *J. Gynecol. Obstet. Biol. Reprod.*, 25(4), 424–425, 1996.
5. Chen, L.H. and Petitti, D.B., Case-control study of passive smoking and the risk of small-for-gestational-age at term, *Am. J. Epidemiol.*, 142(2), 158–165, 1995.
6. Haglund, B., Cnattingius, S., and Otterblad-Olausson, P., Sudden infant death syndrome in Sweden,1983-1990: season at death, age at death and maternal smoking, *Am. J. Epidemiol.*, 142(6), 619–624, 1995.
7. Anderson, H.R. and Cook, D.G., Passive smoking and sudden infant death syndrome: review of the epidemiological evidence, *Thorax*, 52(11), 1003–1009, 1997.
8. Dybing, E. and Sanner, T., Passive smoking, sudden infant death syndrome (SIDS) and childhood infections, *Hum. Exp. Toxicol.*, 18(4), 202–205, 1999.
9. Di Benedetto, G., Passive smoking in childhood, *J. R. Soc. Health*, 115, 13–16, 1995.
10. Charlton, A., Children and passive smoking: a review, *J. Fam. Pract.*, 38(3), 267–277, 1995.
11. Kurz, H., Frischer, T., Huber, W.D., and Gotz, M., Gesundheitsschaden durch Passivrauchen bei Kindern (Adverse health effects in children caused by passive smoking), *Wien. Med. Wochenschr.*, 144(22–23), 531-534, 1994.
12. Strachan, D.P. and Cook, D.G., Health effects of passive smoking. 6. Parental smoking and childhood asthma: longitudinal and case-control studies, *Thorax*, 53(3), 204–212, 1998.
13. Strachan, D.P. and Cook, D.G., Health effects of passive smoking. 1. Parental smoking and lower respiratory illness in infancy and early childhood, *Thorax*, 52(10), 905–914, 1997.

14. Margolis, P.A., Kayyes, L.L., Greenberg, R.A., Bauman, K.E., and LaVange, L.M., Urinary cotinine and parent history (questionnaire) as indicators of passive smoking and predictors of lower respiratory illness in infants, *Pediatr. Pulmonol.*, 23(6), 417–423, 1997.

15. Gasparoni, A., Autelli, M., Ravagni-Probizer, M.F., Bartoli, A., Regazzi-Bonora, M., Chirico, G., and Rondini, G., Effect of passive smoking on thyroid function in infants, *Eur. J. Endocrinol.*, 138(4), 379–382, 1998.

16. Shabib, S.M., Cutz, E., and Sherman, P.M., Passive smoking is a risk factor for esophagitis in children, *J. Pediatrics*, 127(3), 435–437, 1995.

17. Lashner, B.A., Shaheen, N.J., Hanauer, S.B., and Kirschner, B.S., Passive smoking is associated with an increased risk of developing inflammatory bowel disease in children, *Am. J. Gastroenterol.*, 88(3), 356–359, 1993.

18. Montgomery, S.M., Pounder, R.E., and Wakefield, A.J., Smoking in adults and passive smoking in children are associated with acute appendicitis (letter), *Lancet*, 353(9150), 379, 1999.

19. Bredfeldt, R.C., Cain, S.R., Schutze, G.E., Holmes, T.M., and McGhee, L.A., Relation between passive tobacco smoke exposure and the development of bacterial meningitis in children, *J. Am. Board Fam. Pract.*, 8(2), 95–98, 1995.

20. Leone, A., Cigarette smoking and health of the heart, *J. R. Soc. Health*, 115(6), 354–355, 1995.

21. Whidden, P., Passive smoking (letter), *Lancet*, 350(9070), 73, 1997.

22. Merletti, F., Richiardi, L., and Boffetta, P., Effetti per la salute del fumo passivo (Health effects of passive smoking), *Med. Lav.*, 89(2), 149–163, 1998.

23. Engeland, A. and Bjorge, T., Passiv royking og kreftrisiko (Passive smoking and risk of cancer), *Tidsskr. Nor. Laegeforen.*, 118(14), 2183–2186, 1998.

24. Tredaniel, J., Zalcman, G., Boffetta, P., and Hirsch, A., Le tabagisme passif. Effets sur la santé. (Passive smoking. Effects on health), *Rev. Prat.*, 43(10), 1230–1234, 1993.

25. Lagrue, G., Branellec, A., and Lebargy, F., La toxicologie du tabac (Toxicology of tobacco), *Rev. Prat.*, 43, 1203–1207, 1993.

26. Byron, K.A., Varigos, G.A., and Wootton, A.M., IL-4 production is increased in cigarette smokers, *Clin. Exp. Immunol.*, 95, 333–336, 1994.

27. Laan, M., Qvarfordt, I., Riise, G.C., Andersson, B.A., Larsson, S., and Linden, A., Increased levels of IL-16 in the airways of tobacco smokers: relationship with peripheral blood T lymphocytes, *Thorax*, 54, 911–916, 1999.

28. Jacob, C.V., Stelzer, G.T., and Wallace, J.H., The influence of cigarette tobacco smoke products on the immune response. The cellular basis of immunosuppression by a water-soluble condensate of tobacco smoke, *Immunology*, 40, 621–627, 1980.

29. Geng, Y., Savage, S.M., Razani-Boroujerdi, S., and Sopori, M.L., Effects of nicotine on the immune response. II. Chronic nicotine treatment induce T cell anergy, *J. Immunol.*, 156, 2384–2390, 1996.

30. U.S. Environmental Protection Agency, Respiratory health effects of passive smoking: lung cancer and other disorders, in Smoking and Tobacco Control, Monograph 4, NIH Publication No 93-3605, U.S. Department of Health and Human Services, Public Health Service, National Institutes of Health, U.S. Environmental Protection Agency, Office of Research and Development, Office of Air and Radiation, Washington, D.C., 1993.

31. Cook, D.G. et al., Passive exposure to tobacco smoke in children aged 5-7 years: individual, family and community factors, *Br. Med. J.*, 308, 384–389, 1994.

32. Holt, P.G. and Keast, D., Environmentally induced changes in immunobiological function: acute and chronic effects of inhalation of tobacco smoke and other atmospheric contaminants in man and experimental animals, *Bacteriol. Rev.*, 41, 205, 1977.

33. Johonson, J.D., Houchhens, D.P., Kluwe, W.M., Craig, D.K., and Fisher, G.L., Effects of mainstream and environmental tobacco smoke on the immune system in animals and humans: a review, *Crit. Rev. Toxicol.*, 20(5), 369–395, 1990.

34. Magnusson, C.G.M., Maternal smoking influences cord serum IgE and IgD levels and increases the risk for subsequent infant allergy, *J. Allergy Clin. Immunol.*, 78, 898–904, 1986.

35. Monafo, V., De Amici, M., Quaglini, S., Pugni, L., Ottolenghi, A., Terracciano, L., and Burgio, G.R., Fumo passivo e livelli sierici di IgE totali in età scolare (Passive smoking and total serum IgE levels in school-age children, *Riv. Ital. Pediatr.*, 21, 675–680, 1995.

36. Ronchetti, R., Macri, F., Ciofetta, G., Indinnimeo, G.L., Cutrera, R., Bonci, E., Antognoni, G., and Martinez, F.D., Increased serum IgE and increased prevalence of eosinophilia in 9-year-old children of smoking parents, *J. Allergy Clin. Immunol.*, 86, 400–407, 1990.

37. Ownby, D.R., Johonson, C.C., and Peterson, E.L., Maternal smoking does not influence cord serum IgE or IgD concentrations, *J. Allergy Clin. Immunol.*, 88, 55–60, 1991.

38. Paganelli, R., Ramadas, D., Layward, L., Harvey, B.A.M., and Soothill, J.F., Maternal smoking and cord blood immunity function, *Clin. Exp. Immunol.*, 36, 256–259, 1979.

39. Cervi, P.L. and Feighery, C., IgG subclasses in foetal cord and maternal serum: associations with infections in infancy and smoking in pregnancy, *J. Clin. Lab Immunol.*, 34(1), 23–30, 1991.

40. Wagner, V., Wagnerova, M., Hakova, L., Zavazal, V., and Wokounova, D., Humoral defending mechanism in children of smoking parents, *Czech. Med.*, 10(2), 70–78, 1987.

41. Kilian, M., Husby, S., Host, A., and Halken, S., Increased proportions of bacteria capable of cleaving IgA1 in the pharinx of infants with atopic disease, *Pediatr. Res.*, 38(2), 182–186, 1995.

42. Sopori, M.L. and Kozak, W., Immunomodulatory effects of cigarette smoke, *J. Neuroimmunol.*, 83(1–2), 148–156, 1998.

43. Miller, L.G. and Ginns, L.C., Reversible defect of immunoregulatory T-cells in cigarette smokers, *Am. Rev. Respir. Dis.*, 125(4), 60, 1982.

44. Tollerud, D.J., Clark, J.W., Morris Brown, L., Neuland, C.Y., Mann, D.L., Pankiw-Trost, L.K., Blattner, W.A., and Hoover, R.N., The effects of cigarette smoking on T cell subsets. A population-based survey of healthy caucasians, *Am. Rev. Respir. Dis.*, 139, 1446–1451, 1989.

45. Freedman, D.S., Flander, W.D., Barboriak, J.J., Malarcher, A.M., and Gates, L., Cigarette smoking and leukocyte subpopulation in men, *Ann. Epidemiol.*, 6, 299–306, 1996.

46. Tanigawa, T., Araki, S., Nakata, A., Kitamura, F., Yasumoto, M., Sakurai, S., and Kiuchi, T., Increase in memory (CD4* CD229+ and CD4+CD45ro+ T and naive (CD4*CD45RA+) T-cell subpopulations in smokers, *Arch. Environ. Health*, 53, 378–383, 1998.

47. Park, L.P., Margolick, J.B., Giorgi, J.V., Ferbas, J., Bauer, K., Kaslow, R.R., and Munoz, A., Influence of HIV-1 infection and cigarette smoking on leukocyte profile in homosexual men, *J. Acquir. Immune Defic. Syndr.*, 5(11), 1124–1130, 1992.

48. Morimoto, K., Miura, K., Sato, M., Kikuchi, H., and Koizumi, A., Induction of sister chromatid exchanges in human lymphocytes I. Cell stage dependency and effect of blood donor's passive smoking, *Jpn. J. Hum. Genet.*, 29, 211, 1984.

49. Laan, M., Qvarfordt, I., Riise, G.C., Andersson, B.A., Larsson, S., and Linden, A., Increased levels of interleukin-16 in the airways of tobacco smokers: relationship with peripheral blood T lymphocytes, *Thorax*, 54, 911–916, 1999.

50. Matulionis, D.H., Kimmel, E., and Diamond, L., Morphologic and physiologic response of lungs to steroid and cigarette smoke: an animal model, *Environ. Res.*, 36, 298, 1985.

51. Matulionis, D.H., Reaction of macrophages to cigarette smoke. II. Immigration of macrophages to the lung, *Arch. Environ. Health*, 34, 298, 1979.

52. Terashima, T., Klut, M.E., English, D., Hards, J., Hogg, J.C., and van Eeden, S.F., Cigarette smoking causes sequestration of polymorphonuclear leukocytes released from the bone marrow in lung microvessels, *Am. J. Respir. Cell Mol. Biol.*, 20(1), 171–177, 1999.

53. Thomas, W.R., Holt, P.G., and Keast, D., Cigarette smoke and phagocyte function: effect of chronic exposure in vivo and acute exposure in vitro, *Infect. Immun.*, 20(2), 468–475, 1978.

54. Edwards, K., Braun, K.M., Evans, G., Sureka, A.O., and Fan, S., Mainstream and sidestream cigarette smoke condensates suppress macrophage responsiveness to interferon gamma, *Hum. Exp. Toxicol.*, 18(4), 233–240, 1999.

55. Kusaka, Y., Kondou, H., and Morimoto, K., Healthy lifestyles are associated with higher natural killer cell activity, *Prev. Med.*, 21(5), 602–615, 1992.

56. Nakachi, K. and Imai, K., Environmental and physiological influences on human natural killer cell activity in relation to good health practices, *Jpn. J. Cancer Res.*, 83(8), 798–805, 1992.

57. Phillips, B., Marshall, M.E., Brown, S., and Thompson, J.S., Effect of smoking on human natural killer cell activity, *Cancer*, 56, 2789–2792, 1985.

58. Tollerud, D.J., Clark, J.W., Morris Brown, L., Neuland, C.Y., Mann, D.L., Pankiw-Trost, L.K., Blattner, W.A., and Hoover, R.N., Association of cigarette smoking with decreased numbers of circulating natural killer cells, *Am. Rev. Respir. Dis.*, 139, 194–198, 1989.

59. Castellazzi, A.M., Maccario, R., Moretta, A., De Amici, M., Gasparoni, A., Chirico, G., and Rondini, G., Effect of active and passive smoking during pregnancy on natural killer-cell activity in infants, *J. Allergy Clin. Immunol.*, 103, 172–173, 1999.

60. Wilson, C.B., Westall, J., Lewis, D.V., Dower, S.K., and Alpert, A.R., Decreased production of interferon-gamma by human neonatal cells, *J. Clin. Invest.*, 7, 860–867, 1986.

61. Bryson, Y.J., Winter, H.S., Gard, S.E., Fisher, T.J., and Stiem, E.S., Deficiency of immune interferon production by leukocytes of normal newborns, *Cell Immunol.*, 55, 191–200, 1980.

62. Wakasugi, N. and Virelizier, J.L., Defective IFN-γ production in the human neonate. I. Dysregulation rather than intrinsic abnormality, *J. Immunol.*, 134(1), 167–171, 1985.

16 Environmental Tobacco Smoke and Lung Function

Shengjun Wang and Mark L. Witten

CONTENTS

16.1 INTRODUCTION

Environmental tobacco smoke (ETS) is an indoor environmental problem and poses a high health risk to humans. The respiratory system may be greatly compromised by direct exposure to ETS. Substantial evidence has accumulated that ETS exposure affects passive smokers' lungs directly by irritation and functional impairment.[1–7] Exposure to ambient concentrations of ETS induces a series of respiratory symptoms such as cough, wheeze, airway inflammation, and airway hyperreactivity. The most vulnerable subjects are newborns, young children, and possibly the fetus while *in utero*. Exposure to the mother's active smoking is a growing health concern with resulting pulmonary problems in children.[8–11] For children under 18 months of age, ETS exposure is thought to be responsible in the U.S. for up to 300,000 cases annually of bronchitis, bronchiolitis, and pneumonia.[2] ETS also contributes overall to chronic obstructive pulmonary diseases (COPD).[2]

ETS is a diluted mixture of cigarette sidestream smoke (SS, ~85%) and mainstream smoke (MS, ~15%). SS is generated at lower burning temperatures and has a different chemical composition when compared with MS. ETS also undergoes rapid aging, a process that may change its physical-chemical properties.[12] Airborne nicotine and particulate matter are important parameters associated with ETS pollution. It has been measured that exposure to airborne nicotine is usually 10 $\mu g/m^3$ or less; however, in badly ventilated rooms or in cars, it may be five- to tenfold higher.[12] Exposure to ETS-related particulate matter ranges from 18 to 64 $\mu g/m^3$ total suspended particles (TSP).[2,12] Surveys have shown that 44% of male and 33% of female nonsmokers expose others to ETS either at home or at work in U.S.[13] Moreover, 88% of nonsmokers had detectable levels of serum cotinine (a metabolic

0-8493-0311-7/00/$0.00+$.50
© 2001 by CRC Press LLC

product of nicotine). The data from European studies suggest that a greater proportion of the population is exposed to higher levels of ETS than in the U.S. It also is estimated that a 20 cigarettes per day smoker, when compared with an 8 h/day ETS exposure nonsmoker, receives only a 1.5- to 4-fold higher dose of certain gaseous phase constituents of cigarette smoke, such as carbon monoxide, formaldehyde, volatile nitrosamines, and benzene.[14]

16.2 HUMAN INVESTIGATIONS

The association between ETS exposure and lung dysfunction was investigated by a number of epidemiological studies from different countries. It has been shown that people exposed to ETS have an increased frequency of respiratory symptoms and a reduction in lung function.[2,15–19] The evidence is particularly strong in newborns, infants, and preschool children. Exposure to ETS causes increased coughing,[20–22] wheezing,[21] sputum production,[21] and respiratory illnesses;[8,20,22,23] decreased forced expiratory volume in 1 s (FEV$_1$),[10] FEV$_1$/forced vital capacity (FVC),[24] and maximal midexpiratory flow rate (FEF$_{25-75}$);[10] and increased airway reactivity.[11,25,26] Such responses would imply a longer lasting and less readily reversed reaction, e.g., chronic congestion of the bronchial mucosa or hypersecretion of mucus.

There are growing supporting data available on the adverse effects of ETS on adult lung function.[2] An early study reported that ETS exposure induced a decrease of expiratory airflow at 25% of vital capacity. A significant increase in residual volume and functional residual capacity was evident, which suggested a small airway spasm.[27] A study of young people found that FEF$_{25-75}$ of a young man 20 years of age who had never smoked and always lived at home would be 800 ml less if both his parents smoked than if they did not. This decrease of FEF$_{25-75}$ was related to an index of cumulative lifetime environmental exposure to tobacco smoke at home, after taking into account the effects of cumulative exposure at work as well as age, height, body size, respiratory pressures, and cooking fuels used at home. In women, the diffusing capacity of the lung decreased in relation to cumulative exposure to tobacco smoke at work, after accounting for the effects of cumulative lifetime exposure at home and the other factors mentioned previously. A large Swiss study in adult nonsmokers (n = 4197) indicated that ETS, in an exposure-related manner after adjusting for various potential confounding variables, was significantly associated with wheezing, shortness of breath, bronchitis symptoms, and asthma, all with significant odds ratios (OR) in the 1.5 to 1.9 range.[15] Using a probability sample survey design, another investigation (n = 400) found a significant OR = 1.9 associated with COPD for nonsmokers exposed at home at a level of more than one pack of cigarettes per day.[16] The results were supported by a large prospective study in the U.S. which found a significant relative risk of 1.72 for symptoms of COPD among nonsmoking adults (n = 3914) with both childhood and adult home exposure to ETS.[17] Among 1033 adults aged 40 to 69 years living in Beijing, China, Xu and Li found significantly lower levels of FEV$_1$ and FVC, after controlling for potential confounders, among subjects exposed to ETS at home or work compared with subjects not exposed.[29] Other controlled studies found that in ETS-sensitive people even short exposure to low concentrations of ETS causes significant impairment of lung function (FEV_1).[18,19]

There is strong evidence that ETS exposure causes changes in immature lung in childhood. Young children exposed to ETS have decreased FEV_1,[9,10] FEV_1/FVC,[24] FEF_{25-75},[10] and $MMEF$[30] and increased airway reactivity.[11,25,26] In a study of 3285 infants in Shanghai, there was a 4.5-fold increase in respiratory hospitalizations in low-birthweight infants during their first 18 months of life if they were raised in a household where more than 20 cigarettes per day were smoked. A longitudinal study showed that exposure to ETS had a dose-response relationship on lung growth (as measured by FEV_1) in school children.[31] Children exposed to ETS in early life have faster rates of lung function decline in adult life. Also, maternal smoking has rather larger effects on child lung mechanics than paternal smoking. The maternal–child effect is due to not only that mothers are physically closer to their children, but also that the growing lung is more vulnerable to ETS challenge. Children whose mothers smoked during their children's first 2 years of life had lower lung function, possibly characterized by airway impairment.

In utero exposure to cigarette smoking influences lifetime lung function after birth. Martinez et al. noted that airway hyperreactivity was present in 70% of children whose mothers smoked during pregnancy, as compared with only 29% of children whose mothers did not smoke during pregnancy.[11] Hanrahan et al. showed that expiratory flow at FRC was 74.3 ml/s in 4-week-old infants born of continuous smokers vs. 150 ml/s in infants born of nonsmokers.[32] The 75 children of mothers who smoked during pregnancy, but not after birth, were found to have an 11% lower FEF_{25-75} than the never-exposed children. A study of 8863 children by Cunningham et al. found that their lung function at 8 to 12 years of age was lower if the mother smoked during pregnancy.[33] These studies suggest that lung function is not only compromised by postnatal exposure to ETS, but also constructively damaged by prenatal exposure to ETS from the mother. A recent Norwegian study of 803 infants demonstrated the effects of prenatal exposure to ETS, measured by tidal flow-volume loops, compliance, and resistance at 2.7 d after birth.[34] There is a possibility that the lung development in children associated with prenatal exposure to tobacco smoking will translate into lung function decline in adults. Such long-lasting reductions, if continually challenged by air pollutants, are likely to impact on the rates of development of COPD.

16.3 ANIMAL STUDIES

There are limited reports of ETS in the toxicology literature on pulmonary function. These studies provide data that exposure to SS adversely affects pulmonary mechanics and airway responsiveness to pharmacological challenges. In C57BL/6 mice exposed to SS of two cigarettes per day over a period of 45 d, SS exposure significantly increased pulmonary resistance (R_L) while it decreased pulmonary dynamic compliance C_{dyn}.[35] Similarly, using the isolated-perfused lung model in Duncan-Hartley guinea pigs exposed daily to SS (6 h/d, 5 d/week) from day 8 to about day 43 of life, SS exposure significantly increased baseline C_{dyn} by 17% and reduced the capsaicin-induced change in R_L and C_{dyn}. It was also observed that a statistically significant 24% decrease in C_{dyn} in rats exposed to diluted SS both *in utero* and postnatally occurred. Moreover, a study showed that rats exposed to SS from day 2 to week 11 of life developed airway hyporesponsiveness to serotonin.

There are multiple mechanisms that have been proposed for the respiratory effects of passive smoking. These include acute respiratory irritation, mucosal edema, decreased mucociliary clearance, chronic cough, bronchial hyperreactivity, and lower respiratory infection. ETS-induced pathophysiological responses were also involved in a number of chemical mediators, including oxidative stress, cytokine, chemokine, and eicosanoid production. However, an underlying mechanism(s) of action cannot be quantitatively linked to health effects in humans at plausible levels of exposure. There is not one chemical mediator responsible for lung dysfunction, but rather a complex interplay exists among multiple preinflammatory mediators.

Recent studies have moved toward exploration of potential mechanisms that underlie etiologies for the increased respiratory symptoms and lung function decline reported in children. One potential mechanism is the role of the pulmonary afferent neural system in the ETS-induced lung dysfunction. C-fibers are known to be stimulated by tobacco smoke[36,37] and by components of ETS such as nicotine,[38] acrolein,[39] and oxidants.[40] The hypothesis studied was that ETS exposure would attenuate the activity of the nerve fibers responsible for the defense reflex, leaving lungs more vulnerable to noxious agents.[41] When bronchopulmonary C-fibers are activated, the nerve impulses travel to the central nervous system, resulting in rapid shallow breathing, cough, and a cholinergically mediated bronchoconstriction. The nerve impulses also cause a local release of tachykinins, which interact with neurokinin (NK) receptors to cause airway mucus secretion, pulmonary permeability, and neural inflammation. In our laboratory, a role of SP in ETS-induced lung dysfunction is currently under study. It is speculated that in the early phase highly sustained SP levels may activate preinflammatory cytokines, eicosanoids, and oxide radical cross-talks due to SS-altered affinity of NK receptors following ETS exposure. In the second phase, we speculate that C-fibers will fail to respond to repeated ETS challenges.

On the basis of the previous brief review, it appears that the adverse effects of ETS on lung function have been characterized. However, actual human environmental atmospheres are not available as required for animal experimentation. The effects and action mechanisms of ETS on lung function are not fully understood. Using Hill's nine criteria for causal association with ETS, specificity and experimental evidence are not fulfilled in relation to lung function. Cessation of all exposure to ETS, achieved by a ban of active smoking, would be the best way to deal with the human health problem.[3] Also, we will have to continue to rely on additional animal studies to develop appropriate therapeutic and preventive strategies that should help to mitigate some of the serious, widespread, and costly health problems associated with cigarette smoke exposure.[3]

REFERENCES

1. U.S. Surgeon General, The Health Consequences on Involuntary Smoking, A Report of the Surgeon General, DHHS (CDC) 87-8398, U.S. Department of Health and Human Services, Washington, D.C., 1986.
2. U.S. Environmental Protection Agency, Respiratory Health Effects of Passive Smoking: Lung Cancer and Other Disorders, EPA/600/6-90/006F, U.S. EPA, Office of Research and Development, Washington, D.C., 1992.
3. Witschi, H., Joad, J.P., and Pinkerton, K.E., The toxicology of environmental tobacco smoke, *Annu. Rev. Pharmacol. Toxicol.,* 37, 29, 1997.
4. Cook, D.G. and Strachan, D.P., Summary of effects of parental smoking on the respiratory health of children and implications for research, *Thorax,* 54, 357, 1999.
5. Cook, D.G., Strachan, D.P., and Carey, I.M., Health effects of passive smoking. 9. Parental smoking and spirometric indices in children, *Thorax,* 53, 884,1998.
6. Cook, D.G. and Strachan, D.P., Health effects of passive smoking. 3. Parental smoking and respiratory symptoms in schoolchildren. *Thorax,* 52, 1081, 1997.
7. Cook, D.G. and Strachan, D.P., Health effects of passive smoking. 7. Parental smoking, bronchial reactivity and peak flow variability in children, *Thorax,* 53, 295, 1998.
8. Wright, A.L., Holberg, C., Martinez, F.D., and Taussig, L.M., Relationship of parental smoking to wheezing and nonwheezing lower respiratory tract illnesses in infancy. Group Health Medical Associates, *J. Pediatr.,* 118, 207, 1991.
9. Tager, I.B., Weiss, S.T., Munoz, A., Rosner, B., and Speizer, F.E, Longitudinal study of the effects of maternal smoking on pulmonary function in children, *N. Engl. J. Med.,* 309, 699, 1983.
10. O'Connor, G.T., Weiss, S.T., Tager, I.B., and Speizer, F.E., The effect of passive smoking on pulmonary function and nonspecific bronchial responsiveness in a population based sample of children and young adults, *Am. Rev. Respir. Dis.,* 135, 800, 1987.
11. Martinez, F.D., Antognoni, G., Macri, F., Bonci, E., Midulla, F., De castro, G., and Ronchetti, R., Parental smoking enhances bronchial responsiveness in nine-year-old children, *Am. Rev. Respir. Dis.,* 138, 518, 1988.
12. Guerin, M.R., Jenkins, R.A., and Tomkins, B.A., *The Chemistry of Environmental Tobacco Smoke: Composition and Measurement, Indoor Air Research Series,* Lewis Publishers, Boca Raton, FL, 1992.
13. Pirkle, J.L., Flegal, K.M., Bernet, J.T., Brody, D.J., Etzel, R.A., and Maurer, K.R., Exposure of the U.S. population to environmental tobacco smoke: the third National Health and Nutrition Examination Survey, 1988 to 1991, *J. Am. Med. Assoc.,* 275, 1233, 1996.
14. Scherer, G., Conze, C., Tricker, A.R., and Adlkofer, F., Uptake of tobacco smoke constituents on exposure to environmental tobacco smoke (ETS), *Clin. Invest.,* 70, 352, 1992.
15. Leuenberger, P., Schwartz, J., Ackermann-Liebrich, U., Blaser, K., Bolognini, G., Bongard, J. P., Brandli, O., Braun, P., Bron, C., and Brutsche, M., Passive smoking exposure in adults and chronic respiratory symptoms (SAPALDIA study), *Am. J. Respir. Crit. Care Med.,* 150, 1222, 1994.
16. Dayal, H.H., Khuder, S., Sharra, R., and Trieff, N., Passive smoking in obstructive respiratory disease in an industrialized urban population, *Environ. Res.,* 65, 161, 1994.
17. Robbins, A., Abbey, D.E., and Lebowitz, M.D., Passive smoking and chronic respiratory disease symptoms in nonsmoking adults, *Int. J. Epidemiol.,* 22, 809, 1993.

18. Jorres, R. and Magnussen, H., Influence of short-term passive-smoking on symptoms, lung mechanics and airway responsiveness in asthmatic subjects and healthy controls, *Eur. Respir. J.,* 5, 936, 1992.

19. Danuser, B., Weber, A., Hartmann, A.L., and Kreuger, H., Effects of bronchoprovocation challenge test with cigarette sidestream smoke on sensitive and healthy adults, *Chest,* 103, 353, 1993.

20. Ekwo, E.E., Weinberger, M.M., Lachenbruch, P.A., and Huntley, W.H., Relationship of parental smoking and gas cooking to respiratory disease in children, *Chest,* 84, 662, 1983.

21. Dodge, R., The effects of indoor pollution on Arizona children, *Arch. Environ. Health,* 37, 151, 1982.

22. Forastiere, F., Corbo, G.M., Michelozzi, P., Pistelli, R., Agabiti, N., Brancato, G., Ciappi, G., and Perucci, C.A., Effects of environmental and passive smoking on the respiratory health of children, *Int. J. Epidemiol.,* 21, 66, 1992.

23. Schulte-Hobein, B., Schwartz-Bickenbach, D., Abt, S., Plum, C., and Nau, H., Cigarette smoke exposure and development of infants throughout the first year of life: influence of passive smoking and nursing on cotinine levels in breast milk and infant's urine, *Acta Paediatr. Scand.,* 81, 550, 1992.

24. Sherrill, D.L., Martinez, F.D., Lebowitz, M.D., Holdaway, M.D., Flannery, E.M., Herbison, G.P., Stanton, W.R., Silva, P.A., and Sears, M.R., Longitudinal effects of passive smoking on pulmonary function in New Zealand children, *Am. Rev. Respir. Dis.,* 145, 1136, 1992.

25. Young, S., Souef, P.N.L., Geelhoed, G.C., Stick, S.M., Turner, K.J., and Landau, L.I., The influence of family history of asthma and parental smoking on airway responsiveness in early infancy, *N. Engl. J. Med.,* 324, 1168, 1991.

26. Frischer, T., Kuehr, J., Meinert, R., Karmaus, W., Barth, R., Hermann-Kunz, E., and Urbanek, R., Maternal smoking in early childhood: a risk factor for bronchial responsiveness to exercise in primary-school children, *J. Pediatr.,* 121, 17, 1992.

27. Pimm, P., Shephard, R.J., and Silverman, F., Physiological effects of acute exposure to cigarette smoke, *Arch. Environ. Health,* 33, 201, 1978.

28. Dayal, H.H., Khuder, S., Sharra, R., and Trieff, N., Passive smoking in obstructive respiratory disease in an industrialized urban population, *Environ. Res.,* 65, 161, 1994.

29. Xu, X. and Li, B., Exposure-response relationship between passive smoking and adult pulmonary function, *Am. J. Respir. Crit. Care Med.,* 151, 41, 1995.

30. Martinez, F.D., Passive smoking and respiratory disorders other than cancer, in *Respiratory Health Effects of Passive Smoking: Lung Cancer and Other Disorders,* Report of the U.S. Environmental Protection Agency, U.S. National Institute of Health, Washington, D.C., 1993, 205–265.

31. Bono, R., Nebiolo, F., Bugiani, M., Meineri, V., Scursatone, E., Piccioni, P., Caria, E., Gilli, G., and Arossa, W., Effects of tobacco smoke exposure on lung growth in adolescents, *J. Expos. Anal. Environ. Epidemiol.,* 8, 335, 1998.

32. Hanrahan, J.P., Tager, I.B., Segal, M.R., Tosteson, T.D., Castile, R.G., Vunakis, H.V., Weiss, S.T., and Speizer, F.E., The effect of maternal smoking during pregnancy on early infant lung function, *Am. Rev. Respir. Dis.,* 145, 1129, 1992.

33. Cunningham, J., Dockery, D.W., and Speizer, F.E., Maternal smoking during pregnancy as a predictor of lung function in children, *Am. J. Epidemiol.,* 139, 1139, 1994.

34. Carlsen, K.L., Jaakkola, J.K., Nafstad, P., and Carlsen, K.-H., In utero exposure to cigarette smoking influences lung function at birth, *Eur. Respir. J.,* 10, 1774, 1997.

35. Wang, S., Young, R.S., Zhang, J., Watson, R.R., and Witten, M.L., Cytokine production of alpha-tocopherol supplementation in mice exposed to sidestream cigarette smoke, *FASEB J.,* 14, A173, 2000.

36. Delay-Goyet, P. and Lundberg, J.M., Cigarette smoke-induced airway oedema is blocked by the NK1 antagonist, CP-96,345, *Eur. J. Pharmacol.,* 203, 157, 1991.

37. Lee, L.Y., Kou, Y.R., Frazier, D.T., Beck, E.R., Pisarri, T.E., Coleridge, H.M., and Coleridge, J.C.G., Stimulation of vagal pulmonary C-fibers by a single breath of cigarette smoke in dogs, *J. Appl. Physiol.,* 66, 2032, 1989.

38. Saria, A., Martling, C.R., Yan, Z., Theodorsson-Norheim, E., Gamse, R., and Lundberg, J.M., Release of multiple tachykinins from capsaicin-sensitive sensory nerves in the lung by bradykinin, histamine, dimethylphenyl piperazinium, and vagal nerve stimulation, *Am. Rev. Respir. Dis.,* 137, 1330, 1988.

39. Lee, B.P., Morton, R.F., and Lee, L.Y., Acute effects of acrolein on breathing: role of vagal bronchopulmonary afferents, *J. Appl. Physiol.,* 72, 1050, 1992.

40. Coleridge, J.C., Coleridge, H.M., Schelegle, E.S., and Green, J.F., Acute inhalation of ozone stimulates bronchial C-fibers and rapidly adapting receptors in dogs, *J. Appl. Physiol.,* 74, 2345, 1993.

41. Joad, J.P., Bric, J.M., and Pinkerton, K.E., Sidestream smoke effects on lung morphology and C-fibers in young guinea pigs, *Toxicol. Appl. Pharmacol.,* 131, 289, 1995.

17 Environmental Tobacco Smoke and Cardiovascular Disease

Jin Zhang and Ronald R. Watson

CONTENTS

17.1 INTRODUCTION

Environmental tobacco smoke (ETS) is an important source of exposure to toxic air contaminants indoors.[1] Virtually everyone in the U.S. is at risk of exposure and harm from ETS or secondhand (passive) smoke. This takes a variety of exposure patterns, including in the workplace, restaurants, bars, home, and fetal exposure via the placenta.[2,3] Epidemiological data suggest that exposure to ETS is associated with high incidence of lung cancer, cardiovascular disease, intrauterine growth retardation, predisposition to chronic lung disease, and Sudden Infant Death Syndrome.[4] Clinical and epidemiological studies,[5] conducted in a variety of locations, found about a 30% increase in risk of death from ischemic heart disease or myocardial infarction among nonsmokers living with smokers. ETS contributes 37,000 heart disease deaths of the total 530,000 annual deaths, making passive smoking the third leading preventable cause of death, after active smoking and alcohol. The American Heart Association has formally concluded that passive smoking is an important risk factor for heart disease in both adults[6] and children.[7]

0-8493-0311-7/00/$0.00+$.50
© 2001 by CRC Press LLC

17.2 TOXICITY OF ETS

ETS is a mixture of sidestream (SS) and mainstream (MS) cigarette smoke.[4] MS constitutes 15% of total ETS. MS is smoke first inhaled by an active smoker and then exhaled. While MS is retained for a few moments in the lung, the smoke is scrubbed of some of its constituents, most notably nicotine and carbon monoxide (CO), as well as much of the particulate matter. ETS contains about 85% SS, the smoke curling between puffs off the end of a lit cigarette, generated at low burning temperatures. Of the thousands of chemicals in cigarette MS and SS, those that contribute to smoking-induced vascular disease include nicotine, carbon monoxide, polycyclic aromatic hydrocarbons (PAHs), and tobacco glycoproteins.[8] Most notably, SS is richer in certain carcinogens than MS.[9–11] Some toxic, carcinogenic, and cocarcinogenic substances such as nicotine, carbon monoxide, carbon dioxide, catechol, benzopyrenes, N-nitrosamines [N-nitrosodimethylamine, N'-nitrosoanabasine, 4-(methylnitrosamino)-1-(3-pyridyl)-1-butanone], and ammonia are enriched in SS.

17.3 ANIMAL MODELS

Experimental studies with passive smoking have successfully duplicated disease conditions in laboratory animals, particularly the effects of SS on heart disease. Exposure to ETS increased myocardial infarct size in a rat model of ischemia and reperfusion.[12] Passive smoking increased experimental atherosclerosis in cholesterol-fed rabbits.[13] Researchers showed increases in plaque development in young cockerels that were exposed to secondhand smoke.[14] SS tobacco smoke exposure produced measurable oxidative DNA damages in Balb/c mouse heart, liver, and lung tissue,[15] such changes should increase the probability of carcinogenesis and premature apoptosis due to SS exposure. The availability of animal models opens the way to fruitful experimental studies on mechanisms to help better understand SS-induced cardiovascular disease (CVD).

17.4 MECHANISMS OF CVD DUE TO ETS EXPOSURE

The relationship between active tobacco smoking and increased risk of coronary artery disease (stable and unstable angina, acute myocardial infarction, or sudden death), cerebrovascular disease (cerebral infarction and cerebral and subarachnoid hemorrhage), peripheral arterial disease (large and small vessel), and aortic aneurysm has been well established in numerous epidemiological and basic science studies.[16] More recently, passive smoking has been shown to represent an important risk factor for CVDs.[17] Passive smoke causes heart damage in much the same mechanisms as active smoking through hematological, neurohormonal, metabolic, hemodynamic, molecular genetic, and biochemical pathways (Figure 17.1). Many components in tobacco smoke play important roles in the process of heart function damage. Activities of key toxins are described later.

17.4.1 CARBON MONOXIDE (CO)

The CO in ETS displaces and competes with oxygen for binding sites on red blood cells.[18] CO causes tissue hypoxia by inhibition of the blood's ability to deliver oxygen

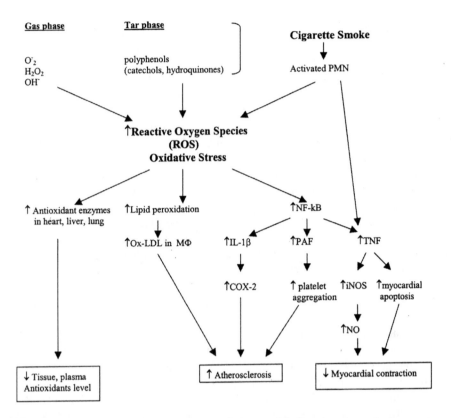

FIGURE 17.1 Cigarette smoke causes heart pathology and dysfunction via producing reactive oxygen species.

to the myocardium. Subsequently, passive smoking reduces the ability of the heart muscle to convert oxygen into the adenosine triphosphate (ATP). Various animal studies have demonstrated that CO interferes with myoglobin, cytochrome P450, and other enzyme functions. It causes lipid peroxidation through neutrophil activation and produces oxidative stress manifested by peroxynitrate deposition in the endothelium. Finally, CO binds to cytochrome-C oxidase, disrupting ATP production and mitochodrial respiration.[19] In a study in rabbits, the activity of the mitochondrial enzyme cytochrome oxidase, which is responsible for ATP synthesis, fell 25% after a single 30-min exposure to secondhand smoke, and the activity continued to decline with the duration of exposure.[20] Additional evidence for limited oxygen utility of passive smoking is that the heart increasingly relies on anaerobic metabolism during passive smoking, which has been indicated by a significant increase in the amount of lactate in venous blood.[21]

17.4.2 Free Radicals

The chemical composition of ETS is very complex. More than 3800 compounds have been identified in cigarette smoke.[22] Included in this list of cigarette smoke compounds are many capable of generating reactive oxygen species (ROS) during metabolism. Major sources of ROS are from both gas and tar phases of cigarette smoke. Oxidative damage to cellular components occurs when the production of ROS overwhelms the cell's antioxidant defenses. This can occur when an organism is exposed to xenobiotic compounds such as those found in ETS.[15] The free radicals entering the body are first trapped by serum aqueous and lipophilic antioxidants, which interact and provide great protection against lipid peroxidation.[15,23] As a result, the antioxidant enzymes in tissues such as the heart, liver, and lung will be activated. Therefore, tissue and plasma antioxidants (vitamins A and C) will be sacrificed against oxidative damage.[24,25] Once the antioxidant barrier is broken, lipid peroxidation can take place and produce oxidative low-density lipoprotein (Ox-LDL). Ox-LDL accumulation in macrophages plays a role in atherosclerosis. Such metabolic changes in lipid profiles due to smoking are generally accepted as a key event of atherosclerosis plaque formation.[26] On the other hand, passive smoking inappropriately activates the neutrophils.[27] In a group of passive smokers with only 3 h of SS, there were significant increases in the circulating leukocyte counts and stimulated neutrophil migration. The activated neutrophils can release oxidants, which may also play a role in tissue damage in passive smokers.[27] Free radicals and oxidants are extremely destructive to the heart muscle cell membrane, as well as other processes within the cell, while producing lipid peroxidation. Low exposure to nicotine or other cigarette smoke constituents significantly worsens reperfusion injury. Rats exposed to secondhand smoke exhibited severely damaged mitochondrial function during the reperfusion injury.[28,29] As mentioned previously, the ability of cardiac mitochondrial cells to convert oxygen into ATP was much more compromised upon oxidative stress.[19] Nuclear DNA is one of the cellular targets of ROS. ROS can react with nuclear DNA, resulting in a number of damaged DNA products.[15] As DNA damage accumulates within a cell, the likelihood of a cytotoxic or mutagenic lesion increases. This may also become one possible etiology of atherosclerosis and coronary heart disease.

ROS also can initiate a series of cellular responses, such as activation of nuclear factor kappa B (NF-κB), which will play an important role in proinflammatory process. NF-κB can induce interleukin-1 (IL-1) beta, platelet-activating factor (PAF), and tumor necrosis factor (TNF).[30] IL-1 beta is well known an inducer of several other proinflammatory enzymes such as inducible cyclooxygenase-2 (COX-2); PAF is shown to benefit platelet aggregation. These are two factors involved in atherogenesis.[31] TNF may decrease myocardial contraction through enhancing inducible nitric oxide synthase (iNOS) and nitric oxide (NO) production or by inducing myocardial apoptosis.[32] Antioxidant nutrients have the potential for modulating inflammatory aspects of immune function through this NF-κB pathway.[30] Thus, increasing the intake of vegetables, fruits, and antioxidant vitamin supplements should lower SS damage due to oxidation in certain degrees.

17.4.3 NICOTINE

Previous findings have shown that nicotine causes significant dose-related myocardial contraction and endothelium-independent vasoconstriction.[33] If these results

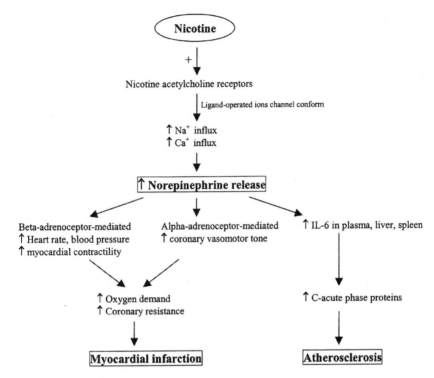

FIGURE 17.2 Nicotine and its role in cardiovascular disease.

can be extrapolated to clinical conditions in humans, then increased myocardial oxygen consumption, secondary to increased contractility and heart rate, and decreased coronary blood flow due to vasoconstriction will cause an imbalance between oxygen demand and supply of the heart and lead to arrhythmia and to myocardial ischemia or infarction, or both.

The mechanism by which nicotine causes myocardial contractility changes was shown to be mainly due to stimulation of sympathetic neutransmission.[34] As nicotine activates nicotine acetylcholine receptors localized on peripheral postganglionic sympathetic nerve endings and the adrenal medulla, binding of nicotine to its receptors leads to a conformational change of the central pore, which results in the influx of sodium and calcium ions. Subsequently, the increase in intracellular Ca^+ triggers the exocytotic release of norepinephrine. Through the release of norepinephrine, nicotine induces a beta-adrenoceptor-mediated increase in heart rate, blood pressure, and myocardial contractility, as well as an alpha-adrenoceptor-mediated increase in coronary vasomotor tone. The resulting simultaneous increase in oxygen demand and coronary resistance has a detrimental effect on the oxygen balance of the heart, which may be a trigger for myocardial infarction. On the other hand, stimulation of norepinephrine can induce IL-6 synthesis in the liver and spleen. IL-6 can induce C-acute phase protein, which is known to be involved in atherosclerosis (Figure 17.2).

Nicotine is also a potential cause of the observed changes in endothelial cells and platelet aggregate ratios after smoking tobacco cigarettes, although other components of cigarette smoke may also be important.[3] Activated platelets can damage the lining of the coronary arteries and facilitate the development of athero-selerotic lesions. Once the platelets are activated due to exposure to ETS, their aggregation is increased and the probability that platelets will adhere at the endo-thelial injury site will be raised. Long-term exposure to passive smoking impairs the arterial endothelial function, probably through impaired endothelial nitric oxide activity,[35] and increases the thickness of the carotid wall.[36] There are population-based data in humans indicating that passive smokers have significantly thicker carotid artery walls, in a dose-response pattern, than subjects who never were exposed to passive smoking.[37] Activated blood platelets have also been shown to increase the likelihood of formation of a thrombus.[38]

17.4.4 CARCINOGENS

The carcinogens in tobacco smoke act as promoters or cofactors, facilitating the development of vascular plaques.[39] Animal studies indicated that PAHs, 7,12-dimethyl-benzanthracene and benzo(a)pyrene, bind preferentially to both LDL and HDL subfragments of cholesterol. The binding may facilitate incorporation of the carcino-genic compounds into the cells lining the coronary arteries and, hence, contribute to both cell injury and hyperplasia in the process of atherosclerosis.[5]

17.5 PREVENTION OF ETS-INDUCED CVD

The most effective way to prevent CVD induced by SS would be elimination of exposure. However, this goal is difficult to achieve because many smokers are addicted to nicotine, continuing the home and community exposure to ETS. Social support and skills training to achieve smoking abstinence can be important elements of smoking cessation. Though smoking inhibition has been done with great success in many public places, much more efforts need to be done in the future to save clean and healthy air for the heart.

Nutritional treatment of people exposed to ETS is theoretically useful as ETS is highly associated with oxidative stress. The antioxidant vitamin C reverses the impairment in endothelium-mediated vasodilatation.[40] High vitamin E consumption also has protective effects against oxidative stress caused by cigarette smoke.[41] From recent studies,[42] an important source of bioflavonoids and related antioxidants are found in an extract found in a pine tree bark. The extract in tissue culture, animals, and humans enhanced endogenous antioxidant systems and increased free radical scavenging. It also showed significantly inhibited tobacco smoke platelet aggrega-tion. Fish oil as a supplement or in the diet may be another available agent for CVD due to ETS.[43] Omega-3-rich fish oils can compensate for a number of adverse impacts of smoking by reducing plasma fibrinogen[44,45] and platelet aggregation,[46] as well as by increasing erythrocyte distensibility.[47] The cardio protective benefits of fish oils may be further complemented by mini-dose aspirin and, in those that have previously experienced a heart attack, beta-blocker therapy.[43] It is universal that healthy eating — meaning reducing our fat intake, reducing our cholesterol, and consuming a high level of fruits and vegetables which are rich in vitamins and minerals — leads to a reduced incidence of CVD. Although these agents have been shown to be theoretically logical, the role of antioxidants to lower the moderate damage due to ETS has been poorly studied and difficult to quantify.

17.6 CONCLUSION

Most studies demonstrate a significant dose-response effect, with greater exposure to ETS associated with greater risk of death from heart disease. In conclusion, ETS may increase the risk of CVD by promoting platelet aggregation and oxidative damage to arterial endothelium and by inducing miocardial apoptosis. Moreover, ETS, in realistic exposures, exerts significant adverse effects on heart disease by reducing the body's ability to deliver and utilize oxygen via depressing cellular respiration at the level of mitochondria. The carcinogens in ETS may initiate and accelerate the development of atherosclerotic plaque. There is evidence that direct and indirect exposure to tobacco smoke increases the heart rate, blood pressure, rate-pressure product, and cardiac output and maximal first derivative of left ventricular pressure.[48] These effects are probably a consequence of adrenergic stimulation by nicotine.

Smoking inhibition is the most important approach for SS-associated CVD prevention. Nutritional treatment would also be advocated as a major health promotion. Treatment or prevention with agents such as aspirin and pycnogenol can reduce platelet aggregation. Increased consumption of fruits and vegetables should lower CVD risk in SS-exposed people. As research on ETS inducing CVD is very limited, more research needs to be done to define risk and to understand the pathological and molecular mechanisms of ETS exposure-related CVD in both animal models and human studies. This may prove very useful in eliminating the underlying causes in order to develop more effective prevention strategies.

ACKNOWLEDGMENT

Research leading to this chapter was supported in part by a grant to Mark Witten from the Arizona Disease Research Control Commission.

REFERENCES

1. Pirkle, J.L., Fiegal, K.M., Bernert, J.T., Brody, D.J., Etzel, R.A., and Maurer, K.R., Exposure of the US population to environmental tobacco smoke: the Third National Health and Nutrition Examination Survey, 1988 to 1991. *JAMA*, 275, 1233–1240, 1996.
2. Thompson, B., ETS exposure in the workplace. Perceptions and reactions by employees in 114 work sites, Working Well Research Group [corrected], (published erratum appears in *J. Occup. Environ. Med.*, 37, 1363, 1995; *J. Occup. Environ. Med.*, 37, 1086–1092, 1995).
3. Office of Environmental Health Hazard Assessment, Health Effects of Exposure to Environmental Tobacco Smoke, Final Draft for Scientific, Public, and SRP Review, California Environmental Protection Agency, Berkeley, February 1997.
4. Witschi, H., Joad, J.P., and Pinkerton, K.E., The toxicity of environmental tobacco smoke, *Annu. Rev. Pharmacol. Toxicol.*, 37, 29–52, 1997.

5. Glantz, S.A. and Parmley, W.W., Passive smoking and heart disease. Epidemiology, physiology, and biochemistry, *Circulation*, 83, 1–12, 1991.

6. Taylor, A.E., Johnson, D.C., and Kazemi, H., Environmental tobacco smoke and cardiovascular disease: a position paper from the Council on Cardiopulmonary and Critical Care, American Heart Association, *Circulation*, 86, 1–4, 1992.

7. Gidding, S., Morgan, W., Perry, C., Isabel-Jones, J., and Bricker, J.T., Active and passive tobacco exposure: a serious pediatric health problem. A statement from the Committee on Atherosclerosis and Hypertension in Children, Council on Cardiovascular Disease in the Young, American Heart Association, *Circulation*, 90, 2581–2590, 1994.

8. Zhu, B.-Q. and Parmley, W.W., Hemodynamic and vascular effects of active and passive smoking, *Am. Heart J.*, 13, 1270–1275, 1995.

9. Hoffmann, D., Adams, J.D., and Brunnemann, K.D., A critical look at N-nitrosamines in environmental tobacco smoke, *Toxicol. Lett.*, 35, 1–8, 1987.

10. Adams, J.D., O'Mara-Adams, K.J., and Hoffmann, D., Toxic and carcinogenic agents in undiluted mainstream smoke and sidestream smoke of different types of cigarettes, *Carcinogenesis*, 8, 729–731, 1987.

11. Guerin, M.R., Jenkins, R.A., and Tomkins, B.A., *The Chemistry of Environmental Tobacco Smoke: Composition and Measurement. Indoor Air Research Series*, Lewis Publishers, Boca Raton, FL, 1992.

12. Zhu, B.-Q., Sun, Y.-P., Sievers, R.E., Glantz, S.A., Parmley, W.W., and Wolfe, C.L., Exposure to environmental tobacco smoke increases myocardial infarct size in rats, *Circulation*, 89, 1282–1290, 1994.

13. Zhu, B.-Q., Sun, Y.-P., Sievers, R.E., Isenberg, W.M., Moorehead, T.J., and Parmley, W.W., Effects of etidronate and lovastatin on the regression of atherosclerosis in cholesterol-fed rabbits, *Cardiology*, 85, 370–377, 1994.

14. Penn, A. and Snyder, C.A., Inhalation of sidestream cigarette smoke accelerates development of arteriosclerotic plaques, *Circulation*, 88, 1820–1825, 1993.

15. Howard, D.J., Briggs, L.A., and Pritsos, C.A., Oxidative DNA damage in mouse heart, liver, and lung tissue due to acute side-stream tobacco smoke exposure, *Arch. Biochem. Biophys.*, 352, 293–297, 1998.

16. Taylor, B.V., Oudit, G.Y., Kalman, P.G., and Liu, P., Clinical and pathophysiological effects of active and passive smoking on the cardiovascular system, *Can. J. Cardiol.*, 14, 1129–1139, 1998.

17. Josefson, D., Passive smoking doubles risk of heart disease, *Br. Med. J.*, 314, 1569, 1997.

18. Moskowitz, W., Mosteller, M., Schieken, R., et al., Lipoprotein and oxygen transport alterations in passive smoking preadolescent children: the MCV Irwin study, *Circulation*, 81, 586–592, 1990.

19. Weaver, L.K., Carbon monoxide poisoning [Review], *Crit. Care Clin.*, 15, 297–317, 1999.

20. Gvozdjak, J., Gvozdjakova, A., Kucharska, J., and Bada, V., The effect of smoking on myocardial metabolism, *Czech. Med.*, 10, 47–53, 1987.

21. McMurrray, R.G., Hicks, L.L., and Thompson, D.L., The effects of passive inhalation of cigarette smoke on exercise performance, *Eur. J. Appl. Physiol.*, 54, 196–200, 1985.

22. Stedman, R.L., The chemical composition of tobacco and tobacco smoke, *Chem. Rev.*, 68, 153–207, 1968.

23. Valkonen, M. and Kuusi, T., Passive smoking induces artherogenic changes in low-density lipoprotein, *Circulation*, 97, 2012–2016, 1998.

24. Smith, J.L. and Hodges, R.E., Serum levels of vitamin C in relation to dietary and supplemental intake of vitamin C in smokers and nonsmokers, *Ann. N.Y. Acad. Sci.*, 489, 144–152, 1987.

25. Stryker, W.S., Kaplan, L.A., Stein, E.A., Stampfer, M.J., Sober, A., and Willett, W.C., The relation of diet, cigarette smoking, and alcohol consumption to plasma beta-carotene and alpha-tocopherol levels, *Am. J. Epidemiol.*, 127, 283–296, 1988.

26. Brown, M.S. and Goldstein, J.L., Lipoprotein metabolism in macrophage: implications for cholesterol deposition in atherosclerosis, *Annu. Rev. Biochem.*, 52, 223–261, 1983.

27. Anderson, R., Theron, A.J., Richards, G.A., Myer, M.S., and van Renberburg, A.J., Passive smoking by humans sensitizes circulating neutrophils, *Am. Rev. Respir. Dis.*, 144, 570–574, 1991.

28. Van Jaarsveld, H., Kuyl, J.M., and Alberts, D.W., Exposure of rats to low concentration of cigarette increases myocardial sensitivity to reperfusion, *Basic Res. Cardiol.*, 87, 393–399, 1992.

29. Van Jaarsveld, H., Kuyl, J.M., and Alberts, D.W., Antioxidant vitamin supplementation of smoke-exposed rats partially protects against myocardial ischaemic/reperfusion injury, *Free Radic. Res. Commun.*, 17(4), 263–269, 1992.

30. Grimble, R.F., Modification of inflammatory aspects of immune function by nutrients, *Nutr. Res.*, 18, 1297–1317, 1998.

31. Bkaily, G. and Dorleans-Juste, P., Cytokine-induced free radical and their roles in myocardial dysfunctions, *Cardiovasc. Res.*, 42, 576–577, 1999.

32. Ferrsri, R., The role of TNF in cardiovascular disease, *Pharmacol. Res.*, 40(2), 97–105, 1999.

33. Lee, T.S. and Hou, X., Nicotine is hazardous to your heart [letter; comment], *Chest*, 109, 584–585, 1996.

34. Haass, M. and Kubler, W., Nicotine and sympathetic neurotransmission, *Cardiovasc. Drugs Ther.*, 10, 657–665, 1996.

35. Celermajer, D.S., Adams, M.R., Clarkson, P., Robinson, J., McCredie, R., Donald, A., and Deanfield, J.E., Passive smoking and impaired endothelium-dependent arterial dilatation in healthy young adults, *N. Engl. J. Med.*, 334, 150–154, 1996.

36. Diez Roux, A.V., Nieto, F.J., Comstorck, G.W., Howard, G., and Szklo, M., The relationship of active and passive smoking to carotid atherosclerosis 12-14 years later, *Prev. Med.*, 24, 48–55, 1995.

37. Howard, G., Burke, G.L., Szklo, M., et al., Active and passive smoking are associated with increased carotid artery wall thickness: the Risk in Communities Study, *Arch. Intern. Med.*, 154, 1277–1282, 1994.

38. Glantz, S.A. and Parmley, W.W., Passive smoking and heart disease: mechanisms and risk, *J. Am. Med. Assoc.*, 273, 1047–1053, 1995.

39. Penn, A., Currie, J., and Snyder, C.A., Inhalation of carbon monoxide does not accelerate arteriosclerosis in cockerels, *Eur. J. Pharamacol.*, 228, 155–164, 1992.

40. Mays, B.W., Freischlag, J.A., Eginton, M.T., Cambria, R.A., Seabrook, G.R., and Towne, J.B., Ascorbic acid prevents cigarette smoke injury to endothelium-dependent arterial relaxation, *J. Surg. Res.*, 84, 35–39, 1999.

41. Pacht, E.R., Kaseki, H., Mohammed, J.R., Cornwell, D.G., and Davis, W.B., Vitamin E in the alveolar fluid of cigarette smokers, *J. Clin. Invest.*, 77, 789–796, 1986.

42. Park, Y.C., Rimbach, G., Saliou, C., Valacchi, G., and Packer, L., Activity of monomeric, dimeric, and trimeric flavonoids on NO production, TNF-alpha secretion, and NF-kappa B-dependent gene expression in RA 264.7 macrophages, *FEBS Lett.*, 465(2–3), 93–97, 2000.

43. McCarty, M.F., Fish oil may be an antidote for the cardiovascular risk of smoking, *Med. Hypoth.*, 46, 337–347, 1996.

44. Hostmark, A.T., Bjerkedal, T., Kierulf, P., et al., Fish oil and plasma fibrinogen, *Br. Med. J.*, 297, 180–181, 1988.

45. Rodack, K., Deck, C., and Huster, G., Dietary supplementation with low-dose fish oils lowers fibrinogen levels: a randomized, double-blind controlled study, *Ann. Intern. Med.*, 111, 757–758, 1989.

46. Goodnight, S.H., Jr., Harris, W.S., and Connor, W.E., The effects of dietary *w* 3 fatty acids on platelet composition and function in man: a prospective, controlled study, *Blood*, 58, 880–885, 1981.

47. Mills, D.E., Galey, W.F., and Dixon, H., Effects of dietary fatty-acids supplementation on fatty-acids composition and deformability of young and old erythrocytes, *Biochim. Biophys. Acta*, 1149, 313–318, 1993.

48. Winniford, M.D., Wheelan, K.R., Kremers, M.S., Ugolini, V., van den Berg, E., Jr., Niggemann, E.H., Jansen, D.E., and Hillis, L.D., Smoking-induced coronary vasoconstriction in patients with atherosclerotic coronary artery disease: evidence for adrenergically mediated alterations in coronary artery tone, *Circulation*, 73, 662–667, 1986.

18 Objective Assessment of Environmental Tobacco Smoke (ETS) Exposure in Pregnancy and Childhood

Daphne Chan, Julia Klein, and Gideon Koren

CONTENTS

18.1 INTRODUCTION

Environmental tobacco smoke (ETS) is one of the most significant indoor air pollutants that people are commonly exposed to. Its potential adverse effects on health have received much attention in the past few decades. Tobacco smoke contains more than 4500 compounds, many of which are known human carcinogens. ETS is a mixture of sidestream and mainstream smoke inhaled by the passive smoker that

0-8493-0311-7/00/$0.00+$.50
© 2001 by CRC Press LLC

differs in the concentration of constituents as a result of differences in combustion kinetics.[1] Whereas mainstream smoke is the minor component (~15%) that is first inhaled and then exhaled by the active smoker, sidestream smoke originating from the tip of the burning cigarette is the dominant component accounting for about 85% of the total content of ETS.[2] Higher concentrations of ammonia, benzene, nicotine, carbon monoxide, and many other carcinogens are found in sidestream smoke because it is neither filtered by the cigarette's filter nor the smoker's lungs.[3] It is therefore reasonable to hypothesize that exposure to ETS may be a larger health hazard compared to active smoking. Epidemiological data provide convincing evidence that children's exposure to ETS is associated with increased rates of lower respiratory illness, otitis media, Sudden Infant Death Syndrome (SIDS), and increased incidence and severity of asthma.[4-6] Smoking in pregnancy has been documented to adversely affect fetal growth and is associated with increased risk of spontaneous abortion, premature birth, perinatal death, and low birthweight. ETS exposure by the non-smoking mother and the health impact on the unborn child, as well as postnatal life, is not well documented and remains controversial.[7,8] The majority of studies in the literature on ETS exposure were qualitative and subjective. Various health risks are linked to the effects of passive smoking without quantifying the extent of the actual exposure, especially in the pediatric population. Therefore, it is necessary to assess objectively both gestational and postnatal ETS exposure with the use of reliable biological markers in order to better define and strengthen the proposed relationship between exposure to ETS and its potential impact on health.

18.2 EPIDEMIOLOGY IN PREGNANCY AND CHILDHOOD

In North America, approximately 13 to 30% of all women 18 years of age or older are regular smokers.[9,10] Despite the well-known potential adverse effects from cigarette smoke, many women continue to smoke while pregnant. While less than 3% of those who attempt to quit are successful,[11] the difficulties of smoking cessation may be related to interindividual variations in nicotine metabolism, previous smoking pattern, concurrent substance intake, as well as other demographic and socioeconomic factors. ETS exposure by the non-smoking mother has not been thoroughly investigated in the past, but it has been estimated that 25 to 30% of non-smoking women are regularly exposed to ETS generated by other active household smokers.[12,13] Many more women may be exposed to ETS at their workplace, but documentation of this exposure is more difficult and remains sparse. Recently, it has been reported that approximately 30 to 50% of adult current smokers reside with children in their homes in the U.S. Smoking is allowed in restricted or all areas of the home in more than 70% of these households.[10] A recent Canadian survey reported that 30% of the pediatric and 34% of the youth population are currently exposed to ETS both in the home and in the community.[14] Passive smoking among children with chronic respiratory diseases is surprisingly high (20 to 60%) considering that these children are predisposed to the harmful effects of the constituents in tobacco smoke.[15,16] The source of ETS in pregnancy and children is not limited to the household, but extends into the workplace as well as the community.[17–20] It is almost impossible to avoid exposure to ETS (especially in the public), and the contribution of the community is difficult to assess. Current statistics available are derived from national and state-specific questionnaires and surveys. However, questionnaires and surveys are not free of limitations. The accuracy of these survey results is based on the respondent's will to provide honest answers and the ability to recall past exposure, which may vary between selected populations (e.g., highly variable especially if parents of a sick child are questioned). Therefore, it is possible that these numbers are vastly underestimated. Given that a substantial proportion of pregnant and pediatric populations are exposed to ETS during critical periods of growth and development, ETS is becoming an increasingly significant public health concern.

18.3 *IN UTERO* EXPOSURE AND PREGNANCY OUTCOME

The focus of this chapter is on the effects of ETS on the fetus as a tertiary passive smoker via the ETS-exposed, non-smoking mother and his/her subsequent development in childhood under additional secondhand smoke exposure. The term *"in utero* ETS exposure" in the following discussion is used in the context of *in utero* exposure experienced by the fetus via the ETS-exposed, non-smoking mother.

Decrement in birthweight (typically a mean decrease of 200g), small for gestational age (SGA), SIDS, and increased risk for spontaneous abortion have been associated with active maternal gestational smoking,[21–26] but these variables have not been thoroughly investigated in mothers who are exposed to secondhand smoke during pregnancy. In fact, controversies exist among the few reports that are published in this area. The risk of spontaneous abortion from ETS exposure has been shown to increase among passive-smoking mothers who have moderate alcohol and heavy caffeine consumption and for those who are exposed for more than 1 h daily.[22,23] The possible confounding effects of concurrent alcohol or caffeine exposure, however, have not been accounted for. While some groups report no increased risk of low birthweight in infants of ETS-exposed women,[27,28] others conclude that infants born from passive-smoking mothers are at risk of low birthweight (mean decrease of 16 to over 200 g)[29–33] after adjusting for confounding parental, neonatal, and environmental variables. Gestational age, however, was not properly controlled in one of these reports and therefore weakens the apparent association.[31] There is evidence supporting a modifying role of maternal age in the association between ETS exposure and adverse pregnancy outcomes (low birthweight and premature birth), but increased risk due to mature age alone cannot be ruled out.[34] Although it is difficult to separate the effects of *in utero* and postnatal ETS exposure (as most of the exposure remains relatively unchanged in the same household), attempts were made to evaluate the effect of maternal and postnatal ETS exposure on growth in childhood, but with limited success. Eskenazi and Bergmann suggested an association between maternal smoking (active and passive) and slower postnatal growth in children at 5 years of age, but failed to show such a relationship after adjusting for birthweight and gestational age.[35] Fried et al., on the other hand, showed that the negative correlation between growth measures at birth and prenatal ETS exposure was overcome by adolescence with no effects from postnatal exposure.[36] In the case of SIDS, it is generally agreed that a dose-response relationship exists between prenatal and postnatal exposure and an increased risk of SIDS. But, it is also uncertain which of the former or the latter is more causative.[24-26] Whether pregnancy outcome measures, fetal and postnatal growth, and the rate of SIDS are predisposed by *in utero* exposure via the non-smoking mother and extenuated by postnatal exposure have yet to be determined. It should be noted that mothers rather than children studied by the aforementioned groups were assessed for ETS exposure quantitatively by means of biological markers such as urinary and hair cotinine. If one is determining the impact of ETS on the fetus and its postnatal health status, it is critical to also objectively quantify the offspring's true exposure in relation to the mother's.

18.4 *IN UTERO* EXPOSURE AND HEALTH CONSEQUENCES IN CHILDHOOD

Whereas pregnancy outcome associated with *in utero* ETS exposure is evident at birth, there are many other health consequences that may be experienced by the ETS-exposed fetus that are only manifested later on in childhood. Since most children born to active- or passive-smoking mothers are healthy at birth, it is difficult to identify and assess the prolonged effects of ETS in adolescence and adulthood without bias. One often draws an association between ETS-exposed children (*in utero* and postnatal) with a particular illness or health outcome in case-control studies that may not be evident in prospective cohort studies. The association hypothesized may or may not be a true one, depending on confounding variables present within each analysis such as age, socioeconomic status, parental education, and adequate sample size. Also, investigating the exposure retrospectively means that it cannot be easily quantified by conventional techniques because of the short half-lives characteristic of most biological markers of tobacco smoke (with the exception of hair analysis that allows investigation of long-term drug history; refer to Section 18.7 on hair analysis). Thus, quantification of the risk for various illnesses or health outcomes as a result of *in utero* ETS exposure is complex.

18.4.1 CHILDHOOD CANCER

There may be a possible link between childhood cancer and *in utero* as well as childhood exposure to ETS. From all epidemiological data available, no strong association has been established between maternal active and passive smoking, postnatal exposure to ETS, and childhood cancer. However, slight increased risks for particular types of tumors cannot be ruled out based on the relative increased risks located in several studies for childhood brain tumors, leukemia, and lymphoma.[37,38] In a recent retrospective study by Filipini et al., an increased risk of central nervous system (CNS) tumors was found for children of non-smoking mothers exposed regularly to tobacco smoke both early and late in pregnancy.[39] Whether this and previous findings will reach significance in large population based studies have yet to be determined in more carefully designed prospective studies that include objective documentation of the actual exposure.

18.4.2 NEUROBEHAVIORAL DEVELOPMENT

It has been proposed that the offsprings of smokers are deficient in stature, cognitive, social, and neurodevelopment, and educational achievement, but confounding variables (e.g., social, demographic, psychological) have not been fully accounted for.[40] ETS exposure can cause subtle changes in children's behavior and neuro-development, as identified by a recent meta-analysis.[41] But whether the impact of *in utero* and postnatal ETS exposure has unique or additive effects on intellectual ability and neurobehavioral development in children still has to be investigated.[41,42] Since it is not known which constituents of ETS are responsible for causing such subtle changes, it is important to assess these exposures objectively to eliminate as many confounding variables as possible.

18.4.3 RESPIRATORY ILLNESS AND LUNG FUNCTION

In contrast to childhood cancer and neurobehavioral development, increased risks for postnatal respiratory symptoms and changes in lung function have been docu-mented for children exposed to ETS prenatally and postnatally. Children born to smoking mothers are more likely to suffer from chronic bronchitis, wheezing, and asthma early in childhood.[43–45] Smoking during pregnancy has been associated with various changes in lung function that may persist well into childhood from the time of birth. Prenatal exposure to tobacco smoke can adversely affect the tidal flow-volume ratios and the compliance of the respiratory system in healthy neonates independently of being SGA.[46] Forced expiratory flow rates are significantly lower in infants exposed prenatally to maternal tobacco smoke, suggesting possible impair-ment of *in utero* airway development and alteration of the lung's elasticity.[47] Abnor-mal lung function that is associated with prenatal exposure and familial predisposi-tion to wheezing is regarded as an important determinant of wheezing and lower respiratory tract infection in the first year of life.[48] Interestingly, several studies found that postnatal ETS exposure is not significantly related to declining lung function.[49–51] Thus, it is proposed that *in utero* exposure to active maternal smoking is indepen-dently associated with decreased postnatal lung function in young children. This suggests that alterations in the airways may occur during early development *in utero*, and that chronic childhood ETS exposure may put these predisposed children at greater risks for persistent altered lung function and recurrent infection.[50,51] Unfor-tunately, very few studies have investigated the potential effects of *in utero* ETS exposure on the respiratory health of young children. Asthma and increased risk of acute respiratory infections have been shown to be significantly associated with exposure to secondhand smoke during pregnancy, suggesting a possible direct effect of sidestream ETS on the child's respiratory health.[52,53] To account for these potential direct adverse effects of ETS on the respiratory system, it is crucial to determine the relationship between the extent of true exposure (by quantifying the ETS exposure in neonates and children) and the severity of respiratory outcome in order to strengthen the apparent causative association presented in the literature.

18.5 POSTNATAL EXPOSURE AND RESPIRATORY HEALTH IN CHILDHOOD

Consistent findings are available from many different countries regarding the association between childhood ETS exposure and occurrence of asthma. It is, however, still controversial whether ETS exposure serves to induce the occurrence of asthma, accelerates its severity, or is a combination of both. More studies seem to support the notion of increased severity of the disease as opposed to its induction. The relationship between parental and household smoking and severity of childhood asthma is likely to be causal, given that the association can be described in a dose-response manner and remains significant when potential confounders are accounted for.[54–57] A very similar trend is observed between childhood ETS exposure and the prevalence of respiratory illness and symptoms[58–60] (such as wheezing, lower respiratory tract infections, and acute chest illness). In contrast, the association between postnatal exposure to ETS and decreased lung function in childhood is not as apparent as that for asthma or respiratory illnesses. While some groups found no difference between the lung function of ETS-exposed and non-exposed children,[61,62] others found a significant correlation between current and past (*in utero*) ETS exposure with deficient lung function in childhood, suggesting a plausible additive effect of previous and current exposure.[63–65] Interindividual variations in the metabolic handling of ETS may play a role in the potential additive nature of subsequent exposures, which should be investigated thoroughly and quantitatively in different pediatric populations. One should appreciate that there are very few studies that correlate directly the actual ETS exposure with adverse health consequences in childhood.

18.6 OBJECTIVE ASSESSMENT OF ETS EXPOSURE

To date, associations between *in utero* and childhood exposure to ETS, pregnancy outcomes, and postnatal health consequences have been established predominantly by means of questionnaires and assessments of health status (e.g., pulmonary function test, occurrence of disease and infection). Very few studies have been conducted to quantify objectively the actual ETS exposure in infants and children. Some studies have assessed maternal exposure to active smoking or ETS quantitatively and have indirectly related the same exposure to the fetus and the child. Since the fetus and the child are different from the mother both pharmacokinetically and pharmacodynamically, efforts should be made to quantify and correlate the offspring's exposure to maternal, paternal, and total household exposure prior to associating the effects of ETS to adverse outcomes in health.

Various methods have been used to estimate total environmental exposure to tobacco smoke. Many of these attempts, however, are limited by the subjective nature of the techniques employed. Data collected from questionnaires may not reflect true household exposure, as many pregnant women and parents of children suffering from certain illnesses tend to underreport their actual daily smoking activities.[16,66,67] Using the number of cigarettes smoked to estimate systemic exposure is complicated by interindividual variations in smoking habits, such as the extent of inhalation and number of puffs per cigarette smoked. Therefore, objective assessment of ETS exposure must account for the concentration of the desirable biological marker in the environment in which the exposure occurs, as well as individual differences in the distribution and elimination of the biological marker quantified.[68,69] Variables such as the size and occupancy of the room and ventilation measures may cause fluctuations in the actual exposure experienced by the passive smoker.[69]

There are various biological markers of tobacco smoke exposure that have been used clinically.[70] These include carbon monoxide, cyanide (which is metabolized to thiocyanate in the human body), adducts to blood cells and albumin, as well as urinary excretion products like nitrosoamines and hydroxyproline. In addition, urinary mutagenicity has also been documented as a potential marker for ETS exposure. However, these chemicals are *non-specific* (i.e., high baseline values in ETS non-exposed passive smokers and significant contribution from other environmental sources) and *insensitive* (i.e., insignificant increment from baseline levels even when ETS exposure is high) markers of ETS exposure.[70] Cotinine, the principle oxidative metabolite of nicotine, is currently the most sensitive and specific biological marker for the objective assessment of ETS exposure that can be quantified in various body matrices (e.g., blood, saliva, urine, and hair).[70,71] It is preferred over nicotine because of its relatively longer elimination half-life (e.g., 19 h compared to 2 h for nicotine) and the absence of ambient contamination (i.e., cotinine enters the systemic circulation after hepatic metabolism of nicotine).[62] Since plasma and urinary cotinine analysis can only reflect recent exposures, a biological marker for chronic exposure to tobacco smoke has to be employed to assess the potential long-term adverse effects from ETS.

18.7 HAIR ANALYSIS IN THE ASSESSMENT OF ETS EXPOSURE

In the past decade, hair analysis has emerged as a sensitive and reliable technique for drug screening and quantification. Neonatal hair is a useful matrix that can provide a long-term history of drug exposure, and its collection is both convenient and non-invasive. Hair starts to grow in the last trimester of pregnancy and is able to provide evidence of intrauterine drug exposure for several months postnatally until the neonatal hair is replaced. Cotinine begins to incorporate into fetal hair as it becomes available in the local bloodstream, perfusing the hair shaft and growing cells in the hair follicle. Rates of incorporation depend on physicochemical factors such as melanin affinity, lipophilicity, and membrane permeability. Given that hair grows at an average of 1 cm/month, the hair is the single body matrix that provides a well-defined history of drug exposure that is unachievable by blood or urine analysis.[73–75]

At the Hospital for Sick Children in Toronto, Canada, we have conducted several studies in which the presence of nicotine and cotinine in hair and urine have been tested as biological markers for (1) *in utero* exposure to active maternal smoking and ETS, (2) ETS exposure in healthy children, and (3) ETS exposure in children with respiratory diseases. These are the first documentations of the relevance of such biomarkers used for the identification and quantification of the true extent of exposure to tobacco smoke in pregnancy and childhood.

18.7.1 PROCEDURE OF HAIR AND URINE ANALYSIS

Hair samples are usually collected by cutting a small amount of hair (10 to 25 shafts) close to the scalp from the posterior vertex (for adults) or the occipital region (for neonates and children). For neonates, the longest hair is usually obtained from the occipital region of the scalp because it may represent the earliest development of scalp hair.[76] Collected samples are stored in individual envelopes at room temperature until the time of analysis in batches. An aliquot of the urine samples with the same numerical identification as the hair sample is stored at −20°C until the time of analysis.

Prior to analysis, the hair samples can be washed with a mild detergent to remove atmospheric contaminants. Samples are then rinsed with distilled water and dried in a warm (37°C) oven overnight. Subsequent studies involving the analysis of newborn and young children's hair have omitted this wash-out step since no difference was found between the concentrations of analytes from washed and unwashed hair. Next, the hair is minced into small segments of 2 to 3 mm with a fine scissors, and approximately 2 to 5 mg of each sample are weighed on an analytical balance. The hair samples are then digested at 50°C overnight in 1 ml of 0.6 N sodium hydroxide, followed by neutralization with 50 to 70 µl of concentrated hydrochloric acid on the next day. Neutral solutions of 100 µl aliquots are analyzed for nicotine and/or cotinine using a radioimmunoassay previously described by Langone et al.[77] For quantification, nicotine and cotinine standards (0.5 to 50 ng/ml and 0.2 to 20 ng/ml, respectively) are used. Results are expressed in nanograms of nicotine per milligram of hair. Urinalysis of cotinine employs the same radioimmunoassay, and results are expressed as nanogram per milliliter and standardized per milligram of creatinine.

18.7.2 THE TORONTO STUDY

In a prospective study, 94 mother–infant pairs were recruited from the nurseries in two Toronto hospitals after informed consent was obtained.[78,79] The mothers were classified as active smokers, passive smokers, and non-smokers based on their self-reported exposure to tobacco smoke on a questionnaire with detailed inquiries about their daily smoking behavior. Hair samples were collected from the mothers and infants shortly after birth and were analyzed for nicotine and cotinine using the radioimmunoassay aforementioned. Active-smoking mothers had hair concentrations of 19.2 ± 4.9 ng/mg and 6.3 ± 4.0 ng/mg for nicotine and cotinine, respectively (mean ± SEM). These concentrations were significantly higher than those found in the non-smoking group of mothers (1.2 ± 0.4 ng/mg for nicotine and 0.3 ± 0.06 ng/mg for cotinine; $p < 0.0001$). Infants of smokers had significantly higher mean concentrations of both hair nicotine and cotinine as compared to those measured in infants from non-smokers (2.4 ± 0.9 vs. 0.4 ± 0.09 for nicotine and 2.8 ± 0.8 vs. 0.26 ± 0.04 for cotinine; p <0.01). Passive-smoking mothers and their infants had significantly higher nicotine and cotinine hair concentrations than non-smoking mothers and infants (3.2 ± 0.8 ng/mg and 0.9 ± 0.3 ng/mg for maternal nicotine and cotinine, respectively; 0.28 ± 0.05 ng/mg and 0.6 ± 0.15 ng/mg for infant nicotine and cotinine, respectively; p <0.01). A significant correlation between maternal and neonatal hair concentrations of nicotine (r = 0.49, p <0.001) or cotinine (r = 0.85, p <0.0001) was found. However, the number of cigarettes smoked daily did not correlate with either maternal or neonatal hair concentrations of nicotine or cotinine in this population. This was the first direct biochemical evidence that mothers and especially infants exposed to ETS are at risk of measurable exposure to constituents of cigarette smoke. The accumulation of cigarette smoke constituents in the hair reflects long-term systemic exposure to these toxins and carcinogens and may be related to increased health risks in the neonate. In fact, when pregnancy and fetal outcomes were analyzed together with the measured concentrations of nicotine and cotinine in maternal and fetal hair, neonates of active smokers were found to have significantly more adverse outcomes as compared to other infants (including lower birthweight, smaller head circumference, shorter length, and more perinatal complications). A dose-response relationship between ETS exposure and hair accumulation of ETS constituents was also evident in this study. Neonatal hair concentrations of nicotine and cotinine in the infants residing in homes with both maternal and other household smokers were threefold higher than those infants residing in homes where their mothers were the only smoker.[79] This and an earlier study conducted on adult smokers[80] suggest that hair analysis is a reliable non-invasive method for the determination and validation of self-reported exposure to active smoking and ETS.

18.7.3 RACIAL DIFFERENCES IN SYSTEMIC EXPOSURE TO ETS

Black children and adults have been documented to be more susceptible to various tobacco smoke-related health hazards for reasons yet to be determined. A study was designed to investigate the racial differences in systemic exposure to cotinine in Black and White children in attempt to provide a pharmacokinetic reasoning behind this apparent increased susceptibility.[81] This observational study included 169 non-smoking children between 2 and 18 years of age who were recruited from a consulting pediatric office in Toronto after informed consent or assent were obtained. A detailed questionnaire was administered to and completed by the parents and the children. Hair and urine samples were collected and analyzed for cotinine as previously described. The number of cigarettes that each child was exposed to correlated significantly with urinary cotinine (r = 0.68, p = 0.0001) and hair cotinine concentrations (r = 0.19, p = 0.02). Urinary and hair cotinine concentrations were also significantly correlated (r = 0.3, p = 0.0005). Despite lower daily smoking activities reported among Black parents as compared to White parents (6.63 ± 3 vs. 12 ± 1.8 cigarettes per day, mean ± SEM, p = 0.2), Black children had higher hair concentrations of cotinine than White children (0.89 ± 0.25 vs. 0.48 ± 0.05 ng/mg, p = 0.05). Since urinary cotinine is the best estimate of total environmental dose of ETS and hair cotinine best reflects cumulative systemic ETS exposure, the ratio of urinary to hair cotinine concentration is an indirect estimate of the total body clearance of nicotine acquired from ETS exposure. Black children were found to have a twofold higher ratio of hair to urine cotinine concentration than White children (0.035 ± 0.01 vs. 0.019 ± 0.002 ng/mg; $p = 0.004$), reflecting higher systemic accumulation under similar exposure. For each cigarette that the Black children were exposed to on a per day basis, the resulting urinary cotinine concentration was twofold higher (14.7 ± 5.2 vs. 6.3 ± 1.2 ng/mg, $p = 0.02$). No significant differences in any of the measures were found between White children with dark or fair hair, suggesting that the apparent difference between Black and White children in the handling of ETS may have a pharmacokinetic and pharmacogenetic basis. The data presented suggest that Black children may be pharmacokinetically different from White children, as reflected by a higher systemic exposure, particularly to nicotine from ETS. In a recent study, Black children whose mothers smoked during pregnancy and postpartum were found to have reduced pulmonary function as compared to White children.[82] Whether this decrease in lung function is related to the pharmacokinetic changes observed in Black children requires further investigation.

18.7.4 FINDINGS IN CHILDREN WITH RESPIRATORY DISEASES

18.7.4.1 Asthma

In the previous sections, data have shown that the prevalence and incidence of asthma correlate directly with parental and/or total household smoking activities. In the two previous studies conducted by us in Toronto, hair concentrations of cotinine have been shown to reflect systemic exposure to tobacco smoke in both children and adults. Since it is not known why some children heavily exposed to ETS do not develop asthma while others do, a study was designed to investigate whether asthmatic children handle nicotine differently from non-asthmatic children exposed to similar degrees of ETS.[83] Seventy-eight asthmatic children and 86 control children between the ages of 2 and 18 years were enrolled in the study while attending a consulting pediatric clinic in Toronto after informed consents or assents were obtained. A questionnaire completed by the parents and children included questions about the daily number of cigarettes the children were exposed to and the identities of the smokers in the household. Urinary and hair samples were collected and analyzed for cotinine as previously described. The asthmatic and control children cohorts were matched for age, gender, ethnic distribution, parental education, and socioeconomic status. Lower urinary cotinine concentrations corrected for creatinine were measured in the asthmatic children (47.1 ± 9.1 vs. 62.5 ± 11.5 ng/mg), which was in agreement with the lower reported daily number of cigarettes by their parents (7.4 ± 1.3 vs. 11.2 ± 2.3/day, $p = 0.14$). Despite lower total environmental exposure, asthmatic children accumulated twofold higher cotinine in their hair than control children (0.696 ± 0.742 vs. 0.386 ± 0.383 ng/mg, $p = 0.0001$). The hair to urinary cotinine concentration ratio was significantly higher in children with asthma than in controls (0.028 ± 0.002 vs. 0.18 ± 0.003 ng/mg, $p = 0.0001$). As suggested by these results, children with asthma experienced twofold higher systemic exposure to nicotine from ETS than normal children under similar degrees of total environmental exposure. This apparent increase in systemic exposure may be attributed to a lower clearance rate. This is the first suggestion of pharmacokinetic predisposition of asthmatic children to ETS. Whether this change in the pharmacokinetic handling of nicotine and potentially other constituents of ETS in children with asthma is directly related to reduced lung function and increased risks of respiratory illnesses has to be thoroughly investigated.

18.7.4.2 Cystic Fibrosis

A pilot study was conducted in a group of children with cystic fibrosis (CF) and their normal healthy siblings. The primary objective was to investigate whether there were pharmacokinetic differences in the handling of nicotine in children with CF as compared to their siblings who were exposed to similar degrees of ETS. Sixty-six children with CF between the ages of 6 months and 7 years were recruited from a Toronto hospital CF clinic. Forty-nine siblings between ages 10 and 17 years participated in the control cohort. Hair and urine samples were collected as previously described after informed consent or assent were obtained. Only 61 patients and 29 siblings were able to provide both urine and hair samples. Data collected from the 29 patient–sibling pairs were analyzed using the Wilcoxon rank test. Statistical analysis of the laboratory data suggested a similar pharmacokinetic predisposition of children with CF to nicotine from ETS as in the asthmatic children previously studied (data shown in Table 18.1). Despite a lower mean environmental exposure to nicotine from ETS, higher systemic exposure to nicotine was found in the CF cohort as reflected by a greater accumulation of cotinine in the hair. Nicotine clearance calculated from the urinary to hair cotinine concentration ratio was threefold lower in the CF cohort. No direct correlation was found between the CF group and their siblings for all the variables analyzed. Demographic and clinical profiles (related to the extent and nature of the ETS exposure) were not available for the present pilot study.

TABLE 18.1
Comparison Between Children with CF and Control Siblings in Systemic Exposure to Nicotine from ETS

Variable	CF children[a]	CF sibling[a]	Significance (*p* value)	Asthmatic children[b]
Hair cotinine (ng/ mg)	0.459 ± 0.482	0.320 ± 0.322	0.489	0.696 ± 0.742
Urinary cotinine (ng/ ml)	62.714 ± 100.823	116.025 ± 181.055	0.015	29.92 ± 43.33
$\dfrac{\text{Urinary cotinine}}{\text{Urinary creatinine}}$ (ng/mg)	49.726 ± 54.382	94.611 ± 135.046	0.5235	47.1 ± 82.1
$\dfrac{\text{Urinary cotinine}}{\text{Hair cotinine}}$	267.36 ± 609.573	664.718 ± 872.306	0.1982	87.16 ± 154.91

[a] Daily exposure data were not available from the CF study.
[b] Asthmatic children were exposed to an average of 7.41 ± 12.10 cigarettes per day.

This is the first suggestion of pharmacokinetic predisposition to nicotine from ETS in children with CF. Under similar environmental exposure, children with CF may be susceptible to higher systemic exposure to constituents of ETS as compared to their healthy siblings. Elevated exposure to the myriad reactive oxygen species in ETS may put the diseased individual at risk for poorer prognosis as a result of increased oxidative burden. It is known that lung dysfunction is a significant determinant of morbidity in individuals with CF. Whether this apparent change in the pharmacokinetics of ETS exposure is related to the pharmacogenetics and pathophysiology of pulmonary dysfunction in the CF disease needs to be investigated in a larger population.

18.8   CONCLUSION

It is evident that not only active maternal smoking causes increased risks of spontaneous abortion, perinatal morbidity, SIDS, low birthweight, and SGA. In fact, much of the available data suggests that *in utero* exposure to ETS by non-smoking pregnant women or postnatal exposure to ETS by the newborns and children are associated with various adverse pregnancy outcomes and health consequences later in childhood. It is becoming clear that exposure to ETS during pregnancy or early in childhood may induce changes in pulmonary function that may become persistent. Susceptibility to recurrent infections and illnesses may also be elevated. More studies have to be conducted in order to quantify objectively the true environmental exposure to tobacco smoke in the pediatric population. Available biological markers such as urinary and hair cotinine measurements are reliable and useful indicators that can be used to assess the extent of ETS exposure. The combination of objective assessment of ETS exposure both during pregnancy and in childhood can strengthen the validity of any associations that are to be drawn between increased ETS exposure and subsequent undesirable health consequences in the pediatric population.

ACKNOWLEDGMENT

This work was supported by a grant from the Medical Research Council of Canada and a Translational Research Grant from The Hospital for Sick Children, Toronto, Canada.

REFERENCES

1. Manuel, J., Double exposure, *Environ. Health Perspect.*, 107, A197, 1999.
2. Witschi, H., Joad, J.P., and Pinkerton, K.E., The toxicology of environmental tobacco smoke, *Annu. Rev. Pharmacol. Toxicol.*, 37, 29, 1997.
3. Eriksen, M.P., LeMaistre, C.A., and Newell, G.R., Health hazards of passive smoking, *Annu. Rev. Public Health*, 9, 47, 1998.
4. Committee on Environmental Health, Environmental tobacco smoke: a hazard to children, *Pediatrics*, 99, 639, 1997.
5. Law, M.R. and Hackshaw, A.K., Environmental tobacco smoke, *Br. Med. Bull.*, 52, 22, 1996.

6. Crawford, W.A., On the health effects of environmental tobacco smoke, *Arch. Environ. Health*, 43, 34, 1988.

7. Charlton, A., Children and passive smoking: a review, *J. Fam. Pract.*, 38, 267, 1994.

8. Koren, G., Fetal toxicology of environmental tobacco smoke, *Curr. Opin. Pediatr.*, 7, 128, 1995.

9. Stachenko, S.J., Reeder, B.A., Lindsay, E., Donovan, C., Lessard, R., and Balram, C., Smoking prevalence and associated risk factors in Canadian adults, *Can. Med. Assoc. J.*, 146, 1989, 1992.

10. Centers for Disease Control and Prevention, State-specific prevalence of cigarette smoking among adults, and children's and adolescents' exposure to environmental tobacco smoke — United States, 1996, *J. Am. Med. Assoc.*, 278, 2056, 1997.

11. Ward, S., Addressing nicotine addiction in women, *J. Nurse Midwif.*, 44, 3, 1999.

12. Smith, N., Austen, J., and Rolles, C.J., Tertiary smoking by the fetus, *Lancet*, 1, 82, 1982.

13. Rubin, D.H., Krasilnikoff, P.A., Leventhal, J.M., et al., Effect of passive smoking on birth weight, *Lancet*, 2, 415, 1986.

14. Leech, J.A., Wilby, K., and McMullen, E., Environmental tobacco smoke exposure patterns: a subanalysis of the Canadian human time-activity pattern survey, *Can. J. Public Health*, 90, 244, 1999.

15. Butz, A.M. and Rosenstein, B.J., Passive smoking among children with chronic respiratory disease, *J. Asthma*, 29, 265, 1992.

16. Kohler, E., Sollich, V., Schuster, R., and Thal, W., Passive smoke exposure in infants and children with respiratory tract diseases, *Hum. Exp. Toxicol.*, 18, 212, 1999.

17. Rebagliato, M., Bolumar, F., and Florey, C.duV., Assessment of exposure to environmental tobacco smoke in nonsmoking pregnant women in different environments of daily living, *Am. J. Epidemiol.*, 142, 525, 1995.

18. Jordaan, E.R. and Potter, P., Environmental tobacco smoke exposure in children: household and community determinants, *Arch. Environ. Health*, 54, 319, 1999.

19. Cook, D.G., Whincup, P.H., Jarvis, M.J., Strachan, D.P., Papacosta, O., and Bryant, A., Passive exposure to tobacco smoke in children aged 5-7 years: individual, family, and community factors, *Br. Med. J.*, 308, 383, 1994.

20. Dell'Orco, V., Forastiere, F., Agabiti, N., Corbo, G.M., Pistelli, R., Pacifici, R., Zuccaro, P., Pizzabiocca, A., Rosa, M., Altieri, I., and Perucci, C.A., Household and community determinants of exposure to involuntary smoking: a study of urinary cotinine in children and adolescents, *Am. J. Epidemiol.*, 142, 419, 1995.

21. Wilcox, A.J., Birth weight and perinatal mortality: the effect of maternal smoking, *Am. J. Epidemiol.*, 137, 1098, 1993.

22. Windham, G.C., von Behren, J., Waller, K., and Fenster, L., Exposure to environmental and mainstream tobacco smoke and risk of spontaneous abortion, *Am. J. Epidemiol.*, 149, 243, 1999.

23. Windham, G.C., Swan, S.H., and Fenster, L., Parental cigarette smoking and the risk of spontaneous abortion, *Am. J. Epidemiol.*, 135, 1394, 1992.

24. Dybing, E. and Sanner, T., Passive smoking, sudden infant death syndrome (SIDS) and childhood infections, *Hum. Exp. Toxicol.*, 18, 202, 1999.

25. Anderson, H.R. and Cook, D.G., Passive smoking and sudden infant death syndrome: review of the epidemiological evidence, *Thorax*, 52, 1003, 1997.

26. Golding, J., Sudden infant death syndrome and parental smoking — a literature review, *Pediatr. Perinat. Epidemiol.*, 11, 57, 1997.

27. Sadler, L., Belanger, K., Saftlas, A., Leaderer, B., Hellenbrand, K., and McSharry, J.-E., Environmental tobacco smoke exposure and small-for gestational-age birth, *Am. J. Epidemiol.*, 150, 695, 1999.

28. Fortier, I., Marcoux, S., and Brisson, J., Passive smoking during pregnancy and the risk of delivering a small-for gestational-age infant, *Am. J. Epidemiol.*, 139, 294, 1994.

29. Windham, G.C., Eaton, A., and Hopkins, B., Evidence for an association between environmental tobacco smoke exposure and birthweight: a meta-analysis and new data, *Paediatr. Perinat. Epidemiol.*, 13, 35, 1999.

30. Peacock, J.L., Cook, D.G., Carey, I.M., Jarvis, M.J., Bryant, A.E., and Anderson, H.R., Maternal cotinine level during pregnancy and birthweight for gestational age, *Int. J. Epidemiol.*, 27, 647, 1998.

31. Eskenazi, B., Prehn, A.W., and Christianson, R.E., Passive and active smoking as measured by serum cotinine: the effect on birthweight, *Am. J. Public Health,* 85, 395, 1995.

32. Rebagliato, M., Florey, C.duV., and Bolumar, F., Exposure to environmental tobacco smoke in nonsmoking pregnant women in relation to birth weight, *Am. J. Epidemiol.*, 142, 531, 1995.

33. Lazzaroni, F., Bonassi, S., Manniello, E., Morcaldi, L., Repetto, E., Ruocco, A., Calvi, A., and Cotellessa, G., Effect of passive smoking during pregnancy on selected perinatal parameters, *Int. J. Epidemiol.*, 19, 960, 1990.

34. Ahluwalia, I.B., Grummer-Strawn, L., and Scanlon, K.S., Exposure to environmental tobacco smoke and birth outcome: increased effects on pregnant women aged 30 years or older, *Am. J. Epidemiol.*, 146, 42, 1997.

35. Eskenazi, B. and Bergmann, J.J., Passive and active maternal smoking during pregnancy, as measured by serum cotinine, and postnatal smoke exposure. I. Effects on physical growth at age 5 years, *Am. J. Epidemiol.*, 142, s10, 1995.

36. Fried, P.A., Watkinson, B., and Gray, R., Growth from birth to early adolescence in offspring prenatally exposed to cigarettes and marijuana, *Neurotox. Teratol.*, 21, 513, 1999.

37. Sasco, A.J. and Vainio, J., From in utero and childhood exposure to parental smoking to childhood cancer: a possible link and the need for action, *Hum. Exp. Toxicol.*, 18, 192, 1999.

38. Boffetta, P., Tredaniel, J., and Greco, A., Risk of childhood cancer and adult lung cancer after childhood exposure to passive smoke: a meta analysis, *Environ. Health Perspect.*, 108, 73, 2000.

39. Filipini, G., Farinotti, M., and Ferrarini, M., Active and passive smoking during pregnancy and risk of central nervous system tumors in children, *Paediatr. Perinat. Epidemiol.*, 14, 78, 2000.

40. Rush, D. and Callahan, K.R., Exposure to passive cigarette smoking and child development, *Ann. N. Y. Acad. Sci.,* 562, 74, 1989.

41. Eskenazi, B. and Castorina, R., Association of prenatal maternal or postnatal child environmental tobacco smoke exposure and neurodevelopment and behavioral problems in children, *Environ. Health Perspect.,* 107, 991, 1999.

42. Eskenazi, B. and Trupin, L.S., Passive and active maternal smoking during pregnancy, as measured by serum cotinine, and postnatal smoke exposure. II. Effects on neurodevelopment at age 5 years, *Am. J. Epidemiol.*, 142, s19, 1995.

43. Karaman, O., Uguz, A., and Uzuner, N., Risk factors in wheezing infants, *Pediatr. Int.*, 41, 147, 1999.

44. Hu, F.B., Persky, V., Flay, B.R., Zelli, A., Cooksey, J., and Richardson, J., Prevalence of asthma and wheezing in the public schoolchildren: association with maternal smoking during pregnancy, *Ann. Allergy Asthma Immunol.*, 79, 80, 1997.

45. Gergen, P.J., Fowler, J.A., Maurer, K.R., Davis, W.W., and Overpeck, M.D., The burden of environmental tobacco smoke exposure on the respiratory health of children 2 months through 5 years of age in the United States: third national health and nutritional examination survey, 1988 to 1994, *Pediatrics*, 102, E8, 1998.

46. Carlsen, K.C.L., Jakkola, J.J.K., Nafstad, P., and Carlsen, K.-H., In utero exposure to cigarette smoking influences lung function at birth, *Eur. Respir. J.*, 10, 1774, 1997.

47. Hanrahan, J.P., Tager, I.B., Segal, M.R., Tosteson, T.D., Castile, R.G., van Vunakis, H., Weiss, S.T., and Speizer, F.E., The effect of maternal smoking during pregnancy on early infant lung function, *Am. Rev. Respir. Dis.*, 145, 1129, 1992.

48. Tager, I.B., Hanrahan, J.P., Tosteson, T.D., Castile, R.G., Brown, R.W., Weiss, S.T., and Speizer, F.E., Lung function, pre- and post-natal smoke exposure, and wheezing in the first year of life, *Am. Rev. Respir. Dis.*, 147, 811, 1993.

49. Tager, I.B., Ngo, L., and Hanrahan, J.P., Maternal smoking during pregnancy: effects on lung function during first 18 months of life, *Am. J. Respir. Crit. Care Med.*, 152, 977, 1995.

50. Cunningham, J., Dockery, D.W., and Speizer, F.E., Maternal smoking during pregnancy as predictor of lung function in children, *Epidemiology*, 139, 1139, 1994.

51. Gilliland, F.D., Berhane, K., McConnell, R., Gauderman, W.J., Vora, H., Rappaport, E.B., Avol, E., and Peters, J.M., Maternal smoking during pregnancy, environmental tobacco smoke exposure and childhood lung function, *Thorax*, 55, 271, 2000.

52. Barber, K., Mussin, E., and Taylor, D.K., Fetal exposure to involuntary maternal smoking and childhood respiratory disease, *Ann. Allergy Asthma Immunol.*, 76, 427, 1996.

53. Jedrychowski, W. and Flak, E., Maternal smoking during pregnancy and postnatal exposure to environmental tobacco smoke as predisposition factors to acute respiratory infections, *Environ. Health Perspect.*, 105, 302, 1997.

54. Strachan, D.P. and Cook, D.G., Parental smoking and childhood asthma: longitudinal and case-control studies, *Thorax*, 53, 204, 1998.

55. Strachan, D.P. and Cook, D.G., Parental smoking and prevalence of respiratory symptoms and asthma in school age children, *Thorax*, 52, 1081, 1997.

56. Cunningham, J., O'Connor, G.T., Dockery, D.W., and Speizer, F.E., Environmental tobacco smoke, wheezing, and asthma in children in 24 communities, *Am. J. Respir. Crit. Care Med.*, 153, 218, 1996.

57. Willers, S., Svenonius, E., and Skarping, G., Passive smoking and childhood asthma, *Allergy*, 46, 330, 1991.

58. Li, J.S.M., Peat, J.K., Xuan, W., and Berry, G., Meta-analysis on the association between environmental tobacco smoke (ETS) exposure and the prevalence of lower respiratory tract infection in early childhood. *Pediatr. Pulmonol.*, 27, 5, 1999.

59. Cook, D.G. and Strachan, D.P., Summary of effects of parental smoking on the respiratory health of children and implications for research, *Thorax*, 54, 357, 1999.

60. Strachan, D.P. and Cook, D.G., Parental smoking and lower respiratory illness in infancy and early childhood, *Thorax*, 52, 905, 1997.

61. Oldigs, M., Jorres, R., and Magnussen, H., Acute effect of passive smoking on lung function and airway responsiveness in asthmatic children, *Pediatr. Pulmonol.*, 10, 123, 1991.

62. Goren, A.I. and Hellmann, S., Respiratory conditions among schoolchildren and their relationship to environmental tobacco smoke and other combustion products, *Arch. Environ. Health*, 50, 112, 1995.

63. Corbo, G.M., Agabiti, N., Forastiere, F., Dell'Orco, V., Pistelli, R., Kriebel, D., Pacifici, R., Zuccaro, P., Ciappi, G., and Perucci, C.A., Lung function in children and adolescents with occasional exposure to environmental tobacco smoke, *Am. J. Respir. Crit. Care Med.*, 154, 695, 1996.

64. Wang, X., Wypij, D., Gold, D.R., Speizer, F.E., Ware, J.H., Ferris, B.G., Jr., and Dockery, D.W., A longitudinal study of the effects of parental smoking on pulmonary function in children 6-18 years, *Am. J. Respir. Crit. Care Med.*, 149, 1420, 1994.

65. Willers, S., Attewell, R., Bensryd, I., Schutz, A., Skarping, G., and Vahter, M., Exposure to environmental tobacco smoke in the household and urinary cotinine excretion, heavy metals retention, and lung function, *Arch. Environ. Health*, 47, 357, 1992.

66. Clark, S.J., Warner, J.O., and Dean, T.P., Passive smoking amongst asthmatic children. Questionnaire or objective assessment?, *Clin. Exp. Allergy*, 24, 276, 1994.

67. Pattishall, E.N., Strope, G.L., Etzel, R.A., Helms, R.W., Haley, N.J., and Denny, F.W., Serum cotinine as a measure of tobacco smoke exposure in children, *Am. J. Dis. Child.*, 139, 1101, 1985.

68. Klein, J. and Koren, G., Hair analysis — a biological marker for passive smoking in pregnancy and childhood, *Hum. Exp. Toxicol.*, 18, 279, 1999.

69. Chan, C.C., Chen, S.C., and Wang, J.D., Relationship between indoor nicotine concentrations, time-activity data, and urine cotinine-creatinine ratios in evaluating children's exposure to environmental tobacco smoke, *Arch. Environ. Health*, 50, 230, 1995.

70. Benowitz, N.L., Biomarkers of environmental tobacco smoke exposure, *Environ. Health Perspect.*, 107(s2), 349, 1999.

71. Phillips, K., Bentley, M.C., Abrar, M., Howard, D.A., and Cook, J., Low level saliva cotinine determination and its application as a biomarker for environmental tobacco smoke exposure, *Hum. Exp. Toxicol.*, 18, 291, 1999.

72. Haufroid, V. and Lison, D., Urinary cotinine as a tobacco-smoke exposure index: a minireview, *Int. Arch. Occup. Environ. Health*, 71, 162, 1998.

73. Klein, J. and Koren, G., Testing for drugs of abuse in the pediatric population, in *Drug Testing Technology*, CRC Press, Boca Raton, FL, 1999, chap. 9.

74. Wennig, R., Potential problems with the interpretation of hair analysis results, *Forensic. Sci. Int.*, 107, 5, 2000.

75. Bailey, B., Klein, J., and Koren, G., Noninvasive methods for drug measurement in pediatrics, *Pediatr. Clin. N. A.*, 44, 15, 1997.

76. Pecoraro, V. and Astore, I.P.L., Measurements of hair growth under physiological conditions, in *Hair and Hair Diseases*, Springer-Verlag, New York, 1990, 237.

77. Langone, J.J., Gjika, H.B., and van Vunakis, H., Nicotine and its metabolites, radioimmunoassays for nicotine and cotinine, *Biochemistry*, 12, 5025, 1973.

78. Eliopoulos, C., Klein, J., Phan, M.K., Knie, B., Greenwald, M., Chitayat, D., and Koren, G., Hair concentrations of nicotine and cotinine in women and their newborn infants, *J. Am. Med. Assoc.*, 271, 621, 1994.

79. Eliopoulos, C., Klein, J., Chitayat, D., Greenwald, M., and Koren, G., Nicotine and cotinine in maternal and neonatal hair as markers of gestational smoking, *Clin. Invest. Med.*, 19, 231, 1996.

80. Eliopoulos, C., Klein, J., and Koren, G., Validation of self-reported smoking by analysis of hair for nicotine and cotinine, *Ther. Drug Monit.*, 18, 532, 1996.

81. Knight, J.M., Eliopoulos, C., Klein, J., Greenwald, M., and Koren, G., Passive smoking in children. Racial differences in systemic exposure to cotinine by hair and urine analysis, *Chest*, 109, 446, 1996.

82. Cunningham, J., Dockery, D.W., Gold, D.R., and Speizer, F.E., Racial differences in the association between maternal smoking during pregnancy and lung function in children, *Am. J. Respir. Crit. Care Med.,* 152, 565, 1995.
83. Knight, J.M., Eliopoulos, C., Klein, J., Greenwald, M., and Koren, G., Pharmacokinetic predisposition to nicotine from environmental tobacco smoke: a risk factor for pediatric asthma, *J. Asthma,* 35, 113, 1998.

19 Cigarette Smoke and Allergic Asthma: Animal Models

Edward G. Barrett and David E. Bice

CONTENTS

19.1 INTRODUCTION

Individuals are routinely exposed to various environmental allergens including house dust mite, animal dander, molds, and pollen. Although most individuals respond to allergen exposure without triggering a debilitating inflammatory response in the lung, some individuals develop an allergic immune response that can cause asthma. The physiological mechanisms that mediate whether an individual responds to an allergen in a "normal" or asthmatic fashion remain poorly understood. However, evidence suggests that the nature of the response to allergen is related to the underlying T-helper (Th) cell cytokine response (reviewed in Reference 1 and 2). Asthma is currently defined as a chronic inflammatory (Th2-mediated) disease of the lung. Normal individuals respond to inhaled allergen through a Th1-mediated cytokine response (interleukin [IL-2] and interferon [IFN]-γ) which is critical in the development of cell-mediated immunity.[3,4] In contrast, asthmatics produce a Th2-mediated cytokine response (IL-4, IL-5, IL-6, IL-9, IL-10, and IL-13) following allergen exposure.[3,4] The strong genetic susceptibility for the development of asthma[5–7] suggests that inheritance influences the polarization of the allergic immune response toward a Th2 phenotype in asthma. However, despite the strong genetic influence in the development of asthma, other risk factors must be involved. Evidence showing that one twin can be asthmatic, while the identical twin is not asthmatic[5,8,9] supports the idea that other risk factors are involved in the development of asthma.

0-8493-0311-7/00/$0.00+$.50
© 2001 by CRC Press LLC

Increasing evidence indicates that smoking by adults is a risk factor for adult onset of asthma (reviewed in Chapter 6).[10] In addition, epidemiological studies have shown that exposure to environmental tobacco smoke during childhood has a dose-related effect on the risk of developing asthma.[11-14] Children of mothers smoking four cigarettes daily have a 14% increase in the prevalence of asthma, but children of mothers smoking 15 or more cigarettes daily have a 49% increased prevalence.[15] In viral infections and cigarette smoke inhalation, concomitant exposure to high levels of potent allergens may be essential for the development of long-term allergic disease.[16,17-20] The odds ratio on smoke exposure for the development of asthma increases when associated with home dampness (odds ratio 1.3) or cat or dog exposure (odds ratio 8.0).[21] Regardless of what triggering factors are responsible for the induction of Th2 immune responses in susceptible individuals, a major concern is that the establishment of Th2 immune responses in the lungs early in life could result in a permanent asthmatic status. Although epidemiological studies indicate that inhaled cigarette smoke induces a Th2 response in the lungs, few experimental studies have been performed to confirm these findings or examine the physiological mechanism.

There are an increasing number of publications in the literature that have utilized animal models to understand how Th2 cytokines and genes influence allergic sensitization and asthma. A complete review of all recent papers is beyond the scope of this chapter; however, selected articles are cited to develop a background from which to examine the use of these models in examining the contribution of cigarette smoke in inducing and/or triggering asthma. Also, several excellent reviews have recently been published examining the use of animal models in asthma research.[2,22]

19.2 ANIMAL MODELS OF ASTHMA

An ideal animal model of asthma should exhibit human characteristics of asthma, including paroxysmal bronchoconstriction, allergen-induced airway responses, late phase bronchoconstriction, airway inflammation including increased eosinophils and Th2 cytokines, increased mucus secretion associated with goblet cell metaplasia and hypertrophy, persistent airway hyperresponsiveness (AHR), and chronic airway remodeling associated with deterioration in lung function. Unfortunately, no single animal model possesses all these characteristics. Although some symptoms comparable to humans have been observed in horses,[23] dogs,[24-26] and cats,[27] animals do not spontaneously develop allergic asthma. Thus, it is extremely important to fully characterize an animal model in terms of which aspect of asthma is being modeled. Numerous animal species have been used to study the pathogenesis of asthma, including primates, sheep, dogs, rabbits, guinea pigs, rats, and mice. Special focus will be placed on rodents in this chapter since they have been used exclusively to examine the interactions of cigarette smoke in asthma.

In the last few years, mice and rats have been used extensively to understand the pulmonary immune response in asthma. Murine models in particular have gained popularity due to the availability of inbred strains, ability to overexpress and knockout specific genes, relatively low cost compared with larger animals, and availability of murine-specific assays/reagents. However, a significant disadvantage of the rodent models has been the inability to induce chronic inflammation associated with airway remodeling and persistent AHR. The reasons why rodents and other animals fail to develop the chronic inflammation seen in humans are unknown. Studies in both rats and mice demonstrate that whereas a single acute challenge with antigen leads to eosinophilic inflammation, mucus cell hyperplasia, an increase in Th2 cytokines, and AHR of variable degrees, repeated exposure to antigen induces tolerance.[28–31] Subsequently, rodent models typically follow a protocol where animals are immunized systemically by intraperitoneal injection with antigen-adjuvant (e.g., alum, *Bordetella pertussis*) and then challenged acutely with antigen to elicit an "allergic" response.

19.3 CIGARETTE SMOKE–ASTHMA ANIMAL MODELS

Although many different animal models of asthma exist, only rodent models have been utilized to examine the relationship between inhalation of cigarette smoke and the exacerbation of asthma. The use of small animals is predicated by the complexity and logistics of generating and exposing animals to controlled amounts of cigarette smoke. Various groups are using guinea pig, rat, and mouse models of asthma in combination with cigarette smoke exposure.

19.3.1 GUINEA PIGS

Guinea pigs are one of the earliest and most used animal models in asthma research, likely because of an easily demonstrated early bronchoconstriction after antigen sensitization and inhalation. Challenge of allergic guinea pigs with antigen produces an acute, histamine-mediated bronchoconstriction, a transient increase in AHR, and sustained lung eosinophilia.[32–34] Under certain experimental conditions, a late phase bronchochonstriction can be measured in the guinea pig.[35] Research with this animal model has helped elucidate the mechanisms associated with the relationship between AHR and airway inflammation. Many studies have examined the recruitment of eosinophils to the lung and their degranulation using the guinea pig model.[36–39] There are several disadvantages in using the guinea pig as a model of asthma. Unlike humans, the immune response to inhaled allergen in the guinea pig is driven by a subclass IgG (IgG$_1$) rather than IgE.[40–42] These animals have a prominent lung eosinophilia independent of sensitization and allergen challenge.[43,44] Also, the frequency, severity, and duration of the late phase response is highly variable. Finally, guinea pigs lack persistent AHR, and it also appears that AHR is mediated primarily by histamine, whereas anti-histamines are relatively ineffective in treating humans.[45]

Initial studies utilizing guinea pigs examined the direct effects of cigarette smoke on AHR independent of allergen sensititization. Acute exposure to cigarette smoke causes AHR in guinea pigs, which resolves within a few hours.[46,47] Although acute cigarette smoke exposure in this animal enhances AHR to histamine, mixed results have been observed with other bronchochonstricors such as acetylcholine.[46–48] Additionally, induction of AHR by acute cigarette smoke exposure is independent of airway inflammation.[49] In contrast, repeated exposure of guinea pigs to cigarette smoke leads to enhanced inflammation, characterized by an influx of neutrophils and eosinophils, and increased AHR that persists for several days after exposure.[50,51] Treatment with a specific platelet-activating factor antagonist prior to repeated cigarette smoke exposure significantly reduces inflammation and AHR, thus suggesting that airway inflammation may be linked to AHR in the repeated exposure model.[50] Other studies have suggested that cigarette smoke induces a biphasic bronchoconstriction in guinea pigs: the first phase is induced by a combination of cholinergic reflex and tachykinins, whereas the second phase involves cyclooxygenase metabolites.[51–53]

Few studies have examined whether the bronchoconstrictive effect of cigarette smoke is enhanced when AHR is induced by allergen sensitization in the guinea pig. A recent study found that ovalbumin sensitization further enhances the broncho-constrictive response to acute cigarette smoke exposure compared to the response in the nonsensitized guinea pig.[54] The animals were sensitized to ovalbumin by repeated inhalation exposures (5 min/d for 2 weeks). Following sensitization, animals were challenged acutely to cigarette smoke. Additionally, the study found that the enhanced AHR was mediated, in part, by endogenous tachykinins evoked by cigarette smoke-induced activation of lung C fibers.[54] To our knowledge, this is the only published study that has examined the effects of cigarette smoke in a sensitized guinea pig model of asthma.

19.3.2 RATS

The Brown Norway (BN) rat, which is a high IgE responder, has been the most common strain of rat used in rat models of asthma.[55] Other rat strains such as Wistar, Fisher, Donryn, Lewis, and Sprague-Dawley have also been used on a limited basis.[56–58] Rat strains such as Lewis and Sprague-Dawley are considered to be low IgE responders. Typically, rats are sensitized by intrapenitoneal or subcutaneous injections of ovalbumin with aluminum hydroxide as an adjuvant. In contrast to guinea pigs, the rat is a weak bronchoconstricor requiring much higher levels of antigen to induce this response. Additionally, bronchoconstriction in rats is primarily mediated by serotonin, whereas allergic bronchoconstriction in the guinea pig is mediated by mast cell-dependent histamine release.[59,60] Many rats exhibit both an early and late phase response after antigen inhalation, where the late phase response can be inhibited by leukotriene antagonists.[58,61,62] Also, most rats exhibit antigen-induced inflammation characterized by an accumulation of neutrophils, lymphocytes, and particularly eosinophils in lung tissue and lavage fluid and the induction of Th2 cytokines.[55,56,61,63] As with guinea pigs, there is some question about the association between airway inflammation and AHR in rats.[30,56,64]

None of the rat models of asthma have been utilized in conjunction with cigarette smoke exposure. However, exposure of nonsensitized rats to cigarette smoke leads to enhanced AHR.[65–67] As with guinea pigs, cigarette smoke-induced AHR in rats is associated with endogenous tachykinins.[68,69] The rat has been used to address the question of whether *in utero* and/or postnatal exposure to cigarette smoke results in impaired lung function and hyperresponsiveness. These studies are aimed at address-ing the mounting evidence that children raised in the homes of smokers have more respiratory ailments such as cough, wheeze, sputum production, asthma, and other respiratory illnesses.[11–14] Following exposure to sidestream smoke *in utero* and

3 weeks postnatally, 8-week-old rats showed increased AHR to methacholine challenge.[70,71] The increase in AHR was not associated with an increase in mast cells or pulmonary neuroendocrine cells, which contain several bronchoconstrictors such as bombesin, endothelin, and serotonin.[71] In addition, those animals that only received *in utero* or postnatal smoke exposure did not develop an increase in AHR.

19.3.3 MICE

In the last 10 years many different murine models of asthma have been developed. Mice have been particularly useful because of the availability of inbred strains, molecular and immunological reagents, and the ability to manipulate specific genes. The capability to evaluate immune response in mice deficient in a specific cell type such as mast cells or lymphocytes[72–74] or in mice deficient for specific cytokines (e.g., IL-4, IL-5, IL-10)[75,76] or other inflammatory mediators is a big advantage. Manipulation of various immune and genetic parameters in the mouse has greatly enhanced our understanding of the underlying immunological and genetic mechanisms associated with asthma.

As with rats, there are significant differences between the various strains of mice and their ability to develop an allergic response.[77] Those strains which do develop an allergic response resemble human asthmatics in several aspects. The TH1/Th2 profiles of sensitized mice are similar to humans.[78] Following immunization and challenge, antigen mice demonstrate peribronchiolar eosinophilia, an influx of CD4+ T cells, and enhanced AHR of varying degrees.[79,80] There are several important disadvantages of the murine model of asthma. As with other rodent models, the murine model is based on an acute antigen challenge without the development of chronic inflammation or significant airway remodeling. As with other rodent models, mice require immunization with antigen-adjuvant and then acute challenge with the antigen to elicit allergic symptoms. Like rats, mice also require high levels of antigen to induce AHR.[82] It has been difficult to demonstrate eosinophilic infiltration of the epithelium with evidence of eosinophil degranulation.[81] Also, demonstration of a late phase response in the mouse has proved problematic,[79] although Cieslewicz et al. has recently developed a model that exhibits a late phase response.[83]

Only one study has been published utilizing a mouse model of asthma in combination with cigarette smoke exposure.[84] In the study, Balb/c mice were either immunized with ovalbumin and then exposed to air or sidestream cigarette smoke or immunized with ovalbumin and then challenged with ovalbumin (60 min) during air or sidestream cigarette smoke exposure. In both exposure scenarios, cigarette smoke elicited a rapid and prolonged exaggerated response with respect to IgE, IgG1, eosinophils, and Th2 cytokines (IL-4 and IL-10). This is the first experimental study to demonstrate that sidestream smoke can enhance the allergic response to inhaled antigen.

The development of tolerance following chronic antigen exposure in the murine asthma models and other rodent models is a particular problem when trying to assess the effects of chronic cigarette smoke exposure. We are utilizing a murine model of asthma using a heterozygous, transgenic mouse (Tg+/–) that has had its germline DNA altered so that 20 to 30% of its T cells express a TCR specific for an epitope of ovalbumin (OVA).[85] An antibody (KJ1-26) recognizes the OVA-TCR, which allows monitoring of antigen-specific T cell traffic during evolution of an anti-OVA pulmonary immune response. The high percentage of responding T cells should significantly improve the sensitivity of assays to detect antigen-specific cytokine responses during the development of pulmonary inflammation and to assess effects on T cell cytokine secretion profiles following various experimental manipulations. This model will allow the examination of chronic effects of inhaled cigarette smoke on the development of pulmonary immunity and AHR to chronic antigen exposure.

We have found (Barrett and Bice, unpublished results) that (1) OVA-TCR-Tg+/– mice develop bronchial inflammation after 2 weeks of aerosolized OVA exposure and reach their maximum following 10 weeks, (2) OVA-TCR+ T cells could be detected in the lungs of aerosolized mice, and (3) there was an increase in the number of OVA-TCR+ cells in the lung associated with lymph nodes in OVA-aerosolized mice compared to air-exposed mice. OVA-TCR-Tg+/– transgenic mice were exposed to aerosolized OVA 6 h/day, 5 days/week (5 mg/m^3). OVA-aerosolized OVA-TCR-Tg+/– transgenic mice developed a peribronchial and perivascular inflammation, and the control air-exposed OVA-TCR-Tg+/– mice did not. Alterations in AHR could be detected after 6 weeks of OVA exposure in the OVA-TCR-Tg+/– mice. Antibodies specific for OVA (IgG1 and IgE) could be detected after 2 weeks of OVA exposure. This model should allow for the combined chronic exposure to cigarette smoke and antigen.

19.4 CONCLUSIONS

Each of the animal models currently used for studying asthma has advantages and disadvantages. The guinea pig model has a robust bronchoconstiction response that is lacking in other rodent models, but very few guinea pig-specific reagents (antibodies molecular probes, etc.) for manipulation are available. The rat has slightly more reagents available and has been useful, particularly the BN strain, in character-izing allergen-induced eosinophilia. By far, however, the mouse model has been used most frequently because of its relatively low cost, abundance of different strains and genetically modified animals, well-studied immune system, and a wide range of reagents. However, none of these rodent models with the exception of the OVA-TCR-Tg+/– mice can be exposed chronically without developing tolerance. Chronic antigen exposures are of particular importance in studying the effects of chronic cigarette smoke exposure on the induction/exacerbation of asthma in children. It is likely that combined chronic exposure to cigarette smoke and antigen during pregnancy and/or early childhood is key factor. Although some other animal models of asthma such as the dog and cat may be more physiologically relevant to human asthma,[22] the cost and complexity of generating cigarette smoke and antigen expo-sures for these large animals would be prohibitive.

Very few studies have been reported in the literature where these animal models of asthma have been utilized in combination with cigarette smoke exposure. The lack of studies in this area is primarily due to the complexity and cost of generating cigarette smoke exposure conditions. Those studies that have examined the combined effects of cigarette smoke and antigen challenge have focused on cigarette smoke-induced exacerbation of acute antigen challenge. New animal models such the OVA-TCR-Tg+/– mouse model need to be developed to examine the combined chronic effects of cigarette smoke and antigen exposure.

REFERENCES

1. Lipscomb, M.F. and Wilder, J.A., Immune dysregulation as a cause for allergic asthma, *Curr. Opin. Pulm. Med.*, 5, 10, 1999.
2. O'byrne, P.M. and Postma, D.S., The many faces of airway inflammation. Asthma and chronic obstructive pulmonary disease, *Am. J. Repir. Crit. Care Med.*, 159, S41, 1999.
3. Mossmann, T.R., Cherwinski, H., Bond, M.W., Gieldin, M.A., and Coffmann, R.L., Two types of murine helper T cell clone. I. Definition according to profiles of lymphokine activities and secreted proteins, *J. Immunol.*, 136, 2348, 1986.
4. Street, N.E. and Mossmann, T.R., Functional diversity of T-lymphocytes due to secretion of different cytokine patterns, *FASEB J.*, 5, 171, 1991.
5. Godfrey, S., Airway inflammation, bronchial reactivity and asthma, *Agents Actions Suppl.*, 40, 109, 1993.
6. Demoly, P., Bousquet, J., Godard, P., and Michel, F.B., The gene or genes of allergic asthma?, *Presse Med.*, 22, 817, 1993.
7. Akasawa, A., Koya, N., and Iikura, Y., Investigation of the genetics in allergic children: the first report — twin study on HLA in allergic disease, *Arerugi*, 40, 428, 1991.
8. Antti-Poika, M., Nordman, H., Koskenvuo, M., Kaprio, J., and Jalava, M., Role of occupational exposure to airway irritants in the development of asthma, *Int. Arch. Occup. Environ. Health*, 64, 195, 1992.
9. Nieminen, M.M., Kaprio, J., and Koskenvuo, M., A population-based study of bronchial asthma in adult twin pairs, *Chest*, 100, 70, 1991.
10. Bodner, C.H., Ross, S., Little, J., Douglas, J.G., Legge J.S., Friend, J.A., and Godden, D.J., Risk factors for adult onset wheeze: a case control study, *Am. J. Respir. Crit. Care Med.*, 157, 35, 1998.
11. Martinez, F.D., Wright, A.L., Taussig, L.M., Holberg, C.J., Halonen, M., and Morgan, W.J., Asthma and wheezing in the first six years of life, *N. Engl. J. Med.*, 332, 133, 1995.
12. Oliveti, J.F., Kercsmar, C.M., and Redline, S., Pre- and perinatal risk facotrs for asthma in inner city African-American children, *Am. J. Epidemiol.*, 143, 570, 1996.
13. Cook, D.J. and Strachan, D.P., Parental smoking and prevalence of respiratory symptoms and asthma in school age children, *Thorax*, 52, 1081, 1997.
14. Stick, S.M., Burton, P.R., Gurrin, L., Sly, P.D., and LeSouef, P.N., Effects of maternal smoking during pregnancy and a family history of asthma on respiratory function in newborn infants, *Lancet*, 348, 1060, 1996.
15. Neuspiel, D.R., Rush, D., Butler, N.R., Golding, J., Bijur, P.E., and Kurzon, M., Parental smoking and post-infancy wheezing in children: a prospective cohort study, *Am. J. Public Health*, 79, 168, 1989.
16. Rizzo, M.C., Arruda, L.K., Chapman, M.D., Fernandez-Caldas, E., Baggio, D., Platts-Mills, T.A., and Naspitz, C.K., IgG and IgE antibody responses to dust mite allergens among children with asthma in Brazil, *Ann. Allergy*, 71, 152, 1993.

17. Gelber, L.E., Seltzer, L.H., Bouzoukis, J.K., Pollart, S.M., Chapman, M.D., and Platts-Mills, T.A., Sensitization and exposure to indoor allergens as risk factors for asthma among patients presenting to hospital, *Am. Rev. Respir. Dis.*, 147, 573, 1993.

18. Sporik, R. and Platts-Mills, T.A.E., Epidemiology of dust-mite-related disease, *Exp. Appl. Acarol.*, 16(1–2), 141, 1992.

19. Call, R.S., Smith, T.F., Morris, E., Chapman, M.D., and Platts-Mills, T.A., Risk factors for asthma in inner city children, *J. Pediatr.*, 121, 862, 1992.

20. Duff, A.L. and Platts-Mills, T.A., Allergens and asthma, *Pediatr. Clin. N. Am.*, 39, 1277, 1992.

21. Sherrill, D.L., Martinez, F.D., Lebowitz, M.D., Holdaway, M.D., Flannery, E.M., Herbison, G.P., Stanton, W.R., Silva, P.A., and Sears, M.R., Longitudinal effects of passive smoking on pulmonary function in New Zealand children, *Am. Rev. Respir. Dis.*, 145, 1136, 1992.

22. Bice, D.E., Seagrave, J.C., and Green, F.H.Y., Animal models of asthma: potential usefulness for studying health effects of inhaled particles, *Inhal. Toxicol.*, 12, 2000.

23. Derksen, F.J., Robinson, N.E., Armstrong, P.J., Stick, J.A., and Slocombe, R.F., Airway reactivity in ponies with recurrent airway obstruction (heaves*), J. Appl. Physiol.*, 58, 598, 1985.

24. Bice, D.E., Williams, A.J., and Muggenburg, B.A., Long-term antibody production in canine lung allografts: implications in pulmonary immunity and asthma, *Am. J. Respir. Cell Mol. Biol.*, 14, 341, 1996.

25. Bice, D.E., Williams, A.J., and Muggenburg, B.A., Long-term antibody production in canine lung allografts: implications in pulmonary immunity and asthma, *Am. J. Respir. Cell Mol. Biol.*, 14(4), 341, 1996.

26. Emala, C., and Hirshman, C.A., Canine models of asthma and hyperresponsiveness, in *Allergy and Allergic Diseases*, Kay, A.B., Ed., Blackwell Science, Oxford, 1997, 1103.

27. Padrid, P., Snook, S., Finucane, T., Shiue, P., Cozzi, P., Solway, J., and Leff, A.R., Persistent airway hyperresponsiveness and histologic alterations after chronic antigen challenge in cats, *Am. J. Respir. Crit. Care Med.*, 151, 184, 1995.

28. Elwood, W., Lotvall, O., Barnes, P.J., and Chung, K.F., Characterization of allergen-induced inflammation and bronchial hyperresponsiveness in sensitized Brown Norway rats, *J. Allergy Clin. Immunol.*, 88, 951, 1991.

29. Holt, P.G. and McMenamin, C., Defence against allergic sensitization in the healthy lung: the role of inhalation tolerance, *Clin. Exp. Allergy*, 19, 255, 1989.

30. Haczku, A., Moqbel, R., Elwood, W., Sun, J., Kay, B., Barnes, P.J., and Chung, K.F., Effects of prolonged repeated exposure to ovalbumin in sensitized Brown Norway rats, *Am. J. Respir. Crit. Care Med.*, 150, 23, 1994.

31. McMenamin, C., Pimm, C., McKersey, M., and Holt, P.G., Regulation of IgE responses to inhaled antigen in mice by antigen-specific γδ T cells, *Science*, 265, 1869, 1994.

32. Yamada, N., Kadowaki, S., and Umezu, K., Development of an animal model of late asthmatic response in guinea pigs and effects of anti-asthmatic drugs, *Prostaglandins*, 43, 507, 1992.

33. Underwood, S., Foster, M., Raeburn, D., Bottoms, S., and Karlsson, J.A., Time-course of antigen-induced airway inflammation in the guinea-pig and its relationship to airway hyperresponsiveness, *Eur. Respir. J.*, 8, 2104, 1995.

34. Itoh, K., Takahashi, E., Mukaiyama, O., Satoh, Y., and Tamaguchi, T., Relationship between airway eosinophilia and airway hyperresponsiveness in a late asthmatic model of guinea pigs, *Int. Arch. Allergy Immunol.*, 109, 86, 1996.

35. Hutson, P.A., Church, M.K., Clay, T.P., Miller, P., and Holgate, S.T., Early and late-phase bronchoconstriction after allergen challenged nonanesthetized guinea pigs, *Am. Rev. Respir. Dis.,* 137, 548, 1988.

36. Rimmer, S.J., Akerman, C.L., Hunt, T.C., Church, M.K., Holgate, S.T., and Shute, J.K., Density profile of bronchoalveolar lavage eosinophils in the guinea pig model of allergen-induced late-phase allergic responses, *Am. J. Respir. Cell Mol. Biol.,* 6, 340, 1992.

37. Van Oosterhout, A.J.M., Ladenius, A.R.C., Savelkoul, H.F.J., Van Ark, I., Delsman, K.C., and Nijkamp, F.P., Effect of anti-IL-5 and IL-5 on airway hyperreactivity and eosinophils in guinea pigs, *Am. Rev. Respir. Dis.,* 147, 548, 1993.

38. Pretolani, M., Ruffie, C., Lapae Silva, J.-R., Joseph, D., Lobb, R.R., and Vargaftig, B.B., Antibody to very late activation antigen 4 prevents antigen-induced bronchial hyperreactivity and cellular filtration in the guinea-pig airways, *J. Exp. Med.,* 180, 795, 1994.

39. Minshall, E. and Sanjar, S., The sensitized guinea pig as a model of allergic asthma, in *Allergy and Allergic Diseases,* Kay, A.B., Ed., Blackwell Science, Oxford, 1997, 1093.

40. Andersson, P., Antigen-induced bronchial anaphylaxis in actively sensitized guinea-pigs. Pattern of response in relation to immunization regimen, *Allergy,* 35, 65, 1980.

41. Desquand, S., Rothhut, B., and Vargaftig, B.B., Role of immunoglobulins G1 and G2 in anaphylactiv shock in the guinea pig, *Int. Arch. Allergy Appl. Immunol.,* 93, 184, 1990.

42. Hsiue, J., Suzuki, S., Hotta, K., Hirota, Y., Mizuno, S., and Suzuki, K., Mite-induced allergic airway inflammation in guinea pigs, *Int. Arch. Allergy Immunol.,* 112, 295, 1997.

43. Banner, K.H., Paul, W., and Page, C.P., Ovalbumin challenge following immunization elicits recruitment of eosinophils but not bronchial hyperresponsiveness in guinea-pigs: time course and relationship to eosinophil activation status, *Pulm. Pharmacol.,* 9, 179, 1996.

44. Rothenberg, M.E., Luster, A.D., Lilly, C.M., Drazen, J.M., and Leder, P., Constitutive and allergen-induced expression of eotaxin mRNA in the guinea pig lung, *J. Exp. Med.,* 181, 1211, 1995.

45. Pretolani, M. and Vargaftig, B.B., From lung hypersensitivity to bronchial hyperreactivity. What can we learn from studies on animal models?, *Biochem. Pharmacol.,* 45, 791, 1993.

46. James, A.L., Dirks, P., Ohtaka, H., Schellenberg, R.R., and Hogg, J.C., Airway responsiveness to intravenous and inhaled acetylcholine in the guinea pig after cigarette smoke, *Am. Rev. Respir. Dis.,* 136, 1158, 1987.

47. Karlsson, J.A., Zackrisson, C., Sjolin, C., and Forsberg, K., Cigarette smoke-induced changes in guinea-pig airway responsiveness to histamine and citric acid, *Acta Physiol. Scand.,* 142, 119, 1991.

48. Omini, C., Hernandez, A., Zuccari, G., Clavenna, G., and Daffonchio, L., Passive cigarette smoke exposure induces airway hyperreactivity to histamine but not to acetylcholine in guinea-pigs, *Pulm. Pharmacol.,* 3, 145, 1990.

49. Nishikawa, M., Ikeda, H., Fukuda, T., Suzuki, S., and Okubo, T., Acute exposure to cigarette smoke induces airway hyperresponsiveness without airway inflammation in guinea pigs. Dose-response characteristics, *Am. Rev. Respir. Dis.,* 142, 177, 1990.

50. Matsumoto, K., Aizawa, H., Inoue, H., Koto, H., Takata, S., Shigyo, M., Nakano, H., and Hara, N., Eosinophilic airway inflammation induced by repeated exposure to cigarette smoke, *Eur. Respir. J.,* 12, 387, 1998.

51. Wu, Z.X. and Lee, L.Y., Airway hyperresponsiveness induced by chronic exposure to cigarette smoke in guinea pigs: role of tachykinins, *J. Appl. Physiol.*, 87, 1621, 1999.

52. Hong, J.L. and Lee, L.Y., Cigarette smoke-induced bronchoconstriction: causative agents and role of thromboxane receptors, *J. Appl. Physiol.*, 81, 2053, 1996.

53. Hong, J.L., Rodger, I.W., and Lee, L.Y., Cigarette smoke-induced bronchoconstriction: cholinergic mechanisms, tachykinins, and cyclooxygenase products, *J. Appl. Physiol.*, 78, 2260, 1995.

54. Wu, Z.X., Zhou, D., Chen, G., and Lee, L.Y., Airway hyperresponsiveness to cigarette smoke in ovalbumin-sensitized guinea pigs, *Am. J. Respir. Crit. Care Med.*, 161, 73, 2000.

55. Schneider, T., van Velzen, D., Moqbel, R., and Issekutz, A.C., Kinetics and quantitation of eosinophil and neutrophil recruitment to allergic lung inflammation in a brown Norway rat model, *Am. J. Respir. Cell Mol. Biol.*, 17, 702, 1997.

56. Kips, J.C., Cuvelier, C.A., and Pauwels, R.A., Effect of acute and chronic antigen inhalation on airway morphology and responsiveness in actively sensitized rats, *Am. Rev. Respir. Dis.*, 145, 1306, 1992.

57. Misawa, M. and Chiba, Y., Repeated antigenic challenge-induced airway hyperresponsiveness and airway inflammation in actively sensitized rats, *Jpn. J. Pharmacol.*, 61, 41, 1993.

58. Renzi, P.M., al Assaad, A.S., Yang, J., Yasruel, Z., and Hamid, Q., Cytokine expression in the presence or absence of late airway responses after antigen challenge of sensitized rats, *Am. J. Respir. Cell Mol. Biol.*, 15, 367, 1996.

59. Dandurand, R.J., Wang, C.G., Laberge, S., Martin, J.G., and Eidelman, D.H., In vitro allergic bronchochonstricition in the brown Norway rat, *Am. J. Respir. Crit. Care Med.*, 149, 1499, 1994.

60. Nagase, T., Fukuchi, Y., Dallaire, M.J., Martin, J.G., and Ludwig, M.S., In vitro airway and tissue response to antigen in sensitized rats. Role of serotonin and leukotriene D4, *Am. J. Respir. Crit. Care Med.*, 152, 81, 1995.

61. Renzi, P.M., Olivenstein, R., and Martin, J.G., Inflammatory cell populations in the airways and parenchyma after antigen challenge in the rat, *Am. Rev. Respir. Dis.*, 147, 967, 1993.

62. Wang, C.G., Du, T., Xu, L.J., and Martin, J.G., Role of leukotriene D4 in allergen-induced increases in airway smooth muscle in the rat, *Am. Rev. Respir. Dis.*, 148, 413, 1993.

63. Haczku, A., Macary, P., Haddad, E.B., Huang, T.J., Kemeny, D.M., Moqbel, R., and Chung, K.F., Expression of Th-2 cytokines interleukin-4 and -5 and of Th-1 cytokine interferon-gamma in ovalbumin-exposed sensitized Brown-Norway rats, *Immunology*, 88, 247, 1996.

64. Bellofiore, S. and Martin, J.G., Antigen challenge of sensitized rats increases airway responsiveness to methacholine, *J. Appl. Physiol.*, 65, 1642, 1988.

65. Fang, L.B., Morton, R.F., Wang, A.L., and Lee, L.Y., Bronchoconstriction and delayed rapid shallow breathing induced by cigarette smoke inhalation in anesthetized rats, *Lung*, 169, 153, 1991.

66. Xu, L.J., Dandurand, R.J., Lei, M., and Eidelman, D.H., Airway hyperresonsiveness in cigarette smoke-exposed rats, *Lung*, 171, 95, 1993.

67. Wright, J.L., Sun, J.P., and Churg, A., Cigarette smoke exposure causes constriction of rat lung, *Eur. Respir. J.*, 14, 1095, 1999.

68. Lundberg, J.M., Alving, K., Karlsson, J.A., Matran, R., and Nilsson, G., Sensory neuropeptide involvement in animal models of airway irritation and of allergen-evoked asthma, *Am. Rev. Respir. Dis.,* 143, 1429, 1991.
69. Joos, G.F. and Pauwels, R.A., The in vivo effect of tachykinins on airway mast cells of the rat, *Am. Rev. Respir. Dis.,* 148, 922, 1993.
70. Joad, J.P., Ji, C., Kott, K.S., Bric, J.M., and Pinkerton, K.E., In utero and postnatal effects of sidestream cigarette smoke exposure on lung function, hyperresponsiveness, and neuroendocrine cells in rats, *Toxicol. Appl. Pharmacol.,* 132, 63, 1995.
71. Joad, J.P., Bric, J.M., Peake, J.L., and Pinkerton, K.E., Perinatal exposure to aged and diluted sidestream cigarette smoke produces airway hyperresponsiveness in older rats, *Toxicol. Appl. Pharmacol.,* 155, 253, 1999.
72. Martin, T.R., Takeishi, T., Katz, H.R., Austen, K.F., Drazen, J.M., and Galli, S.J., Mast cell activation enhances airway responsiveness to methacholine in the mouse, *J. Clin. Invest.,* 91, 1176, 1993.
73. Kung, T.T., Stelts, D., Zurcher, J.A., Jones, H., Umland, S.P., Kreutner, W., Egan, R.W., and Chapman, R.W., Mast cells modulate allergic pulmonary eosinophils in mice, *Am. J. Respir. Cell Mol. Biol.,* 12, 404, 1995.
74. Duez, C., Kips, J., Pestel, J., Tournoy, K., Tonnel, A.B., and Pauwels, R., House dust mite-induced airway changes in hu-SCID mice, *Am. J. Respir. Crit. Care Med.,* 161, 200, 2000.
75. Hogan, S.P., Matthaei, K.I., Young, J.M., Koskinen, A., Young, I.G., and Foster, P.S., A novel T cell-regulated mechanism modulating allergen-induced airways hyper-reactivity in BALB/c mice independently of IL-4 and IL-5, *J. Immunol.,* 161, 1501, 1998.
76. Yang, X., Wang, S., Fan, Y., and Han, X., IL-10 deficiency prevents IL-5 over-production and eosinophilic inflammation in a murine model of asthma-like reaction, *Eur. J. Immunol.,* 30, 382, 2000.
77. Brewer, J.P., Kisselgof, A.B., and Martin, T.R., Genetic variability in pulmonary physiological, cellular, and antibody responses to antigen in mice, *Am. J. Respir. Crit. Care Med.,* 160, 1150, 1999.
78. Mossman, T.R., Cherwinski, H., Bond, M.W., Giedlin, M.A., and Coffman, R.L., Two types of murine helper T cell clone. I. Definition according to profiles of lymphokine activities and secreted proteins, *J. Immunol.,* 136, 2348, 1986.
79. Hessel, E.M., Van Oosterhout, A.J., Hofstra, C.L., De Bie, J.J., Garssen, J., Van Loveren, H., Verheyen, A.K., Savelkoul, H.F., and Nijkamp, F.P., Bronchoconstriction and airway hyperresponsiveness after ovalbumin inhalation in sensitized mice, *Eur. J. Pharmacol.,* 293, 401, 1995.
80. Garlisi, C.G., Falcone, A., Hey, J.A., Paster, T.M., Fernandez, X., Rizzo, C.A., Minnicozzi, M., Jones, H., Billah, M.M., Egan, R.W., and Umland, S.P., Airway eosinophils, T cells, Th-2 type cytokine mRNA, and hyperreactivity in response to aerosol challenge of allergic mice with previously established pulmonary inflammation, *Am. J. Respir. Cell Mol. Biol.,* 17, 642, 1997.
81. Stelts, D., Egan, R.W., Falcone, A., Garlisi, C.G., Gleich, G.J., Kreutner, W., Kung, T.T., Nahrebne, D.K., Chapman, R.W., and Minnicozzi, M., Eosinophils retain their granule major basic protein in a murine model of allergic pulmonary inflammation, *Am. J. Respir. Cell Mol. Biol.,* 18, 463, 1998.
82. Martin, T.R., Gerard, N.P., Galli, S.J., and Drazen, J.M., Pulmonary responses to bronchoconstrictor agonists in the mouse, *J. Appl. Physiol.,* 64, 2318, 1988.

83. Cieslewicz, G., Tomkinson, A., Adler, A., Duez, C., Schwarze, J., Takeda, K., Larson, K.A., Lee, J.J., Irvin, C.G., and Gelfand, E.W., The late, but not early, asthmatic response is dependent on IL-5 and correlates with eosinophil infiltration, *J. Clin. Invest.*, 104, 301, 1999.

84. Seymour, B.W.P., Pinkerton, K.E., Friebertshauser, K.E., Coffman, R.L., and Gershwin, L.J., Second-hand smoke is an adjuvant for T helper-2 responses in a murine model of allergy, *J. Immunol.*, 159, 6169, 1997.

85. Murphy, K.M., Heimberger, A.B., and Loh, D.Y., Induction by antigen of intrathymic apoptosis of CD4+CD8+TCRlo thymocytes in vivo, *Science*, 250, 1720, 1990.

20 Ventilation Control of Environmental Tobacco Smoke

D. Jeff Burton

CONTENTS

0-8493-0311-7/00/$0.00+$.50
© 2001 by CRC Press LLC

20.1   INTRODUCTION

Ventilation is required for every human occupancy and is routinely used to provide thermal comfort and to control airborne contaminants. The three major categories of ventilation used for airborne contaminant control include (1) general or dilution ventilation, (2) displacement ventilation, and (3) local source or local exhaust ventilation. (Terms, units, and definitions are found at the end of the chapter.)

20.1.1   Dilution Ventilation

Natural dilution ventilation (air movement induced by wind and temperature differences) has been used for centuries to control smoke from fireplaces, stoves, and tobacco products. Mechanical dilution ventilation has been used widely since the 1890s to control indoor air contaminants including environmental tobacco smoke (ETS).

Although most public and private buildings (excluding residences) now use smoking bans or designated smoking areas as the primary control of ETS, dilution ventilation is used to control smoke where the ban is not 100% effective. Dedicated dilution ventilation systems also provide the primary control of ETS in designated smoking areas. Residential buildings and much of the hospitality industry, however (restaurants, cafes, bars, bowling alleys, casinos, taverns, dance clubs, hotel lobbies, time-share condos), continue to allow smoking. Hospitality industry patrons and employees (perhaps in the millions of people) continue to be exposed to ETS at varying levels depending on the controls installed.[14]

20.1.2   Displacement Ventilation

Displacement ventilation is an approach that discourages the turbulent mixing found in traditional dilution ventilation. Air is introduced into the space at one location, migrates across the space carrying air contaminants with it, and is exhausted at a remote location from the inlet. Other terms for this type of ventilation include "plug flow" and "pump flow."

20.1.3   Local Exhaust Ventilation

The science and engineering for local exhaust ventilation control of point sources of emission are well established.[4-6] Although existing local exhaust ventilation technology has rarely been applied to tobacco smoking, it has been successfully applied in shops and factories for many decades and can be adapted to the control of tobacco smoke sources.

20.1.4 LIMITATIONS OF VENTILATION

The use of ventilation for the control of ETS is limited, however, by the lack of a clear definition of what it is supposed to accomplish — Provide for occupant satisfaction? To what degree? For health protection? For which health effects and to what degree? For what airborne contaminants or markers? To what acceptable concentrations? Because there are few (if any) officially recognized standards or markers for acceptable airborne ETS concentrations and exposure levels, it is difficult to quantify the appropriate ventilation rates. Nevertheless, ventilation can be used to significantly reduce airborne levels of ETS markers and odors. When suitable and reasonable ETS exposure limits or markers are established, ventilation will prove capable of helping reach these goals.

20.1.5 VENTILATION USED WITH OTHER FORMS OF CONTROL

Ventilation is almost always used in concert with other forms of smoke control, e.g., smoking bans, use of designated smoking areas, and limitations on smoking times. When smokers are required to use designated smoking rooms or areas, for example, specially designed and dedicated ventilation systems are used to dilute the air, isolate the designated area from other parts of the building, and to clean the air prior to its release from the smoking area. Where smoking times are restricted, the limited amount of smoke released can reduce the total amount of dilution or local exhaust ventilation required.

20.2 VENTILATION CONTROL DESIGN CRITERIA

For successful control of ETS, the choice of ventilation must consider *emission sources* (location, strengths, times, smoke behavior), the *air behavior* in the occupied space (direction of flow, mixing potential, velocity), and the *behavior of smokers* (exhalation behavior, smoking time, willingness to cooperate). Three major types of ventilation are suited to the control of ETS: dilution or general ventilation (GV), displacement ventilation (DV), and local exhaust ventilation (LEV).

20.2.1 EMISSION SOURCES

Smoke from tobacco products consists of smoke emitted directly to the air at the burn site ("sidestream smoke") and that exhaled from a smoker ("secondhand smoke"). In a typical situation where cigarettes are being smoked, about half the smoke arises from each source.[3] In this chapter, both are considered to be major constituents of ETS. ETS consists of gases, vapors, and aerosols of very small size that behave like the air they reside in (i.e., they do not settle quickly.) Controlling the contaminated air will control the smoke.

   Sidestream smoke is usually a fixed-point source of smoke, while secondhand smoke is likely to be diffused in the space where smokers reside. In some cases, the smoker can also be considered a point source if the exhaled smoke can be controlled near the smoker's mouth or if the smoker blows the smoke directionally toward an air exhaust opening.

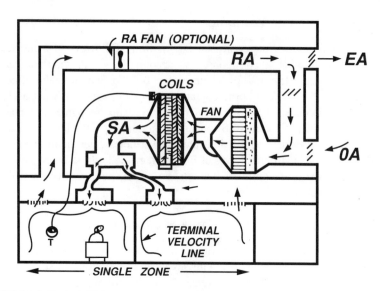

FIGURE 20.1 Typical HVAC air handling system showing terminology.

Another source (mostly of odor complaints) is the vapor emitted from smoke tars deposited on surfaces in the occupied space and in the ventilation system itself. Indeed, many people object to tar vapor odors as much as the smoke itself. Cigarette tars adsorb onto surfaces in the building, including the interior of return air ducts and plenums, air cleaners, coils, and other equipment. These tars slowly desorb, producing offensive odors sometimes for months after smoking has ceased. It has been shown that control of tobacco tar odors typically requires as much as 50 to 300% more outdoor air as compared to spaces where smoking is prohibited.[7] Deodorants and air deodorizers are not much help. Many deodorizers utilize the same irritants or pungent materials as the active ingredient (e.g., ozone, formaldehyde). It appears that these chemicals cause olfactory fatigue or have an anesthetic action. The deliberate addition of additional irritants to the ventilation air seems ironic, if not counterproductive.[8]

20.2.2 AIR BEHAVIOR

Most human occupancies are provided with dilution or general ventilation in accordance with local building codes and American Society of Heating, Refrigeration, and Air Conditioning Engineers (ASHRAE) standards. For example, ASHRAE 62-1999 recommends the delivery of 20 cubic feet per minute (cfm) per person of fresh outside air in typical office spaces and 60 cfm per person in smoking lounges.[1] Most common GV systems create turbulent dilution of airborne contaminants such as ETS, body odors, and effluents from materials and equipment in the space. Supply air (SA) is typically introduced at the ceiling, where it flows along the ceiling and down the wall and then mixes with room air contaminants and is exhausted from the room in return air (RA) grilles mounted in the ceiling (Figure 20.1). Smoke-contaminated air is typically warmer than the surrounding air and will (at least initially) flow up toward the ceiling where it mixes with incoming air and is diluted.

20.2.3 Smoker Behavior

The smoker can have a significant impact on the control of ETS regardless of the control strategy employed. Many smokers today, for example, will exhale vertically to avoid exposure for others nearby and to aid in the dilution of the plume. If the smoker can be enticed to cooperate with the ventilation control system, it has a much better chance of being successful. For example, an air exhaust hood placed along a table top or bar can only be successful if the smoker is willing to blow the smoke toward the exhaust opening. In the limited published data available, it has been reported that smokers were generally willing to cooperate with ventilation smoke controls.[3]

Other smoker behaviors that impact the effectiveness of ventilation controls include (1) the smoking rate (typically one puff per minute at 2 s per puff and eight full puffs per cigarette),[3] (2) the smoker's handling of the burning tobacco product (does it rest primarily in the ashtray, in the hand, or in the mouth), and (3) the smoker's willingness to adhere to any "smoking rules," (smoking in designated rooms or areas, leaving cigarettes in an exhausted ashtray when not inhaling, blowing toward an exhaust hood, etc.)

20.3 DILUTION VENTILATION

Dilution ventilation has been the mainstay of the control of airborne concentrations of ETS for many decades. Indeed, dilution ventilation is the most common type of ventilation in almost all occupancies and has an important roll even when other forms of ventilation are used for smoke control.

20.3.1 ASHRAE 62-1999, Ventilation for Acceptable Air Quality

The most widely used dilution ventilation standard for human occupancies is ASHRAE 62-1999, Ventilation for Acceptable Indoor Air Quality.[1] This standard, against which most indoor air quality performance is measured, was last completely revised by the American Society of Heating, Refrigeration and Air Conditioning Engineers (ASHRAE) in 1989. A new version proposed in 1996 — ASHRAE 62R — met considerable resistance and was tabled. The 1989 standard was subsequently placed on "continuous maintenance," which means the standard is constantly changed and updated, item by item, through "addenda." The date was changed to 1999.[9]

20.3.1.1 Changes in Specific Recommendations for Smoking

The 1989 standard provided for a "moderate amount" of smoking. Addendum "e" recently removed from the ventilation rates table a statement that ventilation rates "accommodate a moderate amount of smoking." Since the standard now makes no mention of smoking, it has been interpreted in two ways. For those against allowing general smoking, it is interpreted as not specifically accommodating smoking except in "smoking lounges." For those supporting smoking, it is seen as not specifically disallowing smoking.

Provisions of the standard which still impact the use of dilution ventilation for tobacco smoke control are listed below by standard paragraph number.

Paragraph 4.0. The standard allows the user to choose one of two methods to comply: the *Ventilation Rate Procedure* (VRP) and the *Indoor Air Quality Procedure* (IAQP). The VRP is based on tabular values of fresh dilution air. See Table 20.1 which is an excerpt of the ventilation rates table provided in the standard. The IAQP is based on actual airborne contaminant concentrations and emission sources. See Sample Calculation 2 in Section 20.3.3 which provides an example of the approach.

Paragraph 4.2. When the user chooses the IAQP, more or less air may be required as compared to the VRP.

Paragraph 4.2. If the VRP is not capable of achieving the minimum air quality required, then the IAQP should be used to achieve minimum requirements.

Paragraph 5.1. When mechanical ventilation is used, provision for air flow measurement should be provided. When natural ventilation is used, sufficient ventilation should be demonstrable.

Paragraph 5.2. Ventilation air should be supplied throughout the occupied zone. This implies delivery to the breathing zone of occupants, as opposed to simply delivering air to the building or space.

Paragraph 5.3. Where *Variable Air Volume Systems (VAV)* are used, and when the supply of air is reduced during times when a space is occupied, indoor air quality should be maintained throughout the occupied zone.

Paragraph 5.6. Where practical, exhaust systems should remove contaminants at the source.

Paragraph 5.7. Where combustion sources are present, dilution air is to be provided.

Paragraph 5.8 and 10. Where necessary, particle filters and gas/vapor scrubbers should be sized and used to maintain air quality.

Paragraph 6. Indoor air should not contain contaminants at concentrations known to impair health or cause discomfort.

Paragraph 6.1. Outdoor air (OA) introduced to the building through the ventilation system should not exceed U.S. EPA National Primary Ambient-Air Quality Standards. If the outdoor air contaminant levels exceed EPA values, the air should be treated. Air known to contain contaminants not on the EPA list (e.g., ETS) should refer to other references for guidance.

Paragraph 6.1.3. A table lists the minimum OA requirements per person for about 100 occupancies, e.g., offices, classrooms, and so forth. An abbreviated sample is shown in Table 20.1. The standard assumes excellent turbulent mixing and good distribution of the outdoor air. It also assumes a certain occupant loading. Where these are not the case, additional OA may be required. (Note: A proposed change, Addendum "n," would change the ventilation rates; see comments in Section 20.3.1.2.)

TABLE 20.1
Sample of ASHRAE 62-1999 Standard Ventilation Rates

Occupancy	Est. max. occupancy per 1000 sf	cfm/person OA[a]	cfm/ft²
		(required ventilation rates)	
Smoking lounge	70	60	—
Office space	7	20	—
Conference room	50	20	—
Retail sales	30	—	0.3
Theater auditorium	150	20	—

[a] OA = outdoor air.

20.3.1.2 Ventilation Rate Changes

In the 1999 version of the provisions of the standard, OA ventilation requirements are mostly based on the number of people in the occupied area. For example, 60 cfm of OA per person is required for typical smoking lounges. Proposed standard revision "addendum n" recognizes that air pollutants also arise from non-people building sources: building materials, furnishings, and the HVAC equipment itself. In order to accommodate all sources, a new calculation approach has been proposed. Most occupancies will have a rate per person plus a rate per square foot. For offices, the proposed values are 6 cfm per person plus 0.07 cfm/ft². New recommendations for smoking lounges or smoking areas were not known at the time of publication.

20.3.2 Designated Smoking Areas (DSAs)

Many buildings that ban general smoking provide a designated smoking room or area for those who do wish to smoke.[13] Where they exist, DSAs should (1) be available to all building users, including visitors; (2) be configured so that no migration of smoke into nonsmoking areas occurs (e.g., provided with floor-to-ceiling walls, be under negative pressure relative to adjacent areas); and (3) be provided with a dedicated ventilation system which does not return ETS to other areas of the building.[7]

There are three basic approaches for specifying a dedicated ventilation system:

1. Allow the building's air handling system (AHS) to supply air to the DSA, but block all returns to the AHS and install a separate exhaust system designed to exhaust more air than supplied. (About 10% more exhausted than supplied air is typical.) This places the DSA under negative pressure and induces any air leakage into the DSA. Typical supply rates are 20 air changes per hour (AC/h) or 60 cfm per potential smoker, whichever is more.[1,8]

   Sample Calculation 1. Suppose a DSA (dimensions 40' × 30' × 10') is designed to support a maximum of 60 smokers at any one time. Using 20 cfm per person, the air handling unit (AHU) should supply no less than 3600 cfm of dilution air and be designed to exhaust around 4000 cfm, with no recirculation of DSA air to the building. Using 20 AC/h, supply air should total 4000 cfm (Equation 1) with total exhaust air (RA) at about 4400 cfm.

$$Q = \frac{AC/h \cdot Vol}{60} = \frac{20 \times 12,000}{60} = 4,000 \text{ cfm} \tag{1}$$

2. Provide a completely separate ventilation system to the DSA which exhausts all of the air to the outside. In this case, the dedicated AHS both supplies and exhausts the air and has no connection with AHSs used in other parts of the building. Again, the system should be designed and operated to keep the DSA under negative pressure with regard to the rest of the building. See Option (1) for design flowrates.
3. As a variation on Option (2), filter and scrub ETS from the air prior to returning it to the DSA. Air should not be returned to the general building AHS. See discussion of air filtration and scrubbing in Section 20.6.

Options (1) and (2) are the most cost-effective approaches in temperate climates. Option (3) requires expensive aerosol filtration and gas/vapor scrubbing.

Table 20.2 provides a list of good practices to follow in a designated smoking area. A detailed discussion of designing DSAs (with sample calculations) is provided in the Appendix at the end of the chapter.

20.3.3 DETERMINING APPROPRIATE DILUTION VENTILATION AIR FLOWRATES BASED ON EMISSIONS AND AIR CONTAMINANT PARAMETERS

For a constant emission source, no sinks, perfect mixing, and a constant air flow, Equation 2 describes the resulting equilibrium concentration in a ventilated space (e.g., chamber, room, building):

$$C = E/Q \tag{2}$$

where

C = concentration in air
E = emission rate of air contaminant or marker
Q = ventilation rate in the same units as E

Figure 20.2, developed by Hal Lavin, a noted IAQ architect, shows the situation for five different emission rates, with all other conditions equal and at steady state conditions.[11] The EF numbers suggest ratios of emission rates based on typical emission factors of 0.1 to 5 mg/m²/h. Note that background concentrations increase significantly at air exchange rates less than 0.75 AC/h. Also, increases above 2 AC/h do not offer linear reductions in concentrations. Figure 20.2 also shows that complete

TABLE 20.2
Good Practices for a Dedicated Dilution Ventilation System Serving a Designated Smoking Area

- The designated smoking area (DSA) should be physically separated from the rest of the building and at all times be maintained under negative pressure as compared to surrounding areas. This will provide movement of air from the nonsmoking area to the smoking area and maintain odor control in the nonsmoking area.
- Where dilution ventilation is used, the smoking area should provide for a high air mixing potential (e.g., no barriers to free circulation within the smoking area, use of fans to mix air, or supply and exhaust registers located to enhance mixing potential). The DSA can also be provided with a displacement ventilation system where cool air is introduced at the floor level and allowed to rise uniformly to return grilles in the ceiling. (See Section 20.4 for details on this approach.)
- The fresh air supply should provide at least 20 AC/h or 60 cfm of OA per smoker, whichever is more at the time of maximum smoker use.
- Supply air should not be recirculated to the DSA unless HEPA and charcoal filtration is provided. It is best to have 100% outdoor air service to the smoking area. This will reduce maintenance and upkeep in the HVAC equipment serving the smoking area.
- A rigorous housekeeping program must be used to keep the smoking area clean. (Smokers will refuse to enter the area after a few weeks without it.)
- Even during times of nonoccupancy, the air handler should continue to supply and exhaust air through the smoking area. This will help reduce tar buildup and maintain odor control. The amount of air supplied and exhausted to the room can be reduced to 4 to 6 AC/h.

Source: From Burton, D.J., *OH&S*, December 1994. With permission.

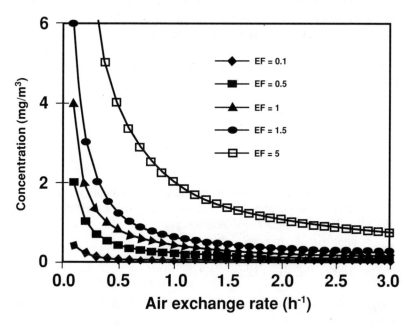

FIGURE 20.2 Concentrations vs. air exchange rates. (From Lavin, H., *IA Bull.*, 3(5), 1996. With permission.)

removal of the contaminant never occurs as long as an emission source is present. It also implies the need for determining an acceptable marker and its allowable concentration in air, e.g., carbon monoxide at 3 ppm or nicotine at 0.01 mg/m³.

Equation 2 is usually modified to incorporate a mixing factor and can be solved for the appropriate dilution volume flowrate when an acceptable concentration is specified.

$$Q_{OA} \approx \frac{q \cdot K_{eff} \cdot 10^6}{Ca} \qquad (2b)$$

where in U.S. (and SI) units Q_{OA} is the volume flowrate of dilution air, cfm (m³/s); q is the emission or generation flowrate of vapor or gas, cfm (m³/s); and Ca is the acceptable exposure concentration, ppm; K_{eff} is a mixing factor to account for incomplete or poor delivery of dilution air (OA) to occupants, unitless.

The mixing of air is sometimes called "ventilation efficiency" or "ventilation effectiveness." Mathematically, it is often stated as

$$K_{eff} = \frac{\text{Actual } Q_{OA} \text{ required to provide minimum } Q_{OA} \text{ to occupied zone}}{\text{Ideal } Q_{OA} \text{ required to provide minimum } Q_{OA} \text{ to occupied zone}}$$

The value of k_{eff} ranges from 1.0 to 1.5 in most indoor occupancies.

Sample Calculation 2. ASHRAE reports that one completely smoked cigarette generates or emits about 55 mg of carbon monoxide.[7] What volume flowrate (Q_{OA}) is required for dilution of carbon monoxide (as a marker for ETS) to an acceptable concentration of 3 ppm above background, if ten people are smoking?

Assume $K_{eff} = 1.25$, STP, Ca = 3 ppm, and 8 min duration per cigarette. The emission rate can be determined by

$$q \approx \frac{0.0244 \cdot G}{MW \cdot t \cdot d} \qquad (3)$$

where

q   = volume of gas or vapor emitted in m³/s, at standard conditions
        (STP in this equation is 70°F at 760 mmHg barometric pressure, dry air)
MW = molecular weight or molecular mass
t   = approximate time, seconds (s)
G   = amount generated, grams (g)
d   = density correction factor, unitless

Plugging these data into Equation 3 and converting cubic meters per second to cfm at STP:

$$q = 0.00209 \text{ scfm}$$

Using Equation 2b,

$$Q_{OA} \approx \frac{0.00209 \cdot 1.25 \cdot 10^6}{3} \approx 870 \text{ cfm or } 87 \text{ cfm per person}$$

This compares favorably with ASHRAE's recommended 60 cfm per smoker.

20.4 DISPLACEMENT VENTILATION

In a 1998 workshop chaired by Steven E. Guffey, University of West Virginia, sponsored by OSHA and ACGIH, displacement ventilation (DV) was determined to be the most cost-effective ventilation approach for control of ETS in the hospitality industry, an occupancy where the pressure to continue to allow general smoking is great.[3] The following discussion generally reflects the findings of the workshop.

DV is a ventilation approach that discourages the turbulent mixing found in traditional dilution ventilation. Air is introduced into the space at one location, migrates across the space carrying air contaminants with it, and is exhausted at a remote location from the inlet.

In the cold-air approach, cold air is released at the floor level and exhausted at the ceiling level. People, burning cigarettes, and electronic devices warm the incoming air creating upward-flowing convection currents that carry the contaminated air to the ceiling area where it is removed through RA grilles. Because the rising plume from both sidestream smoke and exhaled secondary smoke is released at temperatures well above 90°F, cold-air DV makes a good choice.

When the ceiling is high (more than 9 ft above the floor), air contaminants near the ceiling are usually well above space occupants. To be successful, DV requires that there be relatively little mixing in the air by people moving about or the use of mixing fans, Casablanca fans, and so forth. It works best when SA grilles are installed at or near the floor. Smoker cooperativeness and a willingness to exhale upward is also useful. When conditions are suitable, cold-air displacement can be used also with ventilated ashtrays, ventilated booths, and other local exhaust or supply strategies.

Good design criteria include the following:

1. When air must travel substantial distances, environmental conditions require low turbulence and careful control of supply air temperatures.
2. Smokers should be "required" to exhale vertically upward.
3. Supply air volume flowrates for large open areas like classrooms, restaurants, and casinos are typically 1 to 3 cfm/ft².[3]

Potential advantages of cold-air DV include the following:

1. Reduced supply air requirement
2. Ability to control both second-hand and sidestream ETS
3. Reduced drafts
4. As much as a three to five times reduction in ETS concentrations for same dilution airflow and power consumption[3]

Potential disadvantages and concerns include the following:

1. DV is unfamiliar to many HVAC engineers
2. Supply air diffusers may take substantial wall space
3. Ducting of supply air to floor level may be difficult
4. Sensitivity to errors in supply air temperature could affect thermal comfort
5. Low ceilings will adversely affect temperature distribution ("cold feet, warm head")
6. Smoke concentrations near the ceiling will be high, potentially exposing persons near the ceiling

Workshop participants suggested that total ETS reductions were likely to be around 90% or more for good displacement conditions. Poor conditions, especially those due to the introduction of turbulence and large eddies, could sharply lower the reductions.[3]

In the supplied-air island approach, an "island" of clean, smoke-free air is created for nonsmokers. Typically, an SA grille provides a low-velocity diffusion of fresh clean air directly above an occupied desk, bar stool, table, or working employee. Widely used in industry, this approach has seen little application in ETS control, but the technology is available (see Reference 10 for more information).

20.5 LOCAL EXHAUST VENTILATION

20.5.1 OVERVIEW

The potential advantages of using local exhaust ventilation (LEV) in addition to dilution ventilation approaches and dedicated ventilated smoking areas include the following: (1) smokers can smoke at more diverse designated smoking sites (installed in such locations as offices or cafeterias), and (2) there can be reduced quantities of replacement dilution air to temper and clean.

There are some disadvantages to the LEV system.

1. It doesn't discourage smoking in the space or building.
2. It requires the cooperation of the smoker.
3. The retrofit of existing buildings would be more difficult than establishing a dedicated smoking area. New buildings could easily be fitted with a smoke exhaust system.

In practice, LEV systems should be integrated into the general dilution ventilation scheme.

In its preamble to a proposed IAQ standard in 1995, OSHA stated: "… local exhaust ventilation would be the preferred and recommended method for controlling occupational exposures to contaminant point sources like ETS."[2] Interestingly, OSHA goes on to say, "A designated smoking area which is enclosed, exhausted directly to the outside, and maintained under negative pressure is sufficient to contain tobacco smoke within the designated area. Such areas could be considered an application of local exhaust ventilation because the contaminant is being exhausted from a confined space without dispersal into the general workspace."

There are three classes of LEV systems: enclosing hoods (the source is within an exhausted enclosure), capture hoods (the emission source is remote from the hood and relies on air passing the source to capture the emission and carry it into the hood), and receiving hoods (warm contaminated air rises into the hood). All three types have application in the control of ETS.

20.5.2 SIDESTREAM SMOKE CONTROLS

In the 1998 workshop sponsored by OSHA and ACGIH, LEV was determined to be a cost-effective approach for control of sidestream ETS using ashtray exhaust systems.[3] These hoods fall into the "enclosing" class of exhaust hoods. The following discussion reflects the findings of the workshop.

Although exhausted ashtrays of various commercial types have been available since before 1990, they have never been widely used. In ductless ashtrays (the air-smoke mixture is cleaned and returned to the room air) maintenance of associated filters is likely to be a problem in restaurants, bars, and other establishments. For that reason, it is important that a "ductless" unit that returns filtered air to occupied spaces have a filter that is effective in removing ETS and is easy to maintain. Ducted ashtrays (the air-smoke mixture is removed from the space to the outside through ducts or hoses) are preferred. Major concerns of any type of exhausted ashtray included the scarcity of space on bar tops, restaurant tables, and desks and potential problems with cleaning the units and the surfaces they obstruct.

The *advantages* of exhausted ashtrays include:

1. Effectiveness in removing sidestream smoke
2. Reduction of the total burden of ETS in the room
3. Low airflow requirements
4. Low airflow, small ducts, low noise
5. Can be moved within a limited range
6. Can be designed to be easily cleaned
7. Can be built into existing equipment

There are also *disadvantages* of an exhausted ashtray:

1. It provides little or no reductions in the contaminants from exhaled secondhand smoke.
2. It must be ducted to outside unless it has its own fan and integral filter.
3. Ducts may be difficult to run and may be unsightly.
4. It must be cleaned frequently.
5. If not filtered at the hood, ducts must be cleaned frequently.
6. Filters inside the hood must be changed frequently.
7. If screening is omitted, the equipment can plug.
8. Increased airflow across the burn zone can hasten the burn rate, reducing available smoke to the user and thus discouraging its use.

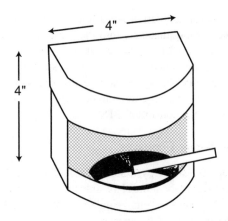

FIGURE 20.3 Ventilated ashtray. (From Guffey, S.E., Proceedings of the Workshop on Ventilation Engineering Controls for Environmental Tobacco Smoke in the Hospitality Industry, ACGIH, Cincinnati, OH, 1998. With permission.)

Design criteria include:

1. Enclose as much as possible.
2. Capture from side or above.
3. Exhaust to outside.
4. Filter locally with at least a particulate filter before sending to ductwork.
5. If ductless, the filters must be effective in removing gases and particulates, have low resistance to flow, and be easy to remove and replace.
6. Make easily detachable from wiring or ducting for maintenance.
7. Provide a fine mesh to protect the filter and prevent trash from entering ducts.
8. Make airflow requirements as low as possible while maintaining 100% collection effectiveness from a cigarette left inside it.
9. The unit must be easily cleanable and must not interfere with cleaning or sanitation surfaces.
10. It must be convenient for smokers to use.

Figure 20.3 shows a typical approach to the design and construction of an exhausted ashtray.

The effectiveness of ventilated ashtrays is affected by the following factors:

1. The fraction of total ETS airborne contaminants of concern that come from sidestream smoke, $F_{sidestream}$
2. The collection effectiveness of the device in capturing and removing ETS as it is generated at the tip of the cigarette, $F_{effective}$
3. The fraction of the total consumption time the cigarette is left inside the ventilated ashtray, F_{time}
4. The degree of difficulty in maintaining and cleaning the devices

Thus, the reduction in ETS emitted from smoking a cigarette can be computed from the first three factors (the fourth would represent degradation in collection effectiveness).

$$\text{Fraction collected} = F_{\text{sidestream}} \times F_{\text{effective}} \times F_{\text{time}} \tag{3}$$

Assuming, for example, that 50% of total ETS comes from sidestream smoke and that the ventilated ashtray collects 99% of effluent while the cigarette was resident and that the burning cigarette would typically remain within the ventilated ashtray 80% of the time, the fraction collected would be about $(0.5 \times 0.99 \times 0.80)$ 40% of the total emitted ETS. This is a significant quantity of the total emitted ETS, and the control equipment is relatively inexpensive to install and operate.

In estimating airflow requirements for ventilated ashtrays, for an enclosed ashtray similar to Figure 20.1, the airflow requirement can be computed as the open area at the face multiplied by the required air velocity at the face. For a face velocity of 100 ft/min, an enclosure with a 4" × 3" opening would require about 8 cfm. If it is necessary to increase the size of the opening to create an easier target for smokers, the airflow requirement increases proportionately to open area.

20.5.3 Secondhand Smoke Controls

20.5.3.1 Enclosing and Capture Hoods

Figures 20.4 and 20.5 suggest potential approaches for the control of exhaled tobacco smoke.[7,12] Figure 20.4 shows conceptual examples of both enclosing and capture hoods. Figure 20.5 shows the conceptual approach for a capture hood placed along a desk, table top, or bar top. All require the cooperation of the smoker.

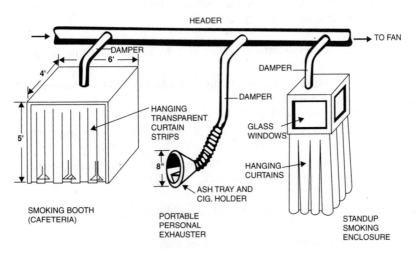

FIGURE 20.4 Conceptual LEV control of secondhand ETS. (From Burton, D.J., *OH&S*, November 1994. With permission.)

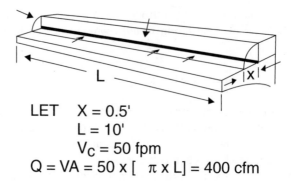

LET   X = 0.5'
      L = 10'
      V_C = 50 fpm
      Q = VA = 50 x [π x L] = 400 cfm

FIGURE 20.5 Conceptual LEV using a slot exhaust hood along flat surfaces like desks and bars with airflow estimation. (From Burton, D.J., *OH&S*, December 1994. With permission.)

Typical hoods would exhaust 200 to 400 cfm. Given the cooperation of the smoker, these devices could capture 90% or more of secondhand smoke generated during smoking.[3]

Before LEV technology can be widely applied, however, someone must demonstrate the technology (i.e., in sample projects), evaluate existing design criteria (for applicability to smoking), estimate realistic costs, and evaluate such systems for unforeseen problems (e.g., would the exhaust system require periodic cleaning?).

20.5.3.2. Receiving Hoods

Figure 20.6 shows a typical receiving or canopy hood over a desk or table to receive warm rising air contaminated by secondhand smoke. A small overhead canopy hood can be made to look attractive and incorporate lighting.

Good design criteria include the following:

1. The hood should not block the line of sight for persons sitting across the table.
2. The hood height, width, and location should be set the so that people do not bump their heads on it.
3. It should require as little airflow as possible.
4. It must be easy to service and clean and should meet all fire and sanitation regulations.
5. It should have low velocities in the ducts to avoid high noise levels.
6. Ashtrays should be placed under the canopy.[3]

Advantages of a canopy hood include the following:

1. It can be aesthetically pleasing if done well.
2. It can control secondhand smoke quite effectively when designed, operated, maintained, and used correctly.

FIGURE 20.6 Conceptual LEV using a receiving canopy hood. (From Guffey, S.E., Proceedings of the Workshop on Ventilation Engineering Controls for Environmental Tobacco Smoke in the Hospital Industry, ACGIH, Cincinnati, OH, 1998. With permission.)

Disadvantages include the following:

1. Installation and operating costs are high.
2. It will cast shadows if it does not double as a lighting fixture.
3. Residues of ETS may create fire and safety hazards.
4. Airflow requirements will be at least 300 cfm.
5. It requires the cooperation of the smoker.

20.6 RECIRCULATION OF ETS-CONTAMINATED RETURN AIR USING FILTERS AND SCRUBBERS

20.6.1 OVERVIEW

In theory, air can be cleaned of ETS contaminants and returned to a designated smoking area or even to the general building AHU. In practice, however, this is rarely done because (1) cost-benefit ratios have not favored filtration/scrubbing and recirculation, (2) successful demonstrations of the technology have rarely been reported in the literature, (3) HVAC designers are not often familiar with the technology, (4) filters and scrubbers have not traditionally been rated for controlling ETS, and (5) energy conservation has been served by simply banning smoking or moving it to a designated area.

Some of these factors are changing. The following discussion provides a brief overview of the existing technology and the changes which are occurring in filter test standards.

20.6.1.1   Recirculation Criteria

Filters have traditionally been used in AHSs for maintaining clean equipment. In recent years, filters and scrubbers have also been designed and built to clean the RA for possible recirculation to occupied spaces. Where it is desired to clean, treat, scrub, filter, and recirculate ETS-contaminated air instead of using fresh OA exclusively for dilution, four types of cleaning should be provided.[9]

$$\rightarrow \text{roughing} \rightarrow \text{bioaerosol} \rightarrow \text{respirable} \rightarrow \text{gas/vapor}$$

Roughing filters or prefilters are used to remove large particles: leaves, bugs, feathers, or paper shreds. Dust spot efficiencies of 25 to 30% or arrestance efficiencies of 80% are appropriate. (See Section 20.6.2 for information on standards.) Roughing filters are used to protect more expensive filters and scrubbers downstream.

A bioaerosol filter is used to remove pollen, mold spores, and bacteria. It is also used to protect filters and scrubbers downstream. Dust spot efficiencies of 85% are appropriate. It must be kept dry or treated with a anti-microbial agent.

Where smoking is allowed or where control of viral particles is required, HEPA filtration is often employed. Widely used in the semiconductor and radioactive materials industries, HEPA filters are available in efficiencies ranging from 95% to over 99.999%. HEPA filters can be very efficient in removing most or all ETS particulates. This filter also protects the gas/vapor scrubber.

Gas/vapor scrubbers (e.g., adsorbers using virgin coconut shell carbon activated to 60% carbon tetrachloride activity) should be capable of removing parts per billion concentrations of VOCs, ozone, NO_x, and SO_2. At the Indoor Air '93 Conference held in Helsinki, papers from the Netherlands describe studies of six commercially available "gas absorption membranes." Water soluble chemical air contaminants (e.g., ammonia, acetone, aldehydes) were scrubbed from the returning airstreams at up to 90% efficiencies and improved "perceived air quality" by up to 50%.[15]

Other potential cleaning mechanisms include

- Electrostatic precipitation (ESP) for particle collection
- Chemisorption for removing gases from the air stream
- Catalytic separation for gases and vapors
- Bipolar ionization for control of microorganisms
- Excitation/acceleration for removal of small particles

The last four cleaning approaches are currently being studied and developed for commercial use.

At high concentrations (percent), ozone reacts with many VOCs to form CO_2 and water. However, it should not be routinely used in HVAC systems or the occupied space in an attempt to neutralize gas or vapor air contaminants. Ozone and its byproducts can be hazardous at concentrations necessary to be effective.

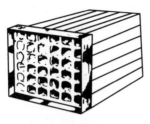

Arrestance   *Dust Spot*   *HEPA*

FIGURE 20.7 HVAC filter types. (From Burton, D.J., *IAQ and HVAC Workbook*, 3rd Ed., IVE, Inc., Salt Lake City, UT, 1999. With permission.)

TABLE 20.3
ASHRAE 52.1 Filter Test Comparisons

	Test	
Filter media	Arrestance	Dust spot
Fine open foams	70–80	15–30
Cellulose mats	80–90	20–35
Wool felt	85–90	25–40
Mats, 5–10 µm, 1/4"	90–95	40–60
Mats, 3–5 µm, 1/2"	60–80	30–40
Mats, 1–4 µm, fibers	>95	80–90
Mats, 0.5–2 µm, glass	—	90–98
Wet-laid glass fibers, HEPA	—	—

Source: From Burton, D.J., *IAQ and HVAC Workbook*, 3rd Ed., IVE, Inc., Salt Lake City, UT, 1999. With permission.)

Filter and scrubbers should normally be located in the mixed air location of the HVAC system, upstream of fans and coils. If filters are not maintained and changed on schedule, the filter itself can be create IAQ problems (i.e., by reducing the air flow through the system and by creating odors itself).

20.6.2 FILTER STANDARDS

There are two major standards for HVAC filters, ASHRAE 52.1 and 52.2 (Figure 20.7). ASHRAE 52.1-1992 provides two tests: the *Weight Arrestance and the Dust Spot Efficiency.* The Arrestance test is a measure of a filter's ability to remove course dust; the Dust Spot test is the removal of fine dust which can cause "dust spots" in the environment served by an HVAC system. Typical comparisons are shown in Table 20.3, by percent removal, for specific filter media (i.e., wool felt may provide 85 to 90% efficiency on the Arrestance test, but only 25 to 40% on the Dust Spot test).

Arrestance filters are recognized as flat, dry, open fiber panel or roll-type filters. Dust Spot filters are often constructed of pleated paper, envelopes, or bags made of

TABLE 20.4
Comparison of Filter Standard Test Results

ASHRAE standard	Filter test designation	Test results	Additional information
	Arrestance	Percent (by weight) of a test dust the filter retains	Low-efficiency filters only
ASHRAE 52.1	Dust Spot efficiency (DSE)	Percent of atmospheric dust removed by filter	Inconsistent results because test uses atmospheric dust
	Dust holding capacity (DHC)	Weight of dust a filter holds at a specified pressure drop	Good information on service life of a filter
	Pressure drop	Filter resistance to airflow, inch w.g.	Lower resistance equals lower energy consumption
	Particle size efficiency (PSE)	Percent retention of test aerosol (KCl) in a filter at specified particle diameters from 0.3 to 10.0 μm	Gives efficiency for various particle sizes; test results consistent and reliable
ASHRAE 52.2	Pressure drop	Filter resistance to airflow, inch w.g.	Lower resistance equals lower energy consumption
	MERV	Categories of performance	Makes filter selection easier

Source: From Burton, D.J., *OH&S*, June 2000. With permission.

filter media. HEPA filters are most often built into wooden boxes. All filters should be kept dry to avoid clogging and the buildup of microorganisms.

Filters for removing ETS aerosols must be of a higher efficiency type (e.g., HEPA filters) and tested under ASHRAE 52.2-1999 or other HEPA filter tests. ASHRAE 52.2-1999 provides minimum particle size efficiency (PSE) data for many filters, including those used for ETS control. The new standard has these unique features:

- It is based on rigorous laboratory testing. This allows reliable comparisons between different filters.
- It reports average minimum efficiencies in three size ranges and at specified preloaded minimum pressure drops across the filter.
- It cannot yet accurately and consistently measure electrostatically charged filter media. This weakness should be corrected in the future. Meanwhile, for these types of filters, don't base selection solely on ASHRAE 52.2.
- It establishes a simplified reporting scheme where all filters are classified into 16 Minimum Efficiency Reporting Value (MERV) groups.

For the foreseeable future, filters will continue to be tested under both standards. Table 20.4 shows a comparison between the three tests supported in both standards. Table 20.5 shows a description of selected MERVs and their approximate relationship

TABLE 20.5
Selected ASHRAE 52.2-1999 Standard MERV Number and Their Characteristic

MERV number[a]	Approximate equivalent ASHRAE 52.1 dust spot efficiency rating	Approximate equivalent ASHRAE 52.1 arrestance rating	Particle Size Efficiency (PSE), (%)			Typical filter type	Typical air contaminant	Typical applications
			(0.3–1 µm, range 1)	(1–3 µm, range 2)	(3–10 µm, range 3)			
1	<20	<65	—	—	<20	Throwaway	Bugs, carpet fibers	Roughing filter
4	<20	75–80	—	—	<20	Electrostatic, washable filters	Pollen, household dust	Residential
7	25–30	>90	—	—	50–70	Pleated, cartridge	Mold spores	Commercial buildings, paint booths
10	50–55%	>95	—	50–65	>85	Box, bag filters	Coal dust, welding fume	Better commercial buildings, hospital labs
13	80–90%	>98	<75	>90	>90	Box, bag filters	Bacteria, copier toner	Best commercial buildings
16	—	—	>95	>95	>95	Box, bag, HEPA filters	All bacteria, most smoke	Smoking lounges
17–20	—	—	99.97+	99.97+	99.97+	HEPA, ULPA	Virus	Cleanrooms

Source: Data from ASHRAE[1] and Burton.[16]

[a] Not all MERV numbers are shown.

to the Arrestance and Dust Spot tests. The last four MERV numbers (17–20) are not yet part of the official standard test, but have been added by ASHRAE for comparison purposes. Although no definitive recommendations have yet been published, filters used for ETS control should probably be MERV number 16 or higher.

20.7   RESEARCH AND DEMONSTRATION PROJECT NEEDS

As identified at the workshop on ventilation of ETS and in other publications, much research and development is needed.[3,8] These data needs include

1. The behavior of tobacco smoke in air at various distances above the smoker
2. The types of systems, materials, and filters that would be necessary to protect the ventilation equipment and allow for recirculation of returning air
3. The burn and smoke production rate when ventilation air flows past the burning tobacco product
4. The behavior of smokers in ventilation controlled environments
5. The effectiveness of various ventilation control approaches in real-world environments
6. The ability of filtration to allow the recirculation of cleaned tobacco smoke-contaminated air
7. The costs of ventilation systems in real-world applications

Demonstration projects are necessary to prove and improve the local exhaust ventilation equipment described in this chapter and in other publications.

20.8   TERMS, UNITS, AND DEFINITIONS

C = Concentration in air

E = Emission rate of air contaminant or marker, any units of "amount/time"

Q = Ventilation rate in the same units as E

Q_{OA} = Volume flowrate of dilution air, cfm (m^3/s)

q = Emission or generation flowrate of vapor or gas, cfm (m^3/s)

Ca = The acceptable exposure concentration, ppm

K_{eff} = An air mixing effectiveness factor to account for incomplete or poor delivery of dilution air, unitless

STP = 70°F at 760 mmHg barometric pressure, dry air (ACGIH definition)

MW = Molecular weight or molecular mass

t = Approximate time, (s)

G = Amount emitted or generated, grams (g)

d = Air density correction factor, unitless

ACH, AC/h = Air changes per hour N = The number of times air is theoretically replaced in a space during 1 h; also, "air change rate"

AHU = Air handling unit; refers to equipment in HVAC systems

AHS = Air handling system; refers to equipment in HVAC systems

ASHRAE = American Society of Heating, Refrigeration, and Air Conditioning Engineers; primary association involved in HVAC design

Dilution ventilation = Air quality control which relies on the dilution of airborne contaminants, usually by using outdoor air (OA); also general ventilation

Effectiveness = A measure of the ventilation system's ability to deliver dilution air to the breathing zone of an occupant or to the occupied zone of a space; also, air mixing efficiency, air change effectiveness

Emission = The release of contaminants into the indoor air; emission rate = mass released per time; emission factor = emission rate per unit of source (e.g., per cigarette)

HEPA = High efficiency particulate arrestance; HEPA filter

HVAC = Heating, ventilating, and air conditioning; air handling system equipment designed and used for controlling temperature, humidity, odor control, and air quality

Industrial ventilation = The equipment or operation associated with the supply and exhaust of air, by natural or mechanical means, to control airborne contaminants in an industrial setting

Local exhaust ventilation (LEV) = A ventilation systems which captures and removes emitted contaminants at or near an emission source

Natural ventilation = The movement of outdoor air into a space through intentionally provided openings, such as windows, doors, or other nonpowered ventilators, or by infiltration; caused by wind and temperature differences

Occupied zone = Usually, the region within an occupied space between the floor and 72 in. above the floor and more than 2 ft from the walls; the region within an occupied space where people inhale

Odor = A quality of gases, vapors, or aerosols which stimulates the olfactory organs; typically unpleasant or objectionable

Outdoor air (OA) = "Fresh" air; outdoor air is usually obtained from outside the building, but exceptions exist (e.g., from an acceptable hallway)

Return air (RA) = Air returned to the fan from the occupied space

Standard air, standard conditions (STP) = Dry air at 70°F, 29.92" Hg; or air at 50% relative humidity, 68°F, and 29.92" Hg; in either definition, the air density $\partial = 0.075$ lb/ft^3.

Supply air (SA) = Air supplied to a space from the HVAC system

Volume flowrate = The quantity of air flowing in cubic feet per minute, cfm, scfm, acfm

Designated smoking areas (DSA) = A space specifically designated, designed, and operated to accommodate smoking

Displacement ventilation (DV) = Air introduced into a space at one location which migrates across the space carrying air contaminants with it and is exhausted at a remote location from the inlet; other terms include "plug flow" and "pump flow"

REFERENCES

1. American Society of Heating, Refrigeration and Air Conditioning Engineers (ASHRAE), ASHRAE 62-1999 — Ventilation for Acceptable Indoor Air Quality, ASHRAE, ASHRAE 55-1992 on Thermal Comfort; ASHRAE 52.1 and 52.2-1992 and 1999 on Filter Testing; ASHRAE 129P on Measuring Air Exchange Effectiveness; contact ASHRAE, 1791 Tullie Circle, NE, Atlanta, GA 30329.
2. DOL, OSHA, *Notice of Proposed Rulemaking on Indoor Air Quality*, 59 FR 15968, U.S. Government Printing Office, Washington, D.C., 1994.
3. Guffey, S.E., *Proceedings of the Workshop on Ventilation Engineering Controls for Environmental Tobacco Smoke in the Hospitality Industry*, ACGIH, Cincinnati, OH, 1998.
4. ACGIH, *Industrial Ventilation Manual*, 23rd Ed., ACGIH, Cincinnati, OH, 1997.
5. Hemeon, W.C.L., *Plant and Process Ventilation*, 3rd Ed., Lewis Publishers, New York, 1999.
6. ASHRAE, *Fundamentals Handbook*, ASHRAE, Atlanta, GA, 1997.
7. Burton, D.J., Engineering technology adjusts workers' exposure to tobacco smoke, *OH&S*, November 1994.
8. Burton, D.J., Proposed smoking ban by OSHA creates need for specialized HVAC controls (Parts 1-4), *OH&S*, December 1994.
9. Burton, D.J., *IAQ and HVAC Workbook*, 3rd Ed., IVE, Inc., Salt Lake City, UT, 1999.
10. Burton, D.J., *Industrial Ventilation Workbook*, 4th Ed., IVE, Inc., Salt Lake City, UT, 1999.
11. Lavin, H., *IA Bull.*, 3(5), 1996.
12. Burton, D.J., Smoking bans create a need for specialized ventilation controls, *OH&S*, August 1990.
13. Turner, S.C., The power of ventilation: an airport smoking area that works, *HPAC*, January 1999.
14. Marino, A. and Ivanovich, M., Accommodating smoking in the hospitality industry, *HPAC*, October 1999.
15. —, Indoor Air '96, Proceedings of the 7th International Conference on Indoor Air Quality and Climate, Nagoya, Japan, Vol. 1–4, 1996.
16. Burton, D.J., New ASHRAE 52.2-1999 standard helps in air filter selection, *OH&S*, June 2000.

APPENDIX

Instructions for providing a dedicated ventilation system to a smoking area.

Step	Action
1	Determine the actual boundaries of the proposed smoking area. (This is the total space where air moves, not an area designated on paper.) Estimate the total room volume of the smoking area in cubic feet, V_{space}.
2	Estimate the maximum number of persons in the smoking area at any one time, Ps.
3	Estimate minimum required fresh outdoor air required ($Q_{OA\text{-}req}$).

$$Q_{OA\text{-}req} = Ps \times (OA, 60 \text{ cfm/person, or as desired.})$$

4   Determine the design (or actual) recirculation percentage at the air handling system, R (e.g., 0.80). Alternatively, determine the design (or actual) OA percentage introduced to the system, N (e.g., 0.20).

5   Estimate the total supply air rate in the smoking area, Q_{SA}.

6   Estimate the design (or actual) outdoor air supplied to the area:

$$Q_{OA\text{-}actual} = Q_{SA} \times (100 - R) = Q_{SA} \times N$$

7   Compare $Q_{OA\text{-}req}$ to $Q_{OA\text{-}actual}$. The latter should equal or exceed the former.

8   Estimate room air changes per hour, N:

$$N = (Q_{SA} \times 60)/V_{space}$$

(also,

$$N = (Q_{OA} \times 60)/V_{space}$$

for outside air only).

9   Check to see that the return air equals or exceeds supply air to the smoking area.

10   Check to see that air from the smoking area is returned directly (not after traveling through other occupied areas).

11   Check the mixing potential in the smoking area (walls, fans, etc.).

12   Recommend or investigate potential alterations if necessary (e.g., moving partitions, installing fans, moving supply/exhaust registers, closing doors and windows).

13   Check to see if the smoking area is under negative pressure (smoke tube, air flows).

14   Establish a comprehensive housekeeping program.

Sample Calculation 3. It is determined that up to five people will occupy a designated smoking lounge, 10' × 10' × 8'. There is a separate, dedicated air supply and exhaust AHU serving the smoking area.

$$Q_{SA} = 500 \text{ acfm}, R = 80\% \ (N = 20\%)$$

Step	Outcome
1	Room volume = $10 \times 10 \times 8 = 800$ ft^3; Room sq ft = 100 ft^2
2	Ps = 5 people
3	$Q_{OA\text{-}req} = 5 \times 60 = 300$ cfm
	(ASHRAE 62-1999 standard interpretation)
4	R = 80% (0.80) or N = 20% (0.20)
5	$Q_{SA} = 500$ cfm
6	$Q_{OA\text{-}actual} = (500)(1.00 - 0.80) = 100$ cfm (or, $500 \times 0.20 = 100$ cfm)
7	$Q_{OA\text{-}actual}$ is not sufficient — it should be 300 cfm (Alterations to the ventilation system will be required to increase the supply of fresh air to the space.)
8	$N = (Q_{SA} \times 60)/V_{space}$
	$= (500 \times 60)/800$
	$= 37.5$ total air changes/h
	also, N = $(300 \times 60)/800 = 22.5$ OA AC/h (Okay)
9-14	As necessary. In this case, the recirculation rate R should be reduced to R = 0.40, which will provide 60 cfm per person of outdoor air.

Index

A

Adolescents
  cardiovascular function, 195
Advertising, 200
Airflow, 150
Allergic alveolitis, 75
Allergy, 69
American Cancer Society, 14
American Society of Heating, Refrigeration, and
  Air Conditioning Engineers (ASHRAE), 281, 352
Amines, 180
Animal models, 129, 310, 337
  asthma, 338
Antioxidants, 45, 59, 198, 244
  supplementation, 50
Asthma, 328
  adult, 81
    exacerbation, 95
    new-onset, 82
  allergic, 337
  occupational, 73
Atherosclerosis, 196
Atopia, 72
Atopy, 71

B

Bans, 152
Bars, 283
Behavior
  air, 352
  smoker, 353
Bias, 3
  publication, 14
Birthweight, 17, 25
  low, 32
  mean, 28
Breast
  lobules, 179
  tissue, 178

C

California State Assembly Bill 13, 92
Cancer
  breast, 177
  childhood, 322
  lung, 8, 13, 17, 109, 129, 131, 249
  nasal sinus, 17

Carbon monoxide, 218, 279, 310, 325
Carcinogenesis, 130
Carcinogens, 109, 178, 187, 314
Cardiovascular disease (*see also* Heart disease),
  61, 309, 315
  prevention, 314
Cardiovascular effects, 217
Cardiovascular function, 195
Centers for Disease Control (CDC), 146
Chemoprevention, 137
Children
  cancer, 322
  cardiovascular function, 195
  exposure reduction, 145
  exposure, 94
  health, 322
  protection, 121
  pulmonary problems, 301
Cholesterol, 59, 63, 198
Chromosomes, 252
Cigarettes, 302
Clinicians, 160
Controls, 10
Conventional medical view, 2
Cooking, 280
Coronary heart disease (CHD), 62
Cotinine, 5, 167, 282
Counseling, 163, 168
Cystic fibrosis, 329

D

Dehydroepiandrosterone (DHEA), 181
Designated smoking areas (DSA), 355
Development
  neurobehavioral, 323
Diagnostics, 3
Diet, 7
DNA, 187, 243

E

Education, 110
  programs, 169
Emission sources, 351
Epithelial cells, 215, 261
  bronchial, 260
  pulmonary, 260
Epithelial, possibly precancerous lesions (EPPL), 4